Blood and Data

UNIVERSITY OF HERTFORDSHIRE
The School of Pharmacy

Blood & Data

*Ethical, Legal and Social Aspects
of Human Genetic Databases*

Published for the International ELSAGEN Conference,
University of Iceland, Reykjavík, 25–28 August 2004

Edited by

Gardar Árnason
Salvör Nordal
Vilhjálmur Árnason

University of Iceland Press
& Centre for Ethics
Reykjavík 2004

Blood and Data — Ethical, Legal and Social Aspects of Human Genetic Databases
Copyright © 2004 by the University of Iceland Press and the Centre for Ethics, University of Iceland, Reykjavík, and for the content of each paper the respective author.

Cover design: Róbert Guillemette
Layout: Brynjólfur Ólason
Printing: Gutenberg

All rights reserved. No part of this publication may be reproduced or used in any form or by any means without written permission of the publishers.

This book is a publication of the ELSAGEN project (Ethical, Legal and Social Aspects of Human Genetic Databases: A European Comparison), which is financed between 2002 and 2004 by the European Commission's 5th Framework Programme, Quality of Life (contract number QLG6-CT-2001-00062).

PRINTED IN ICELAND

ISBN 9979-54-593-3

Table of Contents

List of Contributors 9
Preface 13

BIOETHICS, BIOTECHNOLOGY, BIOSOCIALITY

1 Microfluidic platforms, genetic databases, and biosociality 17
   Michael D. Mehta, Jennifer D. Poudrier

2 Privacy, property, and social relations in bioinformatics research 23
   Anne Gatensby

3 "Interbreeding within the Icelandic population is high compared
   with that of mice or fruit-flies" 29
   Gardar Árnason

GENETICS IN ASIA

4 Socio-genetic marginalization in Asia. A plea for a comparative
   approach to the relationship between genomics, governance,
   and social-genetic identity 39
   Margaret Sleeboom

5 A study on the ethical, legal, and social aspects of the Chinese
   genetic database in Taiwan 45
   Wan-Chiung Cheng, Wan-Ping Li

6 The Singapore human polymorphism/mutation database:
   Our experience with setting up a country-specific database 51
   Ene-Choo Tan, Marie Loh

CONSENT IN MEDICAL RESEARCH

7 The controversy on consent in the Icelandic database case
   and narrow bioethics 67
   Sigrídur Thorgeirsdóttir

8 Toward a tiered approach to consent in biomedical research 79
   Peter Lucas

9 The wolf in sheep's clothing: Informed consent forms as
   commercial contracts 85
   Gerard Porter

10   Gift or duty? A normative discussion for participation in human
     genetic databases research 95
     Nadja K. Kanellopoulou

CONSENT, BIOBANKS, AND GENETIC DATABASES

11   Broad consent — the only option for population genetic databases? 103
     Jane Kaye

12   Databases and informed consent: Can broad consent legitimate
     research? 111
     Sigurdur Kristinsson

13   What is wrong with using anonymized data and tissue for
     research purposes? 121
     Ants Nomper

14   Informed consent for donating biosamples in medical research
     — legal requirements in Iceland 127
     Hördur Helgi Helgason

15   Why we should not relax ethical rules in the age of genetics 135
     Tuija Takala

THE PUBLIC DISCOURSE ON GENETICS AND DATABASES

16   Long-term trends in public sensitivities about genetic
     identification: 1973–2002 143
     Martin W. Bauer

17   Making genes commonly meaningful: Implications of national
     self-images on human genetic databases 161
     Piia Tammpuu

18   Analyzing multiple discourses in the establishment of genetic
     databases 167
     Wendy Marsden

19   Genetic databases and public trust 175
     Mairi Levitt, Sue Weldon

20   "Public databases and privat(ized) property?" A UK study of public
     perceptions of privacy in relation to population based human
     genetic databases 181
     Sue Weldon, Mairi Levitt

21   Becoming masters of our genes: Public acceptance of the Estonian
     Genome Project 187
     Külliki Korts

22   "We don't have that many secrets" — the lay perspective on
     privacy and genetic data 193
     Anna Birna Almarsdóttir, Janine Morgall Traulsen, Ingunn Björnsdóttir

| 23 | Do regulations address concerns? 201
Matti Häyry

VALUES, KNOWLEDGE, AND DIGNITY

| 24 | Interests, values, and genetic databases 211
Ann Bruce, Joyce Tait

| 25 | Genetic databases and what the rat won't do: What is
dignity at law? 217
Mark Cutter

| 26 | Human dignity and technology 223
Daniel Statman

AUTONOMY AND PRIVACY

| 27 | The collection and management of confidential genetic data.
An evaluation of deCODE genetics based on the principle
of autonomy 231
Pascal Schwarz

| 28 | Do biobanks promote paternalism? On the loss of autonomy
in the quest for individual independence 237
Bjørn Hofmann

| 29 | Monozygotic autonomy and genetic privacy 243
Anne Maria Skrikerud

| 30 | Privacy in public 249
Salvör Nordal

GENETIC SCREENING, EUGENICS, AND COMMODIFICATION

| 31 | Genetic screening, prospective parenthood, and the internal
perspective 257
Peter Herissone-Kelly

| 32 | The ugly curve — genetic screening into the 21st century 263
Asterios Tsioumanis, Konstadinos Mattas, Elsa Tsioumani

| 33 | Categorizing genes: Commodifying people? 269
Donald Bruce

GENETIC MEDICINE AND GENETIC RESEARCH

| 34 | Interpretation of genetic data for medical and public
health uses 277
Paul A. Schulte

| 35 | Perceptions of risk and human genetic databases: Consent
and confidentiality policies 283
Timothy Caulfield

36 Individual boundaries and the impact of genetic databases upon
    collectivist cultures: Molecular, cognitive and philosophical views 291
    Janet K. Brewer

RESEARCH BIOBANKS

37 Mapping the language of research-biobanks and health registries:
    From traditional biobanking to research biobanking.
    A project presentation 299
    *Jan Helge Solbakk, Søren Holm, Paula Lobato de Faria, Jennifer Harris,
    Anne Cambon-Thomsen, Marit Halvorsen, Camilla Stoltenberg, Roger
    Strand, Bjørn Hofmann, Anne Maria Skrikerud, Jan Reinert Karlsen*

38 UK Biobank: Social and political landscapes 307
    Helen Busby

39 Cell line research with UK Biobank. Why the new British biobank
    is not just another population genetic database 313
    Sebastian Sethe

BENEFIT-SHARING, JUSTICE, AND DISCRIMINATION

40 Benefit-sharing and public trust in genetic research 323
    Graeme Laurie, Kathryn G. Hunter

41 Biobanks and the "social" in social justice 333
    Sarah Wilson

42 Policy for human genetic resources as compared to
    environmental genetic resources 339
    Karin Erika Bengtsson

43 Problems with targeting law reform at genetic discrimination 347
    Mark J. Taylor

List of Contributors

Anna Birna Almarsdóttir
  Faculty of Pharmacy, University of Iceland, Reykjavík, Iceland.

Gardar Árnason
  Centre for Professional Ethics, University of Central Lancashire, Preston, England.

Martin W. Bauer
  Department of Social Psychology, London School of Economics, London, England.

Karin Erika Bengtsson
  Department of Theology, Uppsala University, Uppsala, Sweden.

Ingunn Björnsdóttir
  Icelandic Pharmaceutical Society, Reykjavík, Iceland.

Janet K. Brewer
  Institute for Indigenous People's Rights, Chicago, U.S.A.

Ann Bruce
  Innogen, ESRC Centre for Social and Economic Research on Innovation in Genomics, University of Edinburgh, Edinburgh, Scotland.

Donald Bruce
  Society, Religion and Technology Project, Church of Scotland, Edinburgh, Scotland.

Helen Busby
  Institute for the Study of Genetics, Biorisks and Society (IGBIS), University of Nottingham, Nottingham, England.

Anne Cambon-Thomsen
  Inserm U558, University of Toulouse, Toulouse, France.

Timothy Caulfield
  Health Law Institute, University of Alberta, Edmonton, Canada.

Wan-Chiung Cheng
  Science and Technology Law Centre, National Taiwan University, Taipei, Taiwan.

Mark Cutter
  ESRC Centre for Economic and Social Aspects of Genomics (CESAGen), Lancaster University, Lancaster, England.

List of Contributors

Anne Gatensby
   McGill University, Montreal, Canada.

Marit Halvorsen
   Department of Private Law, Faculty of Law, University of Oslo, Oslo, Norway.

Jennifer Harris
   Division for Epidemiology, The Norwegian Institute of Public Health, Oslo, Norway.

Matti Häyry
   Centre for Professional Ethics, University of Central Lancashire, Preston, England.

Hördur Helgi Helgason
   The Legal Farm, Hartford VT, U.S.A.

Peter Herissone-Kelly
   Centre for Professional Ethics, University of Central Lancashire, Preston, England.

Bjørn Hofmann
   The Center for Medical Ethics, University of Oslo, Oslo, Norway.

Søren Holm
   Cardiff Institute of Society, Health and Ethics, Cardiff University, Cardiff, Wales.

Kathryn G. Hunter
   Arts and Humanities Research Board (AHRB) Research Centre for Studies in Intellectual Property and Technology Law, School of Law, University of Edinburgh, Scotland.

Nadja K. Kanellopoulou
   Arts and Humanities Research Board (AHRB) Research Centre for Studies in Intellectual Property and Technology Law, School of Law, University of Edinburgh, Edinburgh, Scotland.

Jan Reinert Karlsen
   Center for Medical Ethics, Faculty of Medicine, University of Oslo, Oslo, Norway.

Jane Kaye
   Oxford Genetic Knowledge Park, University of Oxford, Oxford, England.

Külliki Korts
   Centre for Ethics, University of Tartu, Tartu, Estonia.

Sigurdur Kristinsson
   University of Akureyri, Akureyri, Iceland.

Graeme Laurie
   Arts and Humanities Research Board (AHRB) Research Centre for Studies in Intellectual Property and Technology Law, School of Law, University of Edinburgh, Edinburgh, Scotland.

Mairi Levitt
   Institute for Environment, Philosophy and Public Policy (IEPPP), Lancaster University, Lancaster, England.

List of Contributors

Wan-Ping Li
  Science and Technology Law Centre, National Taipei University, Taipei, Taiwan.

Paula Lobato de Faria
  National School of Public Health and Faculty of Law, New University of Lisbon, Lisbon, Portugal.

Marie Loh
  Bioinformatics Institute, National University of Singapore, Singapore.

Peter Lucas
  Centre for Professional Ethics, University of Central Lancashire, Preston, England.

Wendy Marsden
  Arts and Humanities Research Board (AHRB) Research Centre for Studies in Intellectual Property and Technology Law, School of Law, University of Edinburgh, Scotland.

Konstadinos Mattas
  Department of Agricultural Economics, Aristotle University, Thessaloniki, Greece.

Michael D. Mehta
  Department of Sociology, University of Saskatchewan, Saskatoon, Canada.

Ants Nomper
  Oxford Genetic Knowledge Park, University of Oxford, Oxford, England.

Salvör Nordal
  Centre for Ethics, University of Iceland, Reykjavík, Iceland.

Gerard Porter
  Arts and Humanities Research Board (AHRB) Research Centre for Studies in Intellectual Property and Technology Law, School of Law, University of Edinburgh, Scotland.

Jennifer D. Poudrier
  Department of Sociology, University of Saskatchewan, Saskatoon, Canada.

Paul A. Schulte
  Education and Information Division, National Institute for Occupational Safety and Health, Cincinatti, Ohio, U.S.A.

Pascal Schwarz
  Institute for Applied Ethics and Medical Ethics (IAEME), Faculty of Medicine, University of Basel, Basel, Switzerland.

Sebastian Sethe
  Sheffield Institute of Biotechnological Law and Ethics (SIBLE), University of Sheffield, Sheffield, England.

Anne Maria Skrikerud
  Center for Medical Ethics, Faculty of Medicine, University of Oslo, Oslo, Norway.

Margaret Sleeboom
  International Institute for Asian Studies (IIAS), Leiden, The Netherlands.

List of Contributors

Jan Helge Solbakk
   Center for Medical Ethics, Faculty of Medicine, University of Oslo, Oslo, Norway.

Daniel Statman
   Department of Philosophy, University of Haifa, Israel.

Camilla Stoltenberg
   Division for Epidemiology, Norwegian Institute of Public Health, Oslo, Norway.

Roger Strand
   Center of the Study of the Sciences and the Humanities, University of Bergen; and the Center for Medical Ethics, Faculty of Medicine, University of Oslo, Norway.

Joyce Tait
   ESRC Centre for Social and Economic Research on Innovation in Genomics (Innogen), University of Edinburgh, Edinburgh, Scotland.

Tuija Takala
   Department of Moral and Social Philosophy, University of Helsinki, Helsinki, Finland.

Piia Tammpuu
   Centre for Ethics, University of Tartu, Tartu, Estonia.

Ene-Choo Tan
   Defence Medical and Environmental Research Institute, DSO National Laboratories, Singapore.

Mark J. Taylor
   Department of Law, University of Sheffield, Sheffield, England.

Sigríður Þorgeirsdóttir
   Department of Philosophy, University of Iceland, Reykjavík, Iceland.

Janine Morgall Traulsen
   Department of Social Pharmacy, Danish University of Pharmaceutical Sciences, Copenhagen, Denmark.

Elsa Tsioumani
   Deptartment of Law, Democritean University of Thrace, Komotini, Greece.

Asterios Tsioumanis
   Department of Agricultural Economics, Aristotle University, Thessaloniki, Greece.

Sue Weldon
   Institute for Environment, Philosophy and Public Policy (IEPPP), Lancaster University, Lancaster, England.

Sarah Wilson
   North West Genetic Knowledge Park (Nowgen), Lancaster University, Lancaster, England.

Preface

THE LATE 1990S was an exciting time for genetics. Automated DNA sequencing techniques were making rapid progress, and the Human Genome Project was completing its map of the human genome in a tight race with the biotechnology firm Celera Genomics. Geneticists, medical researchers, entrepreneurs, and investors were looking for ways to cash in on the rapid developments in genetics. In 1997, Iceland's deCODE genetics Inc. announced its plans to establish a Health Sector Database, containing health data collected from medical records of almost all 290,000 Icelanders as well as from existing medical records of deceased Icelanders. The Act on Health Sector Database, passed in December 1998, allowed that data from HSD could be connected to data from two other databases, one containing genealogical information for the population, going centuries back, and the other containing genetic data. One of the main uses of such a combined database is to find genes associated with various diseases, in particular common diseases such as cardiovascular diseases and cancer.

The Estonian Genome Project followed in the wake of the Icelandic project, promising to collect health and genetic data of about one million Estonian inhabitants. This databank is to be used for scientific research, public health research and medical treatment. A third project, UK Biobank, is now underway. UK Biobank plans to collect data from about 500,000 people, and would not only consider the genetic basis of common diseases, but also life style factors. A fourth project, the Medical Biobank of Umeå in Sweden, has received much less attention, but it is the only one of the four that has actually been established (already in 1998). However, the company operating the biobank, UmanGenomics, has now practically ceased operating. Despite the initial enthusiasm and optimism, the other projects have made little progress. UK Biobank has not started collecting samples or data yet, and it is not clear when it will start. The Icelandic and Estonian plans have ground to a halt: neither database has been established yet and their future is uncertain. Other population-based databases are being planned, within Europe as well as without, but most on a smaller scale. In addition to them, many smaller biobanks exist, as well as gene banks and mutation databases, which have various degrees of resemblance to the population-based genetic databases.

There is no single term used for the sort of databases, which the four projects have sought to establish. The Swedish and UK projects call them biobanks, emphasizing the collection of biosamples; in Estonia it is called a genome project, emphasizing the collection of genetic data; and in Iceland the focus has been on the collection of health data in the Health Sector Database. All projects plan to collect genetic and health data (and some also genealogical data) from a large number of people in order to provide a resource for research in human genetics. In the title of this volume we have used the term "human genetic database" for these combined databases, with reference to their purpose as a resource for research in human genetics, and to distinguish them from smaller biobanks and gene banks. We have not, however, imposed that terminology on the papers in this volume, which variously refer to human genetic databases as biobanks, gene banks, or population genetic databases.

The plans to establish human genetic databases raise complex questions about their ethical, legal and social implications, many of which are shared by smaller biobanks and gene banks. The bioethics project ELSAGEN (Ethical, legal and social aspects of human genetic databases: A European comparison) was started to address these issues. The project is financed by the European Commission within the 5th Framework Programme from 2002 to 2004. It consists of research groups from University of Iceland, Tartu University and HETA Law Offices in Estonia, Lund University in Sweden, and University of Lancaster, University of Central Lancashire and Oxford University in England. One of the main events organized within the project is the International ELSAGEN Conference on Ethical, Legal and Social Aspects of Human Genetic Databases, held in Reykjavík, Iceland, on August 25–28, 2004. The papers in this volume were contributed to the conference. Many of these papers are written by authors outside of the ELSAGEN group and address issues related to other databases than are in focus of that research whose results will be published in a separate book.

We would like to thank the authors for their contributions and cooperation during the preparation of this volume. For their assistance, we thank Lena Halldenius, Matti Häyry, Hörður H. Helgason, Peter Herissone-Kelly, Jane Kaye, Sigurdur Kristinsson, Margit Sutrop, Tuija Takala, and Sue Weldon.

Gardar Árnason
Salvör Nordal
Vilhjálmur Árnason

Bioethics, biotechnology, biosociality

1

Microfluidic platforms, genetic databases, and biosociality

by Michael D. Mehta and Jennifer D. Poudrier

Introduction

THIS SHORT PAPER examines how the development of portable, low-cost devices (known as microfluidic platforms) may affect how human genetic data are collected and analyzed. By integrating several diagnostic functions onto a hand-held device, microfluidic platforms (MFPs) are likely to have profound impacts on health care. In theory, small samples can be analyzed at the point-of-care, and results available in minutes to health care providers and others. To achieve these goals, MFPs will probably require integration with human genetic databases through wireless and wired networks. To understand the implications of this technology, we will explore the intersection of microfluidic platforms with human genetic databases using Paul Rabinow's concept of "biosociality."

Microfluidic platforms

Microfluidic platforms are microsystems that incorporate microelectronic and biological components onto a hand-held device. Such devices are currently under development, and within 5–10 years are expected to revolutionize how health care providers monitor human health. By placing the capabilities of a large diagnostic lab in a platform the size of a wrist watch, point-of care applications including genetic testing, pharmacogenetic testing, and assessment of viral load (e.g., to assess the effectiveness of a vaccine or degree of infectivity) may become commonplace in medical clinics, pharmacies, and elsewhere. To assist in making diagnoses, MFPs will require real-time data acquisition and integration with human genetic and other kinds of databases.

With a team of scientists, medical practitioners, and social scientists, the authors of this paper are involved in the development of MFPs for medical, forensic, environmental, agricultural, and national security applications.[1] Our team's goal is to develop an open system architecture device that will integrate several functions at the microscale

---

[1] Mehta shares a $1.5 Million CDN grant from the Canadian Institutes of Health Research entitled "Novel Platforms for Genetic Analysis."

on a microfluidic chip. Such chips are pieces of glass or plastic which have several finely etched channels running through them. With the assistance of microscale pumps and other miniaturized devices on the chip, the handheld device (into which the chip is inserted) uses electrical forces to separate components from a sample, and on-board optical technologies for analyzing constituent parts. For medical applications, miniaturized devices for performing on-board polymerase chain reaction (PCR) of DNA samples are being developed to run on these chips. It is anticipated that these devices can conduct several genetic tests in parallel, and use single nucleotide polymorphism (SNP) analysis for identifying responders and non-responders to particular pharmacological agents. The non-medical applications for this technology include genetic identification at crime scenes, water quality monitoring for pathogens, identifying genetically modified plants in the field, and providing data for threat assessments at airports on suspicious materials. In short, MFPs will usher in a new genetic age that will reshape society in profound ways. Their portability, anticipated low cost, and flexibility require that we consider how such technologies renegotiate the often invisible contract between nature and culture.

Biosociality

Paul Rabinow's (1999) concept of "biosociality" refers to a transformative condition under which both nature and scientific work in the life sciences become increasingly revealed as cultural practice.

> If sociobiology is culture constructed on the basis of a metaphor of nature, then in biosociality, nature will be modeled on culture understood as practice. Nature will be known and remade through technique and will finally become artificial, just as culture becomes natural. Were such a project to be brought to fruition, it would stand as the basis for overcoming the nature/culture split (Rabinow 1999, 411).

Biosociality addresses the emerging trends in the conceptual, metaphorical, and discursive boundaries between science and culture. Rabinow argues that biological theories and metaphors, like sociobiology, have become increasingly popular to describe concepts like life, health, culture, and community. For instance, the use of language to describe DNA as the "book of life" promotes a view of life that is maintained, ordered, and understood according to biological principles alone (Kay 1999; Petersen and Bunton 2002). However, with biosociality science loses much of its cultural authority since nature, natural artifacts, and the natural sciences are exposed as socio-cultural practices.

Biosociality and MFPs

The development of MFPs for human genetic testing represents a turning point within the nature/culture divide. As with many genetic issues, MFPs have embedded within them scientific and cultural artifacts that are manifested in debates such as nature versus nurture. On the one hand, such devices have the potential to intensify existing tendencies to explain behavior, health, personality, etc. in strictly biological terms. The very process of looking for genes, SNPs, viral loading, and other biological markers is reductionist in nature, and likely to stimulate the development of a range of techno-scientific solutions like gene-based therapy, germ line therapy, etc. On the other hand, MFPs have the potential to reveal in exquisite detail the genetic variation in the human genome. As developments in genomics, proteomics, and metabolomics progress, MFPs may usher in a new cultural appreciation for difference, and reveal many of the flaws and attribution errors associated with defining health in binary terms (e.g., healthy versus unhealthy). It is at this juncture that MFPs converge with biosociality. To understand how such convergence occurs, we turn our attention to some key elements of biosociality; namely, risk management, identity formation, exposing the cultural practices of science, and the creation of a post-disciplinary society.

Risk management

Rabinow (1999) contends that contemporary genetic science is based on a discursive power. Drawing primarily upon Foucault's (1978) conceptions of knowledge/power and biopower,[2] Rabinow argues that the discourse of genetic science is aimed at regulating the body, and is currently disguised through the language of risk management rather than surveillance and control, and as benevolent intervention and cure rather than as discipline. Currently, the impact of genetic testing technologies on populations as a whole is limited. Few individuals have undergone genetic testing, and most tests conducted are for carrier identification (e.g., Cystic Fibrosis, Tay-Sachs Disease, and Sickle Cell Trait), prenatal diagnosis (e.g., Down Syndrome), newborn screening (e.g., Phenylketonuria), and late onset disorders (e.g., Huntington's Disease). The development of MFPs may change this by making available a wider range of tests to a larger segment of the population.

Under the guise of individual monitoring and population health surveillance, MFPs have the potential to increase the medicalization of health care by placing a premium on testing and personal risk management. From a strictly rational perspective, such testing is associated with improving diagnosis, minimizing adverse drug effects and improving their efficacy, and identifying individuals who may be particularly susceptible to environmental and occupational exposures to chemicals, radiation, and other hazards. Although laudable on some levels, such testing may lead to systemic genetic discrimination that could create a genetic divide that parallels other divides in our society (e.g., digital, informational) (Nelkin and Tancredi 1994). Since health is strongly correlated with socio-cultural factors and inequalities (e.g., age, gender, wealth, ethnicity, level of marginalization), a society where MFPs are used without critical regard runs the risk of entrenching a form of biological (genetic) determinism (Nelkin and Tancredi 1994), and the geneticization of medical knowledge and health care practice (Hedgecoe 1999;

---

2 Biopower refers to Foucault's analysis of the way that power/knowledge discourses serve to rationalize and order bodies at two intersecting levels: the body of society (further theorized with the notion of "governmentality") and the body of the individual (further theorized with the concept of "technologies of the self"). Biopower highlights the social construction of knowledge, in the first instance, but also identifies the way that populations and individual selves are interpellated by and ordered through power/knowledge systems.

Lippman 1991; ten Have 2001). Furthermore, the integration of MFPs with human genetic databases magnifies this tendency significantly due to the simultaneity of decentralizing and centralizing a range of power effects. By their very nature, MFPs are tools of decentralized data collection and local analysis. Instead of biological samples being sent to centralized laboratories for analysis, MFPs bring the lab to the clinic and other points-of-concern. Simultaneously, MFPs will require access to comprehensive human genetic databases. The discrepancy between the ubiquity of MFPs as medical diagnostic tools and access to databases that could prove expensive to access (especially if owned by the private sector), opens up new avenues for analyzing the power effects of this technology, and our changing understanding of the relationship between nature and culture (biosociality).

Identity formation and risk factors

Rabinow suggests that new cultural formations will emerge out of biology. He uses the development of new support groups as an example, whereby individuals who are "at-risk" of developing a particular kind of disorder may come together (due on their encoded status as "at-risk") to form a new social group or network based upon a cultural construction of biological categories. This is especially salient where there are significant gaps between diagnosis and therapies. For Rabinow (1999, 411), the Human Genome Project exemplifies this cultural transformation such that the "new genetics will cease to be a biological metaphor for modern society and will instead become a circulation network for identity terms and restriction loci, around which and through which a truly new type of auto-production will emerge, which I call biosociality…"

MFPs are likely to accelerate, at least initially, the tendency for new social groups to form based along at-risk states, even if the risk is only probabilistic in nature. As such, risk is not a result of specific dangers posed by the immediate presence of an individual or a group, but rather of the combination of interpersonal factors that make a risk probable. Prevention then is surveillance not of the individual but of likely occurrences of diseases, anomalies, and deviant behaviors that place a premium on monitoring populations instead of individuals. Although MFPs are used at the level of the individual, and have an intrinsic capacity to focus biopower at this level, it is in the use of this technology for generalizing from individuals through genetic databases that the subtler effects of biosociality begin to emerge.

Exposing the cultural practices of science

The concept of biosociality relates to an emerging condition whereby biological explanations of society will be less dominant, and where the cultural and social metaphors that define science will emerge. In other words, science will model itself after social values, thereby exposing the cultural practices of science. For example, Rabinow contends that what is now considered natural will become exposed as a cultural construct, or artificial. This tendency to view science as culture is being driven in part by particular technologies that highlight the nature of scientific uncertainty. For instance, MFPs are the result of interdisciplinary research in microbiology, genetics, microfluidics, microscience and nanoscience, engineering, photonics, bioinformatics, and several other areas. The crossing of disciplinary boundaries to create and use this technology reveals several gaps in general scientific understanding, and particular gaps in understanding how to address multiple kinds of uncertainty (e.g., laminar flow in liquids, quantum effects at the nanoscale, genetic markers, etc.). Handling these uncertainties, both individually as scientists and as teams of scientists, reveals the cultural practices of individual scientific disciplines to scrutiny, thereby leading toward biosociality. On another level, the choice and configuration of databases for referencing the results flowing from MFPs reveals other kinds of uncertainties that also expose science to increased scrutiny and demythologizing forces.

In search of a post-disciplinary society

Rabinow appears to embrace the prospects of biosociality because he contends that in biosociality, in an era of artificiality, the boundary between nature and culture will be ruptured and problematized. He contends that a post-disciplinary society will privilege neither nature nor culture, and hopes that both will increasingly be evaluated by ethical standards. By showing how science is artificial, and based upon the creation of artificial "natural" things, science will be exposed as culturally loaded rather than as neutral. For Rabinow (1996), this condition depends upon an anthropologizing of the West where the taken for granted objectivity of science is exposed as cultural practice, and where genetic science in particular is shown as discursive, ideological, and based upon well-entrenched knowledge/power systems. MFPs will accelerate the trend towards problematizing the split between nature and culture. In the same way that various genome projects have shown the high degree of similarity between the genomes of mice and humans (thereby weakening our collective perception of superiority to so-called lower life forms), MFPs will rupture the currently sanctioned comparisons that institutions and individuals make between the able-bodied and disabled, and the healthy and unhealthy. On the surface this appears to be a paradox since MFPs are likely to increase, in the early years, the tendency toward geneticization. However, the use of this technology in conjunction with human genetic databases will ultimately expose the distinctions between culture and science as untenable, and dangerous if held onto too tightly.

Conclusion

These new diagnostic tools have the potential to intensify existing social, ethical, and legal dilemmas on a scale like never before. For example, the speed and precision of MFPs will make the development of thoughtful genetic information policies all the more essential. As policy makers continue to struggle with defining the clinical utility of existing genetic profiling procedures, MFPs will likely allow for the collecting, storing, and sharing of even more information more quickly. Initially, the use of this technology is likely to accelerate concerns about genetic discrimination, geneticization, and the increasing scientization of culture. However, as MFPs, and the databases they rely on, become commonplace, several contradictions between nature and culture will be exposed. Our intent is to continue to draw upon the concept of biosociality to more fully understand and empirically analyze specific intersecting sites of nature/culture and power/knowledge in the production of MFPs, genetic databases, and their respective publics.

References

Foucault, M. 1978. *The history of sexuality: An introduction*. New York: Pantheon Books.

Kay, L. 1999. In the beginning was the word: The genetic code and the book of life. In *The science studies reader*, ed. M. Biagioli, 224–233. New York and London: Routledge.

Hedgecoe, A. 1999. Reconstructing geneticization: A research manifesto. *Health Law Journal* 7:5–18.

Lippman, A. 1991. Prenatal genetic testing and screening: Constructing needs and reinforcing inequities. *American Journal of Law and Medicine* 17:15–50.

Nelkin, D., and L. Tancredi. 1994. *Dangerous diagnostics: The social power of biological information*. Chicago: University of Chicago Press.

Petersen, A., and R. Bunton. 2002. *The new genetics and the public's health*. London and New York: Routledge.

Rabinow, P. 1996. *Essays in the anthropology of reason*. Princeton, NJ: Princeton University Press.

Rabinow, P. 1999. Artificiality and Enlightenment: From sociobiology to biosociality. In *The science studies reader*, ed. M. Biagioli, 407–416. New York and London: Routledge.

ten Have, H. 2001. Genetics and culture: The geneticization thesis. *Medicine, Health Care and Philosophy* 4:295–304.

2

Privacy, property, and social relations in bioinformatics research

by Anne Gatensby

IT WILL NOT have escaped most people's notice that genetic-based research and biotechnologies have quickly captured the public imagination when it comes to the hopes and fears, promises and anxieties of the new biological order that is emerging in the space between science as intellectual pursuit and science as commercial enterprise. Much of this discourse was unleashed when it was announced that a consortium of scientists funded by both state and private institutions was to map the human genome, the entire complement of genetic material that makes up the typical human being. The Human Genome Initiative, as the project came to be known, was particularly significant because it enabled its backers to deploy narratives of moral imperative about health and illness as a means of cultivating widespread popular support for the undertaking, while at the same time downplaying concerns over the commodification of nature.

Yet it is the commodification of nature that is at the heart of an emerging new field of inquiry known as bioinformatics. In general, public access to genetic databases has been the guiding principle within the bioinformatics community. However, there also exist a number of databases that are privately owned that require complex and expensive licensing agreements in order for researchers to gain access to them. While these databases often contain the same sequencing data that public access databases contain, their owners have not failed to complete significant and lucrative rights of access deals, typically worth millions of dollars and mostly with large pharmaceutical firms.

Why is this so? On the surface, it would seem that it would make rational business sense to support the public database system and save millions of dollars in licensing and access fees. And indeed, many of them do support the public system, both with funding and sharing of data. But they also see it as in their interests to support a parallel private system, because having access to information that competitors might not have access to gives firms a jump-start in the race to patent new and novel sequences that might prove over time to have commercial value. Because the first group to discover novel mutations has the right to patent the sequence, at least in the United States, and exploit that patent for commercial gain, private firms are effectively covering all the bases.

Additionally, private database holders see the value of bioinformatics networks not in the sequences themselves, but in the software tools, computer power, and associated information links that they can provide more readily than publicly funded sites. Indeed, the maintenance of private databases that hold the same data as public ones is also justified on the grounds that at some future time, the desire and the will of governments to support a public system might wane, thus positioning the private ventures in an excellent position to monopolize the entire bioinformatic sector.

As one might expect, such contingencies have given rise to a number of debates over what constitutes privacy, property, ethical practice, public interest, and the boundaries between human and non-human bodies. Privacy considerations are of most concern when the information that is held in biomaterials is abstracted from the public and is placed into the dematerialized zone of the network. This is of particular concern for individuals whose genetic and other medical records cease to be shielded by the historically established social relations of confidentiality between medical professionals and patients and move out into public or proprietary spaces.

In general, privacy and property debates tend to adopt individualistic ethical and legalistic frameworks that try to balance the rights of individual patients to maintain confidentiality over personal information with the desires of private institutions to exercise proprietary control over biological data. The questions that tend to be asked relate to the rights and rationales—usually ethical or legal—that the state, or any other group or individual that has the power to informatize biological material, has with regard to their desires to transform the private medical history and autonomy of the individual into a marketable commodity. For example, if the state maintains and supports a public health system that ensures affordable and universal health care, as is the case with many industrialized democracies, does that mean that it then owns the information that it has collected and is free to sell it to the highest bidder without popular consent? Or, do norms of democratic practice and ethical public interest require that the state act in the role of a steward, and require democratic debate and public consultation before it transfers common goods to private hands?

As Hilary Rose (2003) has questioned in her analysis of the Icelandic Health Sector Database, will the social values that underpin the principles of privacy and confidentiality come to be deemed less important than the imperative of commercial gain as bioinformatics becomes more and more taken for granted as the repository of biological knowledge? As yet, social researchers have been relatively slow in reacting to the emergence of bioinformatics as a new regime of biological inquiry, at least in relation to what has developed within legal and bioethical discourses, and are unable to provide satisfactory answers to such questions. There are several reasons for this, I think. First, researchers in social studies of science and technology tend to want to understand the technoscience that they investigate at more than just a superficial level. They want to understand some of the technical details of the discipline in question before they make judgments about its construction, both to avoid the charge of scientific ignorance bandied about by combatants in the science wars, and because such knowledge facilitates a deeper reading of the social relations that circulate within and around technoscientific practices. While many social researchers already have some educational background in science that can be built upon, others have to expend a great deal of time and effort researching the technical details of their subject matter on top of trying to become sufficiently conversant in the theories and methods of the social sciences. This becomes a triple burden in the case of bioinformatics because it draws not just from one disciplinary sector, but two—molecular biology and computer science—each with their own complements of epistemologies, assumptions, practices, and methodologies. Also, because we have inherited a historical division between the scientific disciplines that still separates the physical sciences from the life sciences, it is still rare for individuals to have a point background in biology and in computing. Thus, when faced with decisions about what kinds of science and technology one can feasibly study in the context of a time-limited research project, the general inclination is to study that which can be more easily investigated than that which requires knowledge in several different realms of thought.

The second reason is that the terrain of critical work in bioinformatics has already been staked out by scholars in other fields, most notably law and ethics. This is not surprising given that both law and ethics in contemporary academic inquiry tend to adopt the discursive frame of individual rights and protections, especially in relation to personal privacy and intellectual property. But what is also interesting is that individuals and groups that promote the development of biotechnology and bioinformatics themselves privilege legal and ethical concerns as the primary sites around which debates should be focused. For example, we see the various state regulatory commissions set up in countries that have made biotechnology a prime sector of economic development almost always incorporating an official bioethics component to their advisory boards and expert committees. This form of state-sponsored bioethics tends not to draw on the full range of ethical discourse that has developed within the larger academy, but tends more to be a new form of bioethics in the making, a bioethics that focuses on whether biotechnologies are inherently dangerous or immoral, but not on how they alter social relations within the workplace or the larger society as a whole. And while there may be some nod to concerns over the social aspects of biotechnologies in official documents of this type, the social most often appears as an invocation rather than as a substantive area of debate.

Carr and Levidow (1997), for example, have shown how in the UK situation this new administrative form of bioethics is constructed out of and by state regulatory agencies to displace and fragment critical debates about the control and commodification of nature. Biotechnology, it seems, has its own particular form of ethics, one that has simultaneously given rise to a new class of professional bioethicists that operate to keep the boundaries between technical issues and social issues discrete, and to advance a hegemonic version of bioethics that is based on traditional individualistic and utilitarian ethical theory. It is this new class of bioethicists, that subsume an undifferentiated notion of the social under the rubric of the ethical, that is called upon when the state or private biotech interests need to demonstrate their willingness to address public concerns about genetic engineering, cloning, GM foods, and the like. The very idea of turning to the expertise that has accumulated within the social sciences, where complex notions of subjectivity, contingency, situated practice, social order, and organization have been developed at both the theoretical and practical levels, and might fruitfully contribute to debates about biotechnology and bioinformatics, is still as yet a relatively unexplored option.

Third, the two most well known instances of bioinformatics are the Human Genome Initiative and the Icelandic Health Sector Database. We care most about the HGI and the Icelandic project because they deal most acutely with human populations and what might be done to them in the name of science and business. Yet, when we look at the whole of bioinformatics, including all the possible databases that currently exist, the human genome is just a mere speck of information—the vast majority of the species that are represented are those that belong to the categories of yeasts, bacteria, viruses, fruit flies, nematodes and other invertebrates, and plants. Social relations between humans still exist in the realm of non-human bioinformatics, it is just that they do not immediately seem to be of dire concern because the issues of privacy, protection, access, distribution, and power are not as obvious as in cases where the commodification of human genetic material is the source of social concern. In this sense, we can see that many existing social scientific analyses of the HGI and Icelandic database have not developed completely independently of legal and ethical analyses, but have been shaped by the dominant logic of bioethical and legal discourse. As a consequence, these discourses of privacy and property tend to pit the interests of affected laypersons against the imperatives of commercial bioinformatics firms that seek to control information for the purposes of accumulation. While this might be a logical starting place in reference to human populations, a more complex social analysis means that we have to start addressing something other than human social relations that already have built-in deferences to the concerns of liberal theory. Only by looking at the larger political economy of genetic databases, including the non-human ones, will we be able to move beyond the narrow frame of privacy and property and move on to the terrain of social relations.

This is not to say, however, that there has not been any significant work in social science around bioinformatics, just that because it is so sparse, only a few connections between the social and the bioinformatic have as yet been made. Hilgartner (1995), for example, has proposed that the new biomolecular databases be thought of as new network-based communication regimes for biology. As sociotechnical systems, they have the capacity to continually reshape the boundaries between public and private knowledge, especially when the issues of system control, maintenance, and ownership are brought into the mix.

Fortun (1998; 2001) has also addressed bioinformatics, particularly as it has emerged out of the Human Genome Initiative and how it was deployed in the case of the Icelandic Health Sector Database. For him, genomics is "about the circulation of materials, information, skills, capital, products, explanatory powers, and therapies." Speed genomics is what was set into motion by the HGI, and he suggests that speed is a useful trope for mapping the social, institutional, and conceptual reconfigurations that are brought about by bioinformatics. Fujimura (1998) has looked at the consequences of constructing new representations and meanings in bioinformatics, particularly as they relate to future imaginaries of what it means to be natural and human.

Keating, Limoges, and Cambrosio (1998) have explored how new reliance on molecular biological instrumentation within laboratories leads to changes in organizational and work structure, creating a new division of labor in biology and introducing the social values associated with post-Fordist production regimes into the lab. Thacker (1998;

2000; 2003a; 2003b) has approached bioinformatics from within a cultural studies frame, noting the shift from government or state-sponsored research to industrial and commercial sectors and what that shift means in terms of cultural conceptions of the body and the machine.

Ideas such as those cited have only begun to be addressed within the social sciences and humanities with any sort of complex understanding of the intersections between history, economy, materiality, and sociality. In what does exist, we tend to see a more constrained set of views that pivots between privacy and property debates that posit an instrumentalist and individualist subject as the main locus around which a rights discourse develops. This range, of course, fails to get to the crucial questions that interest many sociologists, such as how the new biological order that is being brought into existence by the introduction of information and computer sciences to the study of life, will affect the parameters of a new social order. When bits of bodies become bits and bytes, we have to modify our understandings of how the social practices of both molecular biology and informatics are altered, and how they, in turn, change the way we interact with each other.

References

Carr, S., and L. Levidow. 1997. How biotechnology regulation sets a risk/ethics boundary. *Agriculture and Human Values* 14:29–43.
Fortun, M. 1998. Projecting speed genomics. In Fortun and Mendelsohn 1998.
Fortun, M. 2001. Breaking the code. *Renssalaer Magazine*. March.
Fortun, M., and E. Mendelsohn, eds. 1998. *The practice of human genetics*. Dordrecht: Kluwer.
Fujimura, J. H. 1998. The practices of producing meaning in bioinformatics. In Fortun and Mendelsohn 1998.
Hilgartner, S. 1995. Biomolecular databases: New communication regimes for biology? *Science Communication* 17:240–263.
Keating, P., C. Limoges, and A. Cambrosio. 1998. The automated laboratory: The generation and replication of work in molecular genetics. In Fortun and Mendelsohn 1998.
Rose, H. 2003. *The commodification of bioinformation: The Icelandic Health Sector Database*. London: Wellcome Trust.
Thacker, E. 1998. Bioinformatics: Materiality and Data between information theory and genetic research. *C-Theory: Theory, Technology and Culture*. 28 October.
Thacker, E. 2000. Redefining bioinformatics: A critical analysis of technoscientific bodies. *Enculturation* 3 (1). http://enculturation.gmu.edu/3_1/thacker.html
Thacker, E. 2003a. Bioinformatics and bio-logics. *Postmodern Culture* 13(2).
Thacker, E. 2003b. Data made flesh: Biotechnology and the discourse of the posthuman. *Cultural Critique* 53:72–97.

3

"Interbreeding within the Icelandic population is high compared with that of mice or fruit-flies"

by Gardar Árnason

Introduction

IN 2000, DECODE genetics Inc. received an exclusive license to construct a Health Sector Database (HSD) for the entire Icelandic population. The HSD is to be connected to two other databases, one containing genealogical data for every Icelander alive and, going centuries back, most of those deceased; and the other containing genetic information. This combination of databases, which I shall simply call the deCODE database, will create "a totally informative population with which to search for drug targets and to model both disease and host-drug interactions" (Gulcher and Stefánsson 1998, 526). Although plans to construct the HSD seem to be on hold, the company has not officially abandoned them.

The purpose of the deCODE database is in great part to provide statistical data for research in human genomics and genetic epidemiology. A further goal is a thorough geneticization of medicine: according to deCODE representatives, the "ultimate goal of the database [is] to usher in an era of preventive health care and individual-based disease management practices based on human genetics" (Gulcher and Stefánsson 1998, 526).

Connecting medical data with genealogical and genetic data will make it possible to find quickly the most likely locations for parts of genotypes linked with such phenotypes as disease symptoms and efficacy of drugs or treatment. For instance, it will be possible to feed the database with encrypted names of individuals suffering from a disease and have it map out clusters of related individuals in a number of pedigrees of varying sizes. The scientist could then pick a pedigree and compare genotypes of healthy and sick individuals from that pedigree, assuming that the genetic factor in the disease is common to all the diseased individuals. If the diseased individuals have a higher frequency of certain genetic differences than the healthy ones, these genetic differences may be a factor in the disease. This would greatly speed up the process of locating genes that are a factor in disease. Genes would not only be linked to diseases, but also to drug efficacy and side-effects, making it possible (in theory at least) to tailor drug treatments according to the genetic profile of the individual.

Roots in eugenics

From the very beginning, proponents of the Icelandic database project have emphasized the value of the Icelandic nation for genetic research. They typically point to three things: Extensive genealogical information, good medical records, and the homogeneity of the nation. The most important of the three is the nation's perceived homogeneity: Icelanders are taken to be more (genetically) alike than other nations. Not only was the population quite isolated for almost 1,000 years, from the tenth century until the Second World War, but it went through three major "bottlenecks," periods when disasters killed off a large part of the population. In the early fifteenth century the plague killed around one third of the approximately 70,000 inhabitants of Iceland, in 1707–9 over one third of the population died of smallpox, and in 1783 Lakagígar, a row of volcanoes in the south of Iceland, erupted, resulting in the death of a quarter of the population. By 1800 Iceland had only about 45,000 inhabitants; on December 31, 2003, there were 290,570 (see *Statistics Iceland*).

These population bottlenecks are supposed to have decreased the variation of genotypes in the population. As the metaphor goes, the Icelandic gene pool is shallow. This is expected to facilitate genetic research, because when genotypes of relatively homogenous individuals are compared there are fewer variations in genotypes (i.e., genetic characteristics) that could account for observed variations in phenotypes (i.e., physical characteristics).

The observation that the Icelandic nation has been subjected to more drastic natural selection than most other nations, and that this somehow makes the nation valuable for science, is not as new as many believe. In the twenties and thirties, Icelandic eugenicists noted what are now called bottlenecks in the history of the nation. This made the nation valuable, not because of homogeneity but because the stock had been improved through "a thousand hardships."

In 1922 Gudmundur Finnbogason, a philosopher at the University of Iceland and a supporter of eugenics, published an article in which he complained that eugenics had not been given much consideration in Iceland. His article was intended to educate the Icelandic public about this new and important science and show what Icelanders could learn from it (see Karlsdóttir 1998a, 427). In his article he discussed favorably the claims made by British eugenicists that poverty, criminality, and misery are inherited characteristics, and then happily pointed out that in Iceland there were no inherited class differences. Quite the contrary, he considered the Icelandic nation as a whole to be of excellent stock. He claimed that the original stock (of Norse and Celtic settlers in the ninth and tenth century) was particularly good, because only the strongest and most diligent men would have traveled this far only to face the dangers and hardships of founding a settlement on a distant island. Since then the stock had only improved, first because almost no foreign blood had mixed with the blood of the original settlers, and second because the hardships over the centuries (eruptions, brutally long and cold winters, plagues, etc.) had weaned out the weakest and so made the stock even stronger (see Karlsdóttir 1998b, 79–80).

During the rise of eugenics in the first three decades of the century, ideas appeared about Iceland's role in the future of eugenics. It was pointed out that Iceland had thorough genealogical records, which would be particularly useful for eugenic research. The philosopher-eugenicist Gudmundur Finnbogason proposed in 1922 that a genealogical institute be established, which would accumulate genealogical information and put together a card catalog, a database of sorts, where all Icelanders would be registered along with information about their health and hereditary characteristics, their biographies and genealogies, as well as samples of their hand writing and voice, and anything else that could serve as the basis for research in genealogy and heredity:

> Scientists would work there, who would want to investigate the heredity of particular characteristics, mental and physical, and those men would go there who would be interested in getting to know the nature of some specific families. There one could see if the race was changing for the better or for the worse. There would be at every time a looking glass, showing how the heart of the nation beats. (Finnbogason 1922, 203–204; my translation.)

Gudmundur Finnbogason described the importance of such an institution in words that could be taken out of the recent debate about the Icelandic Database Project: "A great task awaits Icelanders. They should become and could become that nation which lays down the widest and most solid foundation for the hereditary research of the future" (Finnbogason 1922, 203; my translation).

Another philosopher-eugenicist, Ágúst H. Bjarnason, discussed these ideas in 1926 and thought they would be rather expensive to realize and therefore not feasible at the time. Until then, he suggested that all Icelanders individually record their genealogy and information about themselves and their relatives "and hide neither virtues nor faults, although they concern close relatives" (Bjarnason 1926, 225–226; see also Karlsdóttir 1998b, 59; my translation).

During the war, in 1943, support for the idea came from Halldór Laxness, Iceland's greatest writer and social critic (he was awarded the Nobel prize for literature in 1955). In his article "Human Life on a Card File," Laxness suggests that information about Icelanders could be registered on cards and organized, and then sold to clients for an appropriate fee. He was aware of efforts by the German Nazis to construct databases with such personal information, and distanced himself from their aims, but not from the basic idea of collecting, organizing, and making use of such data:

> I am told that the German secret police has a card file containing information about millions of men from all over the world, with comments on their origin, behavior, views, and character, as well as biographical information—all for the purpose of being able to walk up to these men and kill them at the opportune moment. Compared to Himmler's card file, which is made for murdering, it would be easy to give these few Icelanders life on a card file. (Laxness [1943] 1980, 196; my translation.)

Plans for a genealogical institution and a database for the Icelandic nation for hereditary research ended there, due to lack of funds and lack of political and scientific interest. In the discussion about this possibility, many of the themes in the present database debate occur: the basic idea of establishing a database with hereditary and health information about all Icelanders, the possibility of selling this information, the idea of observing an entire population—how the heart of the nation beats—the value of the extensive genealogical records, as well as the eugenic idea of a strong, pure, and unspoiled race, a homogenous population in the terms of modern genetics, which has only become purer and stronger through the hardships of the centuries, the population bottlenecks in the language of modern human genetics.

Model organisms

The eugenicists believed that the Icelandic "stock" was valuable for scientific research, because of its noble Nordic and Celtic origin and merciless but ultimately beneficent circumstances through its history. Modern geneticists consider the nation valuable because of its homogeneity, which also points to the nation's Nordic-Celtic origin (small founding population) and its often horrific history in Iceland (i.e. population bottlenecks). Homogeneity also points to laboratory animals. Laboratory animals, in particular laboratory mice and rats, are bred specifically for the laboratories and do not exist in the wild. They are model organisms. Experiments require the elimination or at least reduction of such distorting factors as the peculiarities of the individual organism. A part of a controlled experiment is to produce an ideal situation, where the phenomenon being investigated can appear in its purest form. The greatest progress in transmission genetics (i.e, the study of heredity) has emerged from the study of one specific life form: *drosophila*, the fruit fly. One of the central problems in the early drosophila studies was to breed and maintain the right stock of flies. Since not just any fruit fly would do for these studies, geneticists shared fruit flies of "the right stock," otherwise the experiments just would not work. As a result, a specific type of drosophila came into existence, bred specifically for genetic studies.

If it seems like an exaggeration or far fetched to compare a nation with such stocks of laboratory animals, let me justify this comparison by quoting a few texts. It is not particularly my comparison at all. The critics, of course, like to picture Icelanders as the new guinea pigs of genetics. On October 3, 1998, the Swiss news paper *Der Tagesanzeiger* published an article with the title "The Icelanders, Our Lab Mice" (Die Isländer, unsere Labormäuse), attacking the database project. A front page article in the *International Herald Tribune* on January 13, 1999, quotes David Banisar of the Electronic Privacy Information Center in Washington as saying: "Turning the population into electronic guinea pigs should serve as a warning." But not only critics made the comparison with laboratory animals: scientific and financial magazines made it too, and with curious lack of concern. The *Scientific American* was one of the most startling, an article in the February issue 1998 with the title "Natural-Born Guinea Pigs: A start-up discovers genes for tremor and psoriasis in the DNA of inbred Icelanders" opens with these words:

> To build a life among the glaciers and volcanoes of Iceland takes a special breed of people. Not just figuratively, either: the 270,000 citizens of this island nation, a great majority of them descended from seventh-century Viking settlers, form one of the most inbred populations in the world (Gibbs 1998).

One investor, Brian Atwood of Brentwood Venture Capital, doubted the usefulness of the Icelandic nation for genetics in an interview with the business magazine *Red Herring* (1998), going further than just making a comparison of Icelanders to laboratory animals: "Despite the high degree of inbreeding within the Icelandic population, Atwood thinks that the amount of *inter*breeding is still high compared with that of mice or fruit flies, which can be bred more narrowly." It is irrelevant that Atwood was speaking as an investor in a rival genomics company and hence quoted as a financial critic of deCODE. The point is that a financial magazine not only compares the nation to lab animals but —with a perfectly straight face—speaks of a nation *as* lab animals, as laboratory humans that can be bred for experiments (although not as narrowly as fruit flies).

deCODE has also presented the Icelandic nation in a way which suggests laboratory animals. In an information brochure for potential investors, deCODE boasts that "deCODE genetics will be in a position to offer its corporate partners access to the Icelandic population for clinical trials of drug candidates" (Mannvernd 1998). Icelanders are represented as laboratory animals, which makes Iceland the laboratory: "The genetic similarity might make Iceland the best laboratory in the world to study genes," says an ABC News story (Gizbert 1999). The walls of the laboratory have been torn down and now it can encompass an island in the Atlantic. In the new world of genetics, the laboratory animals are no longer locked in cages in the laboratory, they are us.

The international media discussed Icelandic homogeneity in terms that should surprise anyone who has walked around in Iceland with open eyes. Specter's *New Yorker* article is fairly typical:

> It's a cliché, but the first thing a visitor to Iceland notices, after the volcanic landscape that lies beneath the approach to Keflavik International Airport, is just how closely Icelanders resemble each other ... Iceland sometimes seems to be inhabited by one enormous family, not one of whose members ever leaves the neighborhood where he was born ... The hereditary instructions for blue eyes and blond hair, which are so prevalent in Iceland, have been passed undiluted through a small gene pool for fifty generations. After a thousand years of plagues, epidemics, earthquakes, and volcanoes finished weeding out the population, what remains ... is a nation of two hundred and seventy thousand of the most genetically similar people on earth. (Specter 1999, 41–42).

Let me give more examples. The journalist John Schwartz writes in the *Washington Post* on January 13, 1999: "The strikingly uniform DNA of Iceland's largely blue-eyed, blond-haired populace is expected to provide an invaluable resource for studying human genetics, leading to fundamental insights into many diseases, proponents say." And a few paragraphs later Schwartz goes on: "Iceland's population presents a tantalizing opportunity for those who study genetics because Icelanders' blond hair and blue eyes reflect one of the most remarkably homogeneous populations in the world." And he adds: "The original blend of 9th century Norse stock and Celtic seamen has been largely unchanged, and that gene pool was further restricted by bouts of plague, famine, and volcanic eruption."

Second example: Novelist Simon Mawer writes in an op-ed article in *The New York Times* on January 23, 1999, with the headline "Iceland, the Nation of Clones": "But since everyone in Iceland is related to everyone else there (all of them are descended from the same few Vikings), the place is, comparatively speaking, a nation of clones." Third example: A news story on the ABC News web site shows a picture of a blond girl (presumably Icelandic), with a caption that reads: "Virtually all of Iceland's 270,000 residents trace their roots to the Vikings. Their homogeneity might make Iceland the best laboratory in the world to study genes" (Gizbert 1999).

It is not important that a minority of Icelanders are blond, nor that brown eyes are quite common, nor that Icelanders are not inbred, nor even that little scientific evidence exists for the claim that the Icelandic population is homogeneous.[1] What is interesting is that this claim, i.e. that Icelanders are homogeneous, is useful for both business and science, and that this claim—and its justification—is practically the same as the one made by eugenicists in Iceland in the thirties and forties: Icelanders are a very special stock because of their Viking origin and "good breeding through a thousand hardships." Although my examples are mostly from the popular media, the references to homogeneity, the origin of Icelanders, the bottlenecks, the nation-wide laboratory, and the value of the population for research, is not something that the popular media adds to its science reports. These references clearly originate with scientists and is neither invented by the media, nor a product of the popularization of science (see for example Stefánsson 2000; Gulcher and Stefánsson 1998).

Icelanders are not the first group to be singled out for genetic research. A similar fate has befallen Ashkenazi Jews, American Mennonite communities, and the population of the Atlantic island Tristan da Cunha, among others. But this is the first time that an entire nation is turned into a constant source for genetic research, a perpetual laboratory population, a nation in a petri-dish.

I have presented two concerns, which are of a political nature, rather than a moral or an ethical. One is that the database project has deep historical roots in eugenics, and the other is that we are being turned into model organisms for scientific research, lab animals that are observed at all times by science. Perhaps we should ask ourselves: is that who we want to be?

---

1 Some genetic research points to very similar genetic diversity in Iceland as in other countries in northern Europe (see Árnason 2003; Árnason, Sigurgíslason, and Benedikz 2000). deCODE research suggests some homogeneity: "the heterozygosity rate [of microsatellite markers] over 300 Genethon markers in Iceland is 0.75 compared with 0.79 in Europe (Icelandic data is unpublished)," but their main arguments for homogeneity are not genetic but historical and linguistic, referring to the small founding population (although no one knows how small it was, likely at least a few thousand and possibly tens of thousands) and subsequent isolation and population "bottlenecks" over the following one thousand years, as well as little change in the Icelandic language from the 13th/14th century (i.e., when the Sagas were written) to this day (Gulcher and Stefánsson 1998, 524).

Acknowledgments

I thank Gottskálk Jensson and Skúli Sigurdsson for reading and commenting on an earlier draft of this paper. This paper was produced as a part of the ELSAGEN project (Ethical, Legal and Social Aspects of Human Genetic Databases: A European Comparison), financed between 2002 and 2004 by the European Commission's 5th Framework Programme, Quality of Life (contract number QLG6-CT-2001-00062). I gratefully acknowledge the support of the European Community. The information provided is the sole responsibility of the author, the Community is not responsible for any use that might be made of data appearing in this publication. I gratefully acknowledge the support of the Icelandic Research Council for the project "The genetic revolution in Iceland." Earlier drafts of this paper were presented at the conference of the Society for Philosophy and Technology (SPT) in Aberdeen, Scotland, on July 10, 2001; and at a workshop of the ELSAGEN project in Tartu, Estonia, on May 14, 2004. I thank the participants of these meetings for helpful comments and discussions.

References

Árnason, E. 2003. Genetic heterogeneity of Icelanders. *Annals of Human Genetics* 67:5.

Árnason, E., H. Sigurgíslason, and E. Benedikz. 2000. Genetic homogeneity of Icelanders: Fact or fiction? *Nature Genetics* 25:373–374.

Bjarnason, Á. H. 1926. *Sidfrædi*. Vol. II. Reykjavík.

Finnbogason, G. 1922. Mannkynbætur. *Andvari* 47:184–202.

Gibbs, W. W. 1998. Natural-born guinea pigs: A start-up discovers genes for tremor and psoriasis in the DNA of inbred Icelanders. *Scientific American* February 28.

Gizbert, R. 1999. Profiling an entire nation. *ABCNews.com*. Available online at: http://www.mannvernd.is/frettir/abc.wnt990218_iceland.html

Gulcher, J., and K. Stefánsson. 1998. Population genomics: Laying the groundwork for genetic disease modeling and targeting. *Clinical Chemistry and Laboratory Medicine* 36:523–527.

Karlsdóttir, U. B. 1998a. Kynbætt af thúsund thrautum. *Skírnir*, 172 (Fall): 420–450.

Karlsdóttir, U. B. 1998b. *Mannkynbætur*. Reykjavík: Sagnfrædistofnun Háskóla Íslands, Háskólaútgáfan.

Laxness, H. [1943] 1980. *Sjálfsagdir hlutir*. 3rd ed. Reykjavík: Helgafell.

Mannvernd. 1998. Kynningarbæklingur Íslenskrar erfdagreiningar. http://www.mannvernd.is/dreifibref/gppr-database.html

Red Herring. 1998. Icelandic startup DeCode takes a homegrown approach to genomics. May 11.

Specter, M. 1999. Decoding Iceland. *The New Yorker* January: 40–51.

Statistics Iceland. Web site. http://www.hagstofa.is

Stefánsson, K. 2000. The Icelandic Healthcare Database. *"Who owns our genes?" Proceedings of an international conference: October 1999, Tallinn, Estonia; organized by the Nordic Committee of Bioethics.* Copenhagen: The Nordic Council of Ministers.

Genetics in Asia

4

Socio-genetic marginalization in Asia

A plea for a comparative approach to the relationship between genomics, governance, and social-genetic identity

by Margaret Sleeboom

ASIA BOASTS THE main economic players of the 21st century. Especially China, India, and Japan will play major roles in the field of science, particularly in applied modern technologies, such as biomedical research. In this paper I argue for the importance of the comparative study of health care strategies and socio-genetic marginalization under different Asian regimes. The concept of socio-genetic marginalization, first, draws attention to the consequences of the practice of relating the social to the (assumed) genetic make-up of people, regardless of the relevance of such connections. Second, it refers to the isolation of social groups and individuals as a consequence of discrimination on the basis of genetic information. Socio-genetic risk groups may not just have to deal with the psychological burden of their fate, but also with feelings of social inaptitude. Third, it refers to the financial uncertainty that occurs when health care becomes too costly for the socio-economically disadvantaged.

A call for empirical research and comparison

The study of socio-genetic marginalization in China, India, and Japan necessitates a methodological approach that takes into account the prevalent national and local politics of conceptualizing ethnic and social groups, while creating a suitable conceptual framework for analyzing it. I regard comparative research most suitable to that purpose, as it enables the researcher to throw light on conceptualizations of marginalization by means of cross-cultural juxtaposition. This approach acknowledges the importance of three socio-cultural themes of (1) Race and ethnicity, (2) Community, kinship and identity, and (3) Governance and the normative understanding of genomics, which are conceptualized through *three empirical research plans* (1–3), and are outlined below:

1. Genomics, population-policies and local traditions

Governments in China, India, Japan, and Europe treat issues of population planning with various levels of importance and apply different political strategies. In China the issue of the new eugenics, the quality of the population, and the one-child birth policy are of great political and human significance (Ministry of Public Health 1994). The one-child policy, widely practiced since the late 1970s, in combination with a preference for males, has led to a lopsided growth of the population (Handwerker 2001). Through legal prohibition the state has tried to interfere against these practices, as yet unsuccessfully (Ministry of Public Health 1994). In India, too, sterilization, infanticide, and prenatal gender discrimination followed by abortion have led to a population imbalance, in which the state tries to interfere (Verma and Bijarnia 2002; Pandit 2003; Gupta 2000).

Japan and Europe, on the other hand, struggle with the problem of population ageing and falling levels of fertility. In Japan there have been many changes on legal abortion since its industrialization, varying from encouraging birth after the Meiji Restoration in 1868 to strengthen Japanese society and the promulgation of the National Eugenic Law in 1940, followed by the Sterilization Law (which forbade lepers, the blind, and the mentally handicapped to have children), to the implementation of the new Maternal Protection Law in 1996 (Amagasa 1996). At present, eugenic population policies are still important (see The Newsletter of the Questioning Eugenic Thought Network [*yoosei shisoo o tou nettouku*]), as the central authorities encourage women to increase the number of their children. Depending on the locality, subsidies are given to women who cooperate in the birth of healthy children by means of IVF (Satoo 1999).

Thus a variety of genetic technologies are available to the state in policies aimed at raising the quality of the population. Such population policies are part of an attempt to "improve" the genetic composition of individuals or entire peoples. In particular, concerns have been voiced that genetics may entail compulsory sterilization or abortion of those found to possess "undesirable" genetic sequences (Handwerker 2002; Dikötter 1998). The 1995 introduction of eugenic legislation in China, for example, supports the systematic "implementation of premarital medical check-ups" on hereditary, venereal, or reproductive disorders, as well as mental disorders, so as to prevent "inferior births" (Ministry of Public Health 1994; Handwerker 2002; Dikötter 1998). A different tendency can be found in the Netherlands, where members of the medical profession observe that the state is obstructing their duty of providing all possible information and alternative treatment to patients by not allowing them to practice pre-embryo-screening and prenatal genetic research. For instance, the government regards a triple-test for tracing children with Down-syndrome as population research, which requires the permission of the minister responsible for health (Breuning and Tibben 2002).

There is a need for the comparison of clashes of state population policies with local traditions in India, China, and Japan. Modern technologies of genetic engineering increasingly allow the government to intervene and regulate the personal lives of individuals in the name of public health, religion, and the national good. Concomitantly, concepts of health and human values in society are likely to be influenced. However, in some cultural environments, such as India, it is the state that tries to put a brake on the prenatal gender selection of its population.

2. *Genetic sampling and vulnerable groupings in genetic sampling sites*

The twofold aims of this project are, first, to understand the socioeconomic and cultural conditions of genetic sampling and banking in isolated areas and among ethnic minorities in different national contexts and, second, to understand the ways in which research populations are defined and mapped by researchers.

The DNA of these *socially* defined groups is the subject of research into evolutionary genetics, the study of human reactions to various pharmaceutical products (pharmacogenetics), and the study of single nucleotide polymorphism (SNPs). Two kinds of issues are central to this research. The first involve the bioethical aspects of sampling and storing DNA. Current bioethical protocols still fail to deal adequately with the specific conditions raised by population-based research, in particular regarding procedures for group decision making and cultural divergence (MST and MPH 1998; ICMR 2000). The second involve problems inherent to the ways geneticists define sample populations in genetic research. Before the sampling of populations begins, estimates are made about the genetic nature of target populations. The contents of these estimates are intimately related to historical processes of ethnic group formation, the intricacies of cultural perception, and political interests.

As part of the Human Genome Diversity Project (HGDP), the sampling and storing of DNA from indigenous populations serves to "avoid the irreversible loss of precious genetic information" and to "reconstruct the history of the world's populations by studying genetic variation to determine patterns of human migration" (Bamshad et al. 2001; Chen Shuzhuo 2000). Referring to indigenous populations as "isolates of historic interest" (IHI), the HGDP planned to immortalize the DNA of disappearing populations for future study (Cavalli-Sforza 1991). The initial conceptualization of the HGDP has been widely criticized for its consideration of indigenous peoples as mere research subjects, with little regard for their continued livelihood (Lock 1994). Partly, the HGDP has been supplanted by the global initiative of creating the Hapmap (see Hapmap websites) and commercial haplotyping. Though the hapmap is still criticized for its potential to encourage the stigmatization of populations with similar SNP patterns, the dispute on the HGDP is still going strong in India, China, and Taiwan (Sleeboom 2003).

In India, China, and Taiwan, the DNA of various minority groups with suspect unique DNA are exploited commercially by research groups abroad and at home. This has caused considerable local, national, and international strain (Xiong Lei and Wang Yan 2001; Chen Shuzhuo 2000). In Yunnan in Southwest China, ethnic DNA of over 25 so-called national minorities is stored in the world's largest ethnic databank in Kunming (*People's Daily*, 22/11/00). Such research is used to support claims on ethnic and national identity and to resolve conflict over national territory. Bio-anthropological research from South India served to provide genetic evidence for the similarity between high caste Indians and Europeans (Bamshad 2001). In Japan and China similar population research takes place with the historical-nationalist aim of "scientifically" rooting the modern nation into venerable historical origins (see Dikötter 1998).

3. *Genomics, sociogenetic identities and health strategies in China, India, and Japan*

Though increased genetic information means a step forward in predicting and curing genetic diseases, policy-makers also attempt to use it strategically to improve human populations and eliminate "defective" phenotypes. In the private sphere, parallel developments are taking place: early prenatal testing has motivated couples with an increased risk of affected offspring to have children, but the diagnosis of disorders has also led to a steep increase in selective abortion. A central question here is if and when we can speak of a link between national health care policy, public debate, and the private sphere. For instance, in China the government started a one-child family-planning programme in the 1970s to ensure sufficient nutrition for all new-borns, which resulted in a substantially decreased birth rate. At the same time, this policy limits individual freedom and autonomy. On the other hand, infant mortality in China by the 1990s had become considerably lower than in India (Gupta 2000). The criteria for the cost-effectiveness of clinical genetics in developing countries are not the same as in wealthy countries, such as Japan, Singapore, and Taiwan. In developing countries the severely handicapped do not usually survive and, if they do, they are not provided with expensive medical care. Consequently, the targets of genetics services are reached on the basis of a different balance sheet. Thus, in developing countries family planning, carrier testing, genetic counseling, and prenatal diagnosis may have a different rationale.

Preventive measures are expected to help people deal with their problems in a quicker, more effective, and less costly way. This is probably true *a fortiori* in countries with limited health care resources, such as China and India (Gupta 2000; Qiu Renzong 1999; Verma and Bijarnia 2002). Prenatal testing makes little sense, however, when tradition takes a reserved stance on contraception and considers abortion unacceptable. In some instances, among several classes in India and in the Chinese countryside, preference for male offspring has led to abuse of prenatal diagnosis (Handwerker 2002; Pandit 2003). Furthermore, an improvement of diagnostic techniques will increase the demand on health care resources, thereby forcing insurers and health care providers to reorganize the ways in which they allocate their resources and requirements of (potential) patients. In poorer countries, such as China, a further polarization regarding access to modern health care facilities may result (Qiu Renzong 1999; Zhu Fangfang 2001). On the other hand, the standardization of practices of prenatal and neonatal testing in a wealthy area such as Hong Kong was suspected to place undue pressure to abort the fetus and to unnecessarily medicalize kinship (Lam 1999). In Japan, hesitation about communicating with the patient about his or her condition complicates this situation.

In India the commonest application of genetic testing is for reproductive events, and developing countries such as India and China also screen neonates for relatively rare conditions such as phenylketonuria (PKU) (Verma and Bijarnia 2002; Yuan 1997), even before proper evaluation of the practice's usefulness has been performed, and before it has been assessed for its priority among other public health measures for those countries. At the same time, genetic services for the control of common disorders like thalassaemia, Down syndrome, neural tube defects, and muscular dystrophies are not integrated into existing primary health care (Verma and Bijarnia 2002; Yuan 1997). Besides the single gene disorders that form the backbone of present genetic counseling service, predispositions to sundry disorders, such as hypertension, atherosclerosis, cancer, type 1 diabetes, and even schizophrenia, can theoretically be screened for. The financial, economic, and ethical implications of such endeavors are enormous. A central concern is thus the socio-political processes of who decides on which tests to use, their implementation, funding, and motives.

Conclusion

A comparison of three large nation-states, comprising approximately half of the world population, seems rather ambitious. Nevertheless, this research proceeds from the idea that a careful comparison of the local details of a common practice (e.g., genetic sampling, genetic counseling) in a number of very different state-environments can say something worthwhile about the relationship between that common practice and the state. Research on the environment of the actual application of science and technology in Asia is rare. Many approaches emphasize native alternative views to "Western science." This approach, however, looks at the practice of the ways in which those traditions are mobilized against or articulated with these technologies.

References

Amagasa, K. 1996. *Yuusei sosaku no akumei* (The nightmare of eugenics). Tokyo: Shakai Hyooronsha.

Bamshad, M., T. Kivisild, W. S. Watkins, M. E. Dixon, C. E. Ricker, B. B. Rao, J. M. Naidu, et al. 2001. Genetic evidence on the origins of Indian caste populations. *Genome Research* 11:994–1004.

Breuning, M., and A. Tibben. 2002. De keuze aan de ouders (It's up to the parents). *Medisch Contact* 57:1780–82.

Cavalli-Sforza, L. L., A. C. Wilson, C. R. Cantor, R. M. Cook-Deegan, and M. C. King. 1991. Call for a worldwide survey of human genetic diversity: A vanishing opportunity for the human genome project. *Genomics* 11:490–491.

Chen Shuzhuo. 2000. *Yuanzhumin renti jiyin yanjiu zhi lunli zhengyi yu lifa baohu* (The ethical debate and legal protection of human genetic research of aborigines). *Shengwu Keji yu Falue Yanjiu Tongxun*, No. 6, October.

Dikötter, F. 1998. *Imperfect conception: Medical knowledge, birth defects and eugenics in China*. New York: Columbia University Press.

Gupta, J. A. 2000. *New reproductive technologies, women's health and autonomy*. New Delhi etc.: Sage.

Handwerker, L. 2002. The politics of making modern babies in China: Reproductive technologies and the "new" eugenics. In *Infertility around the globe*, ed. M. C. Inhorn and F. van Balen. Berkeley: UCP.

Hapmap: http://www.ncbi.nlm.NIH.gov/SNP/overview.html; http://www.biol.tsukuba.ac.jp/~macer/hapmap.htm; http://www.genome.gov/Pages/Research/HapMap

ICMR. 2000. Ethical guidelines for biomedical research on human subjects. *ICMR Bulletin*, 30, 10: 107–116.

Lam, S. T. 1999. Abortion, sterilization and genetic control – the Hong Kong perspective. In *Chinese scientists and responsibility. Ethical issues of human genetics in Chinese and international contexts. Proceedings of the "First international and interdisciplinary symposium on aspects of medical ethics in China: Initiating the debate," Hamburg, April 9–12, 1998*. Edited by O. Döring. Hamburg: Mitteilungen des Instituts für Asienkunde Nr. 314.

Lock, M. 1994. Interrogating the human diversity genome project. *Social Science & Medicine* 39:603–6.

Ministry of Public Health. 1994. *Zhonghua Renmin Gongheguo Muying Baodianfa* (Law of the People's Republic of China on Maternal and Infant Health Care). Beijing: Ministry of Public Health of the PRC.

Ministry of Science and Technology and the Ministry of Public Health (MST & MPH 1998) (*Zhonghua Renmin Gongheguo kexue jishubu, weishengbu*). 1998. *Renlei yichuan ziyuan guanli zhanxing banfa* (interim measures for the administration of human genetic resources). Beijing: MPH.

Pandit, S. 2003. Hindu bioethics, the concept of dharma and female infanticide in India. In *Genomics in Asia*, ed. M. Sleeboom. London: Kegan Paul.

Qiu Renzong. 1999. Medical ethics in China: status quo and main issues. In *Chinese scientists and responsibility. Ethical issues of human genetics in Chinese and international contexts. Proceedings of the "First international and interdisciplinary symposium on aspects of medical ethics in China: Initiating the debate," Hamburg, April 9–12, 1998*, ed. O. Döring. Hamburg: Mitteilungen des Instituts für Asienkunde Nr. 314.

Satoo, K. 1999. *Shusseizen Shindan* (Prenatal diagnosis). Tokyo: Yuhikaku Sensho.

Sleeboom, M. 2003. Sampling policies of isolates of historical interest in the PRC and the ROC. Paper presented at the International Conference of Asian Studies, August 18–22, in Singapore.

Verma, I. C., and S. Bijarnia. 2002. The burden of genetic disorders in India and a framework for community control. *Community Genetics* 5:192–196.

Xiong Lei and Wang Yan. 2001. *Ling rensheng yi de guoji jiyin hezuo yanjiu xiangmu* (Suspicious international collaboration in joint genetic research projects). *Liaowang*, 13, 26 March.

Yuan, L. F. 1997. Clinical genetic services in the big cities of China. In *Proceedings of 1st Hong Kong Medical Genetics Conference 25–27 January 1997*, ed. S. Lam et al. Hong Kong: The Hong Kong Society of Medical Ethics.

Zhu Fangfang. 2001. *"Anquanwang" he "jianzhenqi"* ("Safety net" and "shock absorber"). Beijing: Zhongguo Guoji Guangbo Chubanshe.

5

A study on the ethical, legal, and social aspects of the Chinese genetic database in Taiwan

by Wan-Chiung Cheng and Wan-Ping Li

I   Introduction

IN 2003, THE Taiwanese Government started a collaborative study to establish a cell bank and a genetic database of Chinese or non-Aboriginal Taiwanese (hereafter "Chinese Genetic Database" or "Super Control Genomic Database"). The aim was to conduct population genetics research on ethnic Chinese-related diseases, especially high blood pressure, diabetes, manic-depressive psychosis, arthritis, and asthma. At this moment, the project has accomplished the genotype analysis of a family suffering from Psoriasis in Taiwan.[1] The project has been undertaken by the Institute of Biomedical Sciences, Academia Sinica.[2]

Since the Chinese Genetic Database can be the largest and the most representative genetic database in the world, including more than 1 billion Chinese people,[3] this paper will introduce this significant project and indicate the relevant controversies, including the problematic informed consent form, and the incompleteness of the privacy protection mechanism.

II   Chinese genetic database in Taiwan: Methodology and expectations

To draw a picture of the Chinese Genetic Database, some fundamental knowledge about Taiwan is required. Accordingly, Part II will demonstrate why the Chinese Genetic Database is unique and significant for ethnic-Chinese population genetics research. It will also illustrate the sources of genealogical data and health records, which are indispensable for the cross-matching method of genetic study.

---

1   See the important research results of Academia Sinica (Academia Sinica b).
2   Academia Sinica was founded in Nanking, Taipei on June 9, 1928 as the highest academic institution in Taiwan (Republic of China) (see Academia Sinica a).
3   This is due to the blood relationship shared by the Chinese and the Taiwanese through their history. See Part II A below for further discussions.

A. Ethnic background: The umbilical cord of non-aboriginal Taiwanese and Chinese

Except for the pureblooded minority aboriginals, around 98% of Taiwan's population is made up of two main ethnic groups. In Taiwanese Han, the names of these groups are Hoklo (Minnan) and Hakka.[4] Based on family pedigree records, Hoklo and Hakka are traditionally considered descendants of migrants of Han from the Central Plain and the areas of the Yellow and Luo Rivers of northern China (see for example Chu 2000). But recent biological data, including data from immunoglobulin, human lymphocyte antigen, glucose-6-phosphate, dehydrogenase mutants, thalassemia mutants, microsatellite, mitochondrial DNA, and other biological studies, all indicate that the Hoklo and Hakka originated mainly from the southern groups of China (see Chu 2000; also Chen 1997). By any means, it would be proper to name non-Aboriginal Taiwanese "Ethnic Chinese." That is why the genetic information gathered by statistical sampling would be representative for more than one billion Chinese people all over the world.

B. Genealogical database resulting from a well-established census register system

To facilitate the governance of Taiwan, a National Census Register System has been very well established since the Japanese Colonization Period (1895–1945) in Taiwan. Taiwanese genealogical data are abundant and complete because of the National Census Register System, as well as the Chinese traditional custom of carefully preserving family pedigree records.

C. Where do the health records come from? — National health insurance medical record databank

There was no centralized health record databank in Taiwan before the implementation of the National Health Insurance in 1995. As of April 2003, there were 21,869,478 individuals enrolled in the NHI with a coverage rate of 96%.[5] Since 1998, the Bureau of National Health Insurance has authorized National Health Research Institutes to establish a National Health Record Databank. So far, it contains the full health records of more than 20 million Taiwanese.

D. How to create a Chinese genetic database?

In order to set up a representative Chinese genetic database, this project has chosen the statistical sampling method. The population of concern is made up of nationals of Taiwan, who have a registered census and were born before Dec. 31, 1981. This project established 6 sampling frames (see column A of table 1 below) in accordance with the administrative districts in Taiwan. Afterwards it chose 34 towns out of the 6 frames (column B of table 1 below) and 3 basic units of city administration (column C of table

---

[4] According to the official statistics, there are 429,534 aboriginals in Taiwan (2003), and the whole population is 22,520,776. See the Yearbook of Taiwan 2003-2004 (Taiwan Government).
[5] See the latest statistical data from the Bureau of National Health Insurance (Taiwan Bureau of National Health Insurance).

1 below) from each town. Finally, from every basic unit of city administration, this project will draw out the same amount of samples (see column D of table 1 below). All the samples are randomly selected using the PPS Sampling Method. They are composed of 6 different age intervals and have an equal sex ratio. There are 278 samples of each age interval and 3,336 in all (column E of table 1 below).

Frames	A Districts	B N. of Towns	C N. of Basic Units of City Administration (3 out of each town)	D N. of Persons	E Total N. of Samples in Each Frame
1	Area 1	8	24	42	1,008
2	Area 2	6	18	25	450
3	Area 3	6	18	35	630
4	Area 4	6	18	31	558
5	Area 5	6	18	30	540
6	Area 6	2	6	25	150
	Sum	34	102		3,336

Table 1: The Distribution of Samples in the Chinese Genetic Database

III The ethical, legal and social aspects of the Chinese genetic database

A. PRIVACY AND PERSONAL AUTONOMY CONCERNS

According to the official statement, the sample provider will be asked to sign an informed consent form which contains a list of questions as follows (Academia Sinica c):

a. The sample provider's age; gender; his/her record of formal schooling; medical history; habits of smoking, drinking, and betel nut chewing; and the original family home of his/her parents and grandparents. In addition, if the sample provider is older than 65, he/she will be asked to do another cognitive ability test.
b. The sample provider will be asked to undergo measurements of blood pressure, lung capacity, and body conditions such as height and weight.
c. The sample provider will be asked to draw a 23 c.c. blood specimen for further examination. In addition, the specimen will be used to extract DNA and culture cell lines for long-term genetic research.

As this shows, an enormous amount of information will be disclosed after the sample provider has signed the consent form. But the similarly important information concerning how the data will be used and protected is quite insufficient. In other words, the sample providers have no idea of where their personal, medical, and biological data will be after signing the form. Thus we believe it is not informed consent and raises even more disputes, as we will discuss in paragraph III. B.

On the other hand, the executive institute explains that it will encrypt the genetic information and the researchers can only access encoded data and will not be able to discover the name or contacting information of the sample providers. But it is still questionable if this privacy protection mechanism is enough, since there is always a safe-keeper who owns the "key" to trace the identity of biological samples or decoded genetic information. For example, if a provider changed his mind and wanted to remove his information from the genetic database, there is always a need to trace the data.[6]

B. Legal issues

Unlike Iceland, Taiwan has not legislated any specific law governing the creation of a health sector or genetic database. But it is believed that the Constitution does provide the legal basis for protection of personal autonomy and privacy (see Art. 22 of the Taiwan Constitution).[7]

According to the "Key Points of Collection and Use of Human Biological Samples for Research Purposes,"[8] the executive institute is supposed to inform the sample providers of the specific purpose of collecting the samples and how long they will be stored (see art. 4, par. 1). Researchers are also required to disclose the potential use of samples by a third party or foreign companies (see art. 4, par. 5), but the Chinese Genetic Database did not comply. However, since the "Key Points of Collection and Use of Human Biological Samples for Research Purposes" is only an administrative regulation or guidance issued by the Department of Health, Executive Yuan (the executive branch of the Republic of China) without prior authorization granted by the Legislative Yuan, the regulation may lack legal effectiveness. Therefore, it is still under dispute whether the violation of "Key Points of Collection and Use of Human Biological Samples for Research Purposes" will bring any legal problems. Furthermore, the regulation itself does not provide for any civil or criminal punishment.

C. Commercialization and benefit-sharing

The licensing policy of decoded genetic information and the benefit-sharing rule of the Chinese Genetic Database are also very unclear. For example, it has been announced that this project will cooperate with the Food Industry Research and Development Institute, and it is expected to share genetic information with local and international biotechnology and pharmaceutical companies.[9] However, the sample providers were not informed of this possibility, nor of the conditions of potential licenses. They are not endowed with any privilege to share the R&D or commercial benefits, either.

---

6 Likewise, Vilhjálmur Árnason (2004) has also discussed the differences between "unidentifiable" and "anonymous" information. Thus, genetic information in the Chinese Genetic Database shall also be defined as "unidentified for research purposes, but can be linked to their sources through the use of a code."
7 Art. 22 of the Taiwan Constitution states: "all other freedoms and rights of the people that are not detrimental to social order or public welfare shall be guaranteed under the Constitution."
8 Promulgated by the Order of Department of Health, Executive Yuan (2002).
9 See the policy declaration of the funding authority, National Science Council of the Executive Yuan (2004).

IV Present condition of this project

As discussed above, due to the serious lack of public debate on the ethical, legal, and social aspects of the creation and maintenance of a genetic database or a health sector database, over one third of the chosen sample providers cannot be contacted now, or have refused to donate a blood specimen. Some chosen sample providers complain that they did not receive sound responses to their inquiries, and some are afraid of potential harms. Furthermore, the sampling method chosen by the Chinese Genetic Database project could result in a crucial failure, since although the researchers may always try to find another equivalent sample, the final data may be biased and no longer representative.[10]

V Conclusion

In brief, as the first large-scale plan to establish a Chinese genetic database gets underway, it is regrettable that the public attention paid to the ethical, legal, and social aspects of the project is not enough at all. Compared with the Icelandic experience in the creation of a health sector database (IHD), Taiwan's attempt to conduct population genetics research gets less attention, yet it is, we believe, much more debatable.

In this paper, we have tried to argue that the consent obtained from sample providers is far from being an informed consent, especially when potential commercialization and the issue of benefit-sharing are taken into concern. Secondly, the privacy protection mechanism for genetic information should not be omitted under the guise of encryption technique. We have not been able to deal with the problem of using personal health information in the cross-matching of phenotypic, genetic, and genealogical data in this paper (see also Cheng 2003, A15). But it is similarly controversial, because the opt-out provision of the National Health Record Databank is neither sound nor practicable. Using these data without personal consent may also violate the "Act on Protection of Processing Personal Information."[11] Hence we hope this paper will draw more attention to the Genetic Database Project in Taiwan and induce more discussions both domestically and internationally.

References

Academia Sinica. a. Homepage. http://www.sinica.edu.tw/as/index.html.
Academia Sinica. b. Research results – web page 11.
    http://www.sinica.edu.tw/info/import-results/11.html
Academia Sinica c. Supercontrol, introduction. http://ncc.sinica.edu.tw/supercontrol/introduction/supercontrol.htm#2
Árnason, V. 2004. Coding and consent: Moral challenges of the database project in Iceland. *Bioethics* 18:27–49.

---

10 It is the so-called "self-selection bias" in statistics.
11 Promulgated by the Order of President Hua Chu Yi No. 5960 on August 11, 1995.

Chen, S. 1997. Observations on Taiwanese residents: Biological relationship via genetic indexes. *Hope (I Wang Tsa Chih)* 19:75–83.

Cheng, W. 2003. Humane thinking on the passionate genetic holy war. Civil Forum in *China Times* July 30.

Chu, J. 2000. Biological relationships of ethnic groups in Taiwan. *EAST ASIA International Forum*, Southern Illinois University at Edwardsville Press. Available online at: http://www.siue.edu/EASTASIA

Department of Health, Executive Yuan. 2002. Yi No. 0910012508. January 2. Chinese version available at: http://websrv.doh.gov.tw/DohGuest/OpenGuest.asp?DocID=146

National Science Council of the Executive Yuan. 2004. Policy declaration. http://www.nsc.gov.tw/head.asp?add_year=2004&tid=5.

Order of President Hua Chu Yi No. 5960 on August 11, 1995. Chinese version available at: http://www.moj.gov.tw/f6_frame.htm

Taiwan Bureau of National Health Insurance. About the program. BNIH Web site http://www.nhi.gov.tw/00english/e_index.htm

Taiwan Government. Yearbook of Taiwan 2003–2004. E-Taiwan Government web site. http://www.gov.tw/EBOOKS/TWANNUAL/

6

The Singapore human polymorphism/mutation database: Our experience with setting up a country-specific database

by Ene-Choo Tan and Marie Loh

Introduction

FOLLOWING THE COMPLETION of the sequencing of the human genome, the focus has shifted to the identification of inter-individual variations. It is believed that a high abundance of polymorphisms will provide more choices for markers and lead to more informative association studies which will facilitate the identification of genes/polymorphisms involved in common complex diseases. As a result of a few large-scale projects to identify variations in the human genome, the number of single nucleotide polymorphisms documented in the public database is now over 4 million. However, most of the polymorphisms in that database are not validated and the frequencies for specific populations are unavailable.

Population-specific differences in genetic mutations/polymorphisms are well documented (Stephens et al. 2001; Dvornyk et al. 2003; Hassan et al. 2003; Lee et al. 2003; Rosenberg et al. 2002; Tan et al. 2003). The existence of population and country-specific databases will be a very valuable resource for studying population diversity, history, and disease susceptibility. For candidate gene association studies on complex traits or phenotypes which are based on comparison of frequencies between groups, the availability of allele frequencies for specific ethnic groups in our population would also be useful for the planning and selection of informative genetic markers and estimation of sample size required to achieve sufficient power.

For Mendelian disorders, the different mutation spectrum and frequencies for the genes in question may also be different for different ethnic groups. Some diseases caused by single gene mutations are primarily found in specific ethnic groups. Thus certain groups have higher risks of certain diseases, such as cystic fibrosis in Caucasians, hemochromatosis in persons of Northern European descent, sickle cell anemia in Negroid populations, and thalassemia for persons of Mediterranean and Southeast Asian descent. The spectrum of mutation for different ethnic groups could be very different due to ancient mutations (examples are Bobadilla et al. 2002 for cystic fibrosis; de Villiers et al. 1999 for hemochromatosis; Kwok et al. 2002 for glucose-6-phosphate dehydrogenase; Waye et al. 2001 for thalassemia), hence the necessity of having a database with all the mutations which have been identified in the population here. It would be useful when designing assays and diagnostic tests to detect the presence of mutations or when confirming a diagnosis for diseases which share similar symptoms. A database containing all published mutations (and also unpublished data made available by individual researchers) identified in Singaporeans for Mendelian disorders would be most useful when designing screening or diagnostic tests.

Ethnic background

Singapore is a small island with a land area of 693 sq km (The World Factbook). It lies near the equator to the South of the Malay Peninsula. The closest neighbors are Malaysia and Indonesia. The city-state although small has a heterogeneous mix of ethnic groups. The population is approximately 4,000,000, comprising 77% Chinese, 14% Malay, 8% Indian, and 1% others (mostly Eurasians) (information from Singapore Infomap). All are descendents of immigrants. The Chinese in Singapore are mostly descendents of those who came from the two Southern provinces of China, Fujian and Kwangtung, in the 1800s and early 1900s. For the Malays, it is estimated that 60–80% originated from Java and surrounding regions. The Indians are mostly second, third, or fourth generation immigrants from various provinces of India. Approximately two thirds are Tamil-speaking and originate from the South of India and Sri Lanka. Most Tamils only marry within the Indian community in Singapore or have spouses who are sourced from a similar background and the same region in India. For the 20 to 30% of Indians who are Muslims, there is significant inter-marriage, and consequent assimilation, with the Malays (Library of Congress Country Studies). Under a scheme which offers permanent residency and citizenship to expatriates with professional qualifications and skills, most of the recent immigrants are ethnic Chinese and Indians born in China, Hong Kong, Malaysia, or the Indian subcontinent.

Data collection

The search for publications on human mutations or polymorphisms identified in Singapore during our data collection stage consisted of both electronic and manual search procedures. A search was first done at PubMed literature database to generate a list of possible publications. Two searches were performed in May 2003 using the following keywords: (i) Singapore and polymorphism and (ii) Singapore and genetic mutation. One hundred and eleven articles were found for the polymorphism search and 87 articles for the mutation search, giving a combined total of 200 articles.

The next step was to determine whether the articles found from the search were indeed useful for our database. We first looked through the articles which had direct links from PubMed and required no subscription. An online search was carried out using the Google search engine (http://www.google.com.sg/), as well as through the National University of Singapore (NUS) Digital Library (http://www.lib.nus.edu.sg/digital/). The university digital library is an online collection of electronic journals from many disciplines that the university library system currently subscribes to. Free access is available to NUS staff and students who would first need to log in with their university account. A total of 25 articles out of the 111 articles from the polymorphism search and 21 out of 89 from the mutation search were obtained, for a total of 46 out of 200 articles. However, after going through the articles, only 21 out of the 25 articles for the polymorphism search and 2 out of the 21 articles for the mutation search were found to be relevant. Hence, we were left with 23 useful publications from our internet search.

Articles were categorized as not relevant for one or more of the following reasons. Firstly, those articles might not be studies on the human species, i.e. they might be Singapore-based studies, but studies of animals or viruses. Thus, they did not fit our criteria of human polymorphism or mutation identified in Singapore or in a Singaporean population. In fact, sometimes, it could simply be the case that the word Singapore appeared in the title, abstract, or affiliation (e.g. the researcher was from a Singapore-based organization or research institute) but the study had nothing to do with polymorphisms or mutations in the Singapore population. Secondly, one essential component of our database is allele, or genotype frequencies. If the article did not have such information, then it was not useful for the database. This was especially true for the earlier papers published in the 1970s when DNA sequencing was not common. Thirdly, the article might not be on new research findings but just be a summary or review of previously published studies.

The percentage of articles from the polymorphism search which were not relevant was reasonable and not unexpected but the higher proportion for the mutation search was somewhat surprising. In fact, the main reason for rejecting an article in the mutation search was that the studies were not on human subjects. An effort was made to review the articles again, but to no avail. The word "mutation" seemed to lead to search results giving more non-human studies, whereas "polymorphism" was more often associated with human studies in the published literature. The explanation could be that new variants identified in animals were usually listed as mutations rather than polymorphisms because of the absence of additional population data from the same species to verify them as polymorphisms.

Another point to note is that only about a quarter of the publications that were listed on our PubMed searches were found from our online search. This was mainly due to two reasons: (i) although most journals do have electronic versions, in most cases paid subscription is required for access to full-text articles. Thus, we were unable to make use of the most direct, and probably most convenient way of accessing the articles without the required membership/subscription. (ii) even for the National University of Singapore (NUS) digital library which had subscription to some of the journals, those articles that were available online were the more recent ones (usually those published within the last 5–10 years) as only the more up-to-date issues were available online. The older articles were only available as printed and bound copies or only abstracts were available.

The next step was to search the university libraries. We first used the Library INtegrated Catalogue (LINC), an online search engine that enables one to search the collections of all the libraries on campus without having to physically visit each of them. From the LINC, we could check whether the journals of interest were available and in which library, as well as to request older items that were no longer on the open shelves but kept in the closed stacks. For the closed stacks items, an email would be sent to us once the journal issue had been retrieved. Via this method, another 111 articles out of the 200 were obtained, out of which 68 were found to be not relevant. Thus, another 43 useful articles were found this way. All were found in the NUS Medical Library. Of these 43 articles, 36 were from the polymorphism search (including 9 closed stacks items) and 7 from the mutation search.

The remaining articles were obtained by approaching the authors or co-authors directly. These articles were mostly those published earlier. Quite a number of the authors were current university staff or ex-staff who still maintained contact with ex-colleagues. An additional 5 relevant articles were obtained this way. In fact, another 6 articles that were not listed in the PubMed search results were found as some authors contributed their own articles that did not appear in our search results.

Next, we went through the remainder and found 10 of the abstracts from PubMed to provide sufficient information for our purpose, even though we were not able to obtain the original full articles. Lastly, we were left with 31 articles that were not available to us.

Hence, using all the above methods (both manual and computerized search and personal contact), we found a total of 89 relevant publications. The number of relevant articles obtained using the various methods as well as the respective percentages are summarized in Table 1.

Method Used	Number of publications	
Computerized search		
(PubMed subscription-free links, NUS digital library, Google search engine)	23	(25.8%)
Manual search		
NUS libraries	43	(48.3%)
Supplied by authors or co-authors	7	(7.9%)
Articles not listed in PubMed search provided by authors	6	(6.7%)
Abstracts with sufficient information used	10	(11.3%)
Total	89	

Table 1: Number of relevant articles obtained via the different methods.

Data analysis

The relevant articles that are used for our database dated from 1972 to 2003. The year of publication and the number of publications from each year are summarized in Table 2. Calendar years with no relevant publications are omitted from the table. The majority of articles are from the late 1990s. In fact, more than 50% of the articles are dated 1998 and later. This is mainly due to the rapid progress in gene technology and changing strategies in complex disease research, especially in the area of DNA sequencing and genotyping in the last few years.

Year	Number of publications
1972	1
1984	1
1985	1
1987	3
1989	4
1990	2
1991	3
1992	6
1993	2
1994	8
1995	8
1996	2
1997	2
1998	6
1999	12
2000	9
2001	6
2002	7
2003	6

Table 2: Summary of articles used in the database by year of publication.

The articles come from 48 different journals. The names of the journals, the number of publications from each and the percentage of articles contributed are listed in Table 3. The journal *Human Heredity* alone accounts for 19.1% of all articles, and the journal *Clinical Genetics*, second in ranking, accounts for only 6.74%. As shown in Figure 1, the 10 journals with the highest number of relevant articles account for approximately 50% of the publications used. Thus, slightly less than 50% of all articles come from the remaining 38 journals. This proportion is expected to increase in future as polymorphism and mutation data are likely to be published in a more diverse selection of journals (such as those catering to specialties like cardiology, asthma, obesity, nutrition, psychology, and toxicology). Hence, scanning of even more obscure journals for our data collection should still be worthwhile, so as not to miss any data which have been produced.

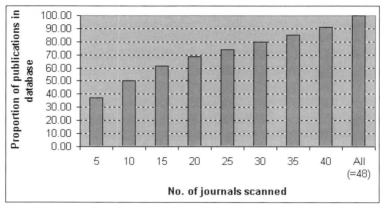

Figure 1: Histogram showing the number of journals scanned, and the proportion of publications in the database they represent. Journals are listed in terms of the number of relevant articles found. For example, the first bar shows that the 5 journals with the highest number of articles account for approximately 37% of all articles.

Journal	Number	%
Human Hered	17	19.10
Clin Genet	6	6.74
Ann Acad Med Singapore	4	4.49
Am J Hum Genet	3	3.37
Br J Cancer	3	3.37
Hum Biol	3	3.37
Pharmacogenetics	3	3.37
Allergy	2	2.25
Ann Hum Biol	2	2.25
Ann Hum Genet	2	2.25
Cancer	2	2.25
Carcinogenesis	2	2.25
Human Immunol	2	2.25
Human Genet	2	2.25
J Clin Endocrinol Metab	2	2.25
Pediatr Res	2	2.25
Am Heart J	1	1.12
Am J Hematol	1	1.12
Am J Phys Anthropol	1	1.12
Anticancer Res	1	1.12
Arterioscler Thromb	1	1.12
Arterioscler Thromb Vasc Biol	1	1.12
Autoimmunity	1	1.12

Blood Coagul Fibrinolysis	1	1.12
Cancer Epidemiol Biomarkers Prev	1	1.12
Cancer Res	1	1.12
Clin Endocrinol	1	1.12
Eur J Clin Invest	1	1.12
Eur J Clin Pharmacol	1	1.12
Eur J Hum Genet	1	1.12
Fertil Steril	1	1.12
Forensic Sci Int	1	1.12
Genet Epidemiol	1	1.12
Horm Res	1	1.12
Hum Reprod	1	1.12
Int J Cancer	1	1.12
International Journal of Cardiology	1	1.12
J Affect Disord	1	1.12
J Forensic Sci	1	1.12
J Med Genet	1	1.12
J Psychiatry Neurosci	1	1.12
J Trop Pediatr	1	1.12
Mol Psychiatry	1	1.12
Nephron	1	1.12
Neuroreport	1	1.12
Psychiatry Res	1	1.12
Schizophr Res	1	1.12
Singapore Microbiologist	1	1.12

Table 3: Summary of articles used in the database by journal of publication.

Database structure and content

The total number of polymorphisms/mutations included in the database is 231 for 91 genes. The gene with the highest number of polymorphisms/mutations in our database is p53, with a total of 33 polymorphisms/mutations. p53 is the well-known tumor-suppressor gene which plays a critical role in the development of cancer in man. The number of articles on p53 also reflects the emphasis on cancer research in Singapore. This is especially high as there were only 4 other genes with more than 10 recorded polymorphisms/mutations. As shown in Table 4, more than 70% of the genes in the database have only one reported polymorphism/mutation.

The data collected were mostly published single-base substitutions in gene regions. Most were single-base substitutions in the coding, introns, and regulatory regions such as promoters of human nuclear genes. There were also some deletions, insertions, duplications, repeat expansions, and variable number of tandem repeats (VNTR) or short

No. of Polymorphisms/Mutations	No. of genes with stated no. of polymorphisms/mutations	%
1	67	73.6
2	9	9.9
3	3	3.3
4	2	2.2
5	2	2.2
6	0	0.0
7	1	1.1
8	1	1.1
9	1	1.1
10	0	0.0
11	0	0.0
12	0	0.0
13	1	1.1
14	0	0.0
15	1	1.1
16	1	1.1
17	0	0.0
18	0	0.0
19	0	0.0
20	1	1.1
21	0	0.0
22	0	0.0
23	0	0.0
24	0	0.0
25	0	0.0
26	0	0.0
27	0	0.0
28	0	0.0
29	0	0.0
30	0	0.0
31	0	0.0
32	0	0.0
33	1	1.1

Table 4: Summary of the number of genes in terms of the number of polymorphisms/mutations.

tandem repeats (STR). The proportion of the different mutation types in the database is presented in Table 5. Single nucleotide polymorphisms (SNPs) account for almost 60% of all listed mutations/polymorphisms. The second highest is insertions/deletions. The remainder is more or less evenly distributed among the other mutation types.

Mutation Type	Number	%
Single Nucleotide Polymorphsims (SNPs)	138	59.7
Insertion/Deletion	49	21.2
Length Polymorphism	12	5.2
Variable Number of Tandem Repeats (VNTRs)	6	2.6
Short Tandem Repeat (STR) Polymorphism	2	0.9
Inversion	1	0.4
Unclassified/Unknown	23	10.0
Total	231	100.0

Table 5: Summary of the proportion of the different types of polymorphism/mutations in the database.

Another point worth noting is the proportion of disease-associated polymorphisms in the database, which accounts for 86.1% (199) of all the polymorphisms/mutations. Only 13.9% (32) are not disease-associated. The high proportion of disease-associated polymorphisms in our database could be due to the fact that, given the limited resources of a small country like Singapore, researchers are more likely to perform studies on genes that already have some data available from other countries which will be looked upon favorably by funding authorities, rather than attempting to find novel polymorphisms/mutations or to embark on non-phenotype related population studies.

Database access

The database is freely accessible to anyone via the World Wide Web at http://web.bii.a-star.edu.sg/~mariel/SHMPD/index.html. Genes are categorized alphabetically. Users navigate by clicking on the first alphabet of the gene name (Figure 2). Every gene in the Singapore human mutation/polymorphism database (SHMPD) is allocated one web page each if the data are available for the Singapore population. Population frequency data for each mutation/polymorphism is presented in a table.

Notations and database organization

Genes are categorized according to the first letter of the gene name. All the genes under each initial are listed on the same page (Figure 3). By clicking on the gene name, all the data available for reported polymorphisms/mutations for the gene are displayed on the same page. Each polymorphism/mutation of the gene is listed only once to prevent any confusion. Different articles/studies reporting on the same mutation of the gene are grouped together under that mutation. Where there is more than one study or publication on that variant, they will be listed in separate tables on the same page.

For each gene, some basic information such as what the gene codes for is provided. For each mutation, details such as disease associated, sample size, gender, and ethnic affiliation of the sample, the alleles studied, as well as their various allelic and genotypic

Figure 2: Screenshot of the homepage of the Singapore Human Mutation and Polymorphism Database.

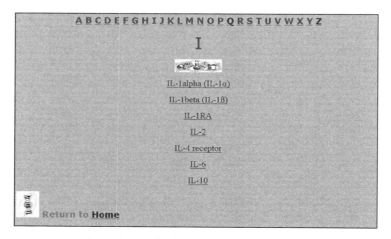

Figure 3: Sample of a web page showing the page with gene names starting with the letter "I".

frequencies are listed when available (Figure 4). Sometimes the articles might only provide this information for the entire sample set but not for each gender or ethnic group. In such cases, we would do the calculations if the information could be deduced from the data provided. All the allele frequencies are reported in 3 decimal places, and up to 1 decimal place for percentages. Values of zero are also shown if the researchers for the study indicate that they did test for the genotype or allele but it was not found. In cases where it is not clear, or if the value zero is not clearly stated, it is omitted. Various links and bookmarks are also available on each page so that one can move freely between the different genes and mutations. Apart from listing the title, journal of publication, and

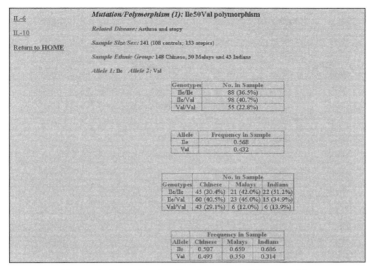

Figure 4: Screenshot showing data captured for one of the interleukin 4 receptor (IL4R) polymorphisms.

the authors of the article, direct links are also made available for all the articles to their abstracts on PubMed (except for one which is not listed by PubMed). OMIM numbers are also provided.

Application and uses

We hope that with the information in the database, genotyping efforts can be rationalized and both research and DNA-based diagnostics can be made more cost-effective, saving both time and effort spent on establishing population frequencies before going on to case-control studies. Researchers working with the ethnic group involved will be able to design their studies and estimate the required sample size based on allele frequencies available; and clinicians with suspected Mendelian cases will also have ready information on the mutational spectrum identified thus far and the laboratory or researcher which has the genotyping assay or test available.

Database submission

New information can be submitted to the database by contacting either of the authors, whose email addresses are provided on the website. We will be putting up a web form for online data submission. In addition, we are working toward auto-inclusion of data by designing search programs which would scan for relevant keywords from newly-published articles.

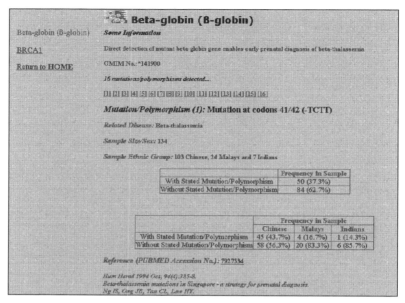

Figure 5: Screenshot showing OMIM number, citation and pubmed link with abstract number.

Conclusion

With online submission and frequent update from PubMed searches, we hope to keep the database current so it will be an important resource for academic, clinical, and commercial users. Besides Southeast Asian countries with the same ethnic background, Western countries such as the United States, Canada, and the United Kingdom which have significant numbers of immigrants of similar ethnic origin will also find the database useful when taking care of such patients.

Acknowledgements

This database was set up as an internship project in partial fulfillment of the requirement for the Master of Science Program in Bioinformatics from the National University of Singapore. The authors thank the many colleagues who have contributed data to our project, as well as users for their feedback and support.

References

Bobadilla, J. L., M. Macek Jr, J. P. Fine, and P. M. Farrell. 2002. Cystic fibrosis: a worldwide analysis of CFTR mutations–correlation with incidence data and application to screening. *Human Mutation* 19:575–606.

de Villiers, J. N., R. Hillermann, L. Loubser, and M. J. Kotze. 1999. Spectrum of mutations in the HFE gene implicated in haemochromatosis and porphyria. *Human Molecular Genetics* 8:1517–1522. Erratum in: *Human Molecular Genetics* 8:1817.

Dvornyk, V., X. H. Liu, H. Shen, S. F. Lei, L. J. Zhao, Q. R. Huang, Y. J. Qin, et al. 2003. Differentiation of Caucasians and Chinese at bone mass candidate genes: implication for ethnic difference of bone mass. *Annals of Human Genetics* 67:216–227.

Hassan, M. I., Y. Aschner, C. H. Manning, J. Xu, and J. L. Aschner. 2003. Racial differences in selected cytokine allelic and genotypic frequencies among healthy, pregnant women in North Carolina. *Cytokine* 21:10–16.

Kwok, C. J., A. C. R. Martin, S. Au, and V. M. S. Lam. 2002. G6PDdb, an integrated database of glucose-6-phosphate dehydrogenase (G6PD) mutations. *Human Mutations* 19:217–224.

Lee, J. K., H. T. Kim, S. M. Cho, K. H. Kim, H. J. Jin, G. M. Ryu, B. Oh, et al. 2003. Characterization of 458 single nucleotide polymorphisms of disease candidate genes in the Korean population. *Journal of Human Genetics* 48:213–216.

Library of Congress. Country Studies. http://memory.loc.gov/frd/cs/

PubMed. Medical Literature Database. http://www.ncbi.nlm.nih.gov/

Rosenberg, N. A., J. K. Pritchard, J. L. Weber, H. M. Cann, K. K. Kidd, L. A. Zhivotovsky, and M. W. Feldman. 2002. Genetic structure of human populations. *Science* 98:2381–2385.

Stephens, J. C., J. A. Schneider, D. A. Tanguay, J. Choi, T. Acharya, S. E. Stanley, R. Jiang, et al. 2001. Haplotype variation and linkage disequilibrium in 313 human genes. *Science* 293:489–493.

Singapore Infomap. Website. http://www.sg

Tan, E. C., C. H. Tan, U. Karupathivan, and E. P. H. Yap. 2003. Mu opioid receptor gene polymorphisms and heroin dependence in Asian populations. *Neuroreport* 14:569–572.

The World Factbook. Singapore. http://www.odci.gov/cia/publications/factbook/geos/sn.html

Waye, J. S., B. Eng, M. Patterson, L. Walker, M. D. Carcao, N. F. Olivieri, and D. H. Chui. 2001. Hemoglobin H (Hb H) disease in Canada: molecular diagnosis and review of 116 cases. *American Journal of Hematology* 68:11–15.

Consent in medical research

7

The controversy on consent in the Icelandic database case and narrow bioethics

by Sigrídur Thorgeirsdóttir

THE ICELANDIC DATABASE controversy, i.e. the controversy about the law passed in parliament in December 1998 (see Icelandic Parliament 1998) allowing the health authorities to hand over the health records of all Icelanders, alive and deceased, to a private company (deCODE) to construct the Health Sector Database (HSD),[1] got a great deal of media coverage in and outside of Iceland, since it seemed to herald a new era of large-scale bio databases. The database was to be coupled with a genealogy and a genotypic database, yielding a "Genealogy Genotype Phenotype Resource" (hereafter referred to as the Icelandic database).

At this point it is unclear whether the database as originally planned will be realized (even though the company has collected blood samples of more than a third of the population and can connect them to genealogical and some medical data).[2] The bioethical issues raised in the controversy remain, as the case poses a challenge to some key bioethical principles. The most important is the question of informed consent that was not deemed mandatory in the database legislation for the handing over of the data. Instead, presumed consent was seen as being sufficient by the lawmakers. The critics contended that this was a violation of the principle of informed consent that has been common practice for participation in, or the giving of health information to, biomedical research. The lawmaker claimed that it was not necessary to abide by the rule of informed consent in this matter, the reason being that the data in the database are supposed to be coded, and thus anonymous and non-personal.[3] The critics, on the other

---

1 See Act on Health Sector Database no. 139/1998: http://eng.heilbrigdisraduneyti.is/laws-and-regulations/nr/659.
2 The company has through cooperation with medical researchers in the public sector indirectly access to much of the phenotypic data needed: "The discoveries we are making are based on the same fundamental idea we have been using so far, i.e. to get as comprehensive information as possible about certain groups in terms of diseases and heredity." Kári Stefánsson, CEO of deCODE, in an interview with the local daily newspaper *Morgunbladid*, on November 8, 2003, p. 12. The company has also marketed a so-called "Clinical Genome Miner" that entails genealogies, genetic and health information.
3 Justification for this was found in a regulation from the Council of Europe on personal and non-personal data, according to which decoded and encrypted data are said to be non-personal data.

hand, maintained that, when linked to genotypic data and family trees, encrypted health data could not be truly anonymous in a population of just over 285,000 people.[4] So, roughly, the database seemed to pose an either-or option between those (among them doctors, bioethicists, data protectionists) who thought that informed consent was mandatory, and those who thought presumed consent was acceptable.

Problems with informed consent

In recent articles on the question of informed consent with regard to database research, serious doubts have been raised about the possibility of requiring informed consent for large-scale databases containing health and/or genetic information. With regard to the Icelandic database, the main stumbling block is the apparent impossibility of informing people adequately about the proposed research they contribute to with their biodata, prior to their consenting to it. It is difficult to foresee how pharmacogenetic, biomedical, or epidemiological database research will develop in the years to come, and therefore it is almost impossible to inform participants in a sufficient way about possible risks or benefits of future research as the principle of informed consent demands. The principle of informed consent therefore seems highly problematic or even inapplicable to secure the ethicality of participation in large-scale databank research, since it would amount to a broad, open consent.

In the Icelandic debate the philosopher Vilhjálmur Árnason has in the light of these difficulties proposed an alternative to requiring informed consent in the database legislation. He calls for an "explicit, written authorization" instead of informed consent (Árnason 2004). Such an authorization would serve, to a degree, to respect the autonomy of patients. Such autonomy is for many the core idea behind informed consent, and the idea that mere presumed consent certainly disrespects. Hjörleifur Finnsson, another Icelandic philosopher, has in a recent article, in view of the database case, criticized the notion of autonomy that is expressed in the demand for a free and non-compulsory informed consent (Finnsson 2003, 194). He claims that the emphasis on individual autonomy and free choice inherent in the principle of informed consent results in a perspective on the issue that overlooks the social and political implications of the database, i.e. what kind of science and health policy is being consented to. In sum: The more bioethics focuses on the individual, the more it loses sight of the bigger picture. Despite Finnsson's lack of belief in making informed consent the main issue for an ethical approach to this case, he criticizes deCODE for having "violated one of the main principles of bioethics," namely the very same principle of informed consent (Finnsson 2003, 191). He thus ends up with the contradiction of criticizing the principle of informed consent *and* demanding that it be respected at the same time without giving any arguments as to *why* (even if it is a problematic principle) and *how* it could be respected.

In the following, I will discuss Árnason's and Finnsson's different criticisms of the applicability of informed consent with regard to the Icelandic database case. Against Árnason, I will list arguments in favor of holding on to informed consent, and trying

---

4 See the website of Mannvernd, Association of Icelanders for Ethics in Science and Medicine, for information on the protest against the database law: http://www.mannvernd.is/english/index.html.

to adapt it to the database conditions. I however agree with Finnsson's criticism of the shortcomings of a narrow bioethical perspective on the case that does not take the social and political implications of the database in question sufficiently into consideration. Therefore, I aim to address the apparent contradiction in Finnsson's analysis, with the intention of listing arguments for why informed consent should have been demanded, and how it can be turned into an effective principle.[5] Rendering the principle of informed consent more effective, requires putting the question of consent into context. A bioethical take on the case therefore requires figuring out how individual consent can become informed not only about the specifics of the research, but also about the broader social and political implications of the kind of research intended with a population-wide database of the kind proposed in Iceland. This analysis will therefore consist in a critique of a reductionist and even naïve kind of bioethics that does not take any critical stand on the science and health, and even economic, politics inherent in the case, but only works within the confines of that science, and for lack of an independent stance never questions the world view of it. It is a kind of bioethics that views itself as mediating between the demands of scientific research and individual rights, and thus adapting to and accommodating the situation. As Christine Overall (1996, 167) writes: "Bioethicists commonly demonstrate great interest in issues of power within the patient-provider relationship, focusing on autonomy and paternalism, informed consent, and control of information; there is relatively little discussion of power inequities … within the larger social sphere."

Is informed consent inapplicable to database research?

Árnason is not alone in his skeptical attitude to the applicability of informed consent for large-scale database research. Onora O'Neill has recently put forward a forceful critique of the principle of informed consent in times of large-scale database research and genetic information. Her goal is to reject "the current tendency to suppose that informed consent procedures are the touchstone of ethically acceptable medical provisions" (O'Neill 2004, 11). In her view this principle, which was developed in the postwar period, was initially directed at ensuring that the patient consents voluntarily and in an informed manner to medical practice as spelled out in the Nuremberg Code and subsequently in the Helsinki Declaration (O'Neill 2004, 4–5). The principle of informed consent later on became instrumental in counteracting medical paternalism in the wake of increased awareness about patients' rights. In the course of that development, the notion of autonomy became increasingly integral to the idea of informed consent.[6]

The model of the doctor-patient relationship that is most often associated with the principle is one of the reasons for questioning its applicability for ensuring the ethicality of patient participation in database research. Even though the one-on-one relationship is a better context in which to inform the patient than the context in which a partici-

---

5 I do not claim that the solution I propose should be paradigmatic for other similar database ventures for they are not identical to the Icelandic one.
6 As most evidently seen in the so-called Kennedy school of bioethics. See Beauchamp and Childress 1994.

pant in databank research can be informed, there most often exists a considerable knowledge gap between the doctor and the patient about the medical facts. This deficiency of the principle of informed consent becomes much more pronounced in cases like the Icelandic database case. Added to the previously mentioned impossibility of foreseeing future research developments, large portions of the population are "actually genetically illiterate" (i.e. have little knowledge of modern genetics), as Søren Holm (2002, 84) points out. Furthermore, even those "with a high level of scientific training … are challenged by the ways in which information is now organized." O'Neill therefore shares with Finnsson a concern about an overemphasis on a person's autonomy, albeit on different grounds. She thus finds it necessary to figure out possibilities of reducing the demands placed on individual consent, and even discarding informed consent procedures altogether "whenever the goods or benefits to be provided are public goods" (O'Neill 2004). She comes to the conclusion that strict data protection and "background institutions that secure decent standards in medical and scientific practice might be used to frame and provide a warrant for the particular procedures for which individual consent was sought." (O'Neill 2001, 701–702). It has however to be emphasized that O'Neill *only* addresses databases constructed for public health purposes, and not—as in the case of the Icelandic database—for purposes of profit.

In view of the Icelandic database, Árnason comes to a similar conclusion to O'Neill's by emphasizing the importance of a science ethics committee. His idea of an explicit written authorization is thus meant to strike a balance between presumed consent deemed sufficient in the database law and informed consent. The authorization is more general and open than informed consent, and does not contain consent to any particular research project. The research protocols regarding the scientific mining of the health data are to be assessed and approved by a Research Ethics Committee that would protect the interests of every participant in the database, living and deceased (Árnason 2004, 45–46).

Whereas O'Neill wants to be realistic about agents, abilities to inform themselves, and criticizes an overemphasis on autonomy in the notion of informed consent, Árnason wants to hold on to the idea of autonomy. His idea is that a written authorization can amount to respecting people's autonomy in this case. He is certainly right that it does respect their autonomy to the degree that it guarantees that people have to give their signature as a sign of allowance to use their health data. But does the signature on the dotted line really respect people's autonomy?

The Kantian origin of the idea of autonomy implies a reasoned form of autonomy and not only a free choice; i.e., individuals are autonomous about their decisions to the extent that they are knowledgeable about and conscious of the implications of their decision or stance on a particular subject matter (O'Neill 2001, 691). The written authorization Árnason requests can hardly be seen as a sign of respect for autonomy in this Kantian sense. It therefore seems that the idea of the authorization is not really fit to protect autonomy. It rather protects a basic form of agency. Furthermore, the written authorization does not put any pressure on the database licensee to inform the patient in any detail about the use of the data. The authorization is thus one-sided compared to the idea of informed consent as a kind of a dialogical or discourse process that puts the responsibility of informing on the researcher or medical practitioner, and allows for reactions to the research plan. This, authorization however seems to be a better solution than presumed consent, as the latter does not call for any action on behalf of the citizens. It is assumed that they are in agreement with the use of their health data in the database.[7]

Presumed consent: "Moral junk"

There are indeed grave disadvantages to presumed consent in the Icelandic case. For one, presumed consent means that the data of children, the dead, and incompetent adults will be automatically included in the database.[8] Can one claim that children, incompetent adults, or the dead "presumptively consent"? Consenting requires agency, and it entails a positive response to a proposition. Despite these logical imperfections of the concept, its defenders in the Icelandic case claim that it works just as well, on account of the opt-out possibility.[9]

The conception of presumed consent has been most widely debated within bioethics in relation to organ donation. Without getting into its use in that field, it is merely important to note that many bioethicists are highly critical of this concept (Veatch and Pitt 1995; Gross 2000). Tom Beauchamp, one of the authors of *Principles of Biomedical Ethics*, says it is best to do away with it, even from the field of organ donation (Beauchamp and Childress 1994). He terms it "moral junk."[10]

What is in my view most disturbing about the application of presumed consent in the database legislation is the fact that the primary goal of the database is the commercial use of it, even though it is to be of some use to public health authorities. It is to be in the hands of a private, for-profit biotech company that plans to sell the products of the database research, be they diagnostic tools, findings of genes linked to the regulation of drug responses, etc. As Merz, McGee, and Sankar (2004) have argued, the process of developing the database and its design vary in significant ways from typical government data collection and analysis activities for public health purposes. Merz, McGee, and Sankar therefore come to the conclusion that, as a result of these differences, the database will probably serve the interests of deCODE more than it serves the interests of the public. A commercial database venture is inclined to give priority to the demands of the shareholders for profitability (see Krimsky 2003). And that

---

7   If they are not, they have the possibility of opting out, which requires that they pick up a form to fill out, and mail it off to the office of the Directorate of Health. (So far 20,000 or 7% of the population did that.)

8   The rights of families to decide about the data of deceased relatives has recently been acknowledged in a decision of the Icelandic Supreme Court (No. 151/2003) from November 27, 2003, in the case of Ragnhildur Gudmundsdóttir versus the State of Iceland. Ms. Gudmundsdóttir objected to the transfer of data belonging to her deceased father to the health record database. This case shows how the Supreme Court has set limits as to how far commercial population genomics and biotechnology can intrude into the private lives of citizens.

9   The mere fact, however, that the opt-out from the HSD has spawned a new database with dissenters kept by the Directorate of Health is one problematic result of the database law.

10 Beauchamp used this term about presumed consent in a discussion we had about the Icelandic database and the controversy on consent.

undermines the claim, Merz, McGee, and Sankar continue, that presumed consent for the collection of health data and the proprietary use of it is ethical. Informed consent therefore should have been required, since the contributors of data are rendered more vulnerable than in a government-run database in the service of public health.[11]

Does informed consent make the individual the main responsible party?

A possible counter-argument against the demand for informed consent is that it may put less pressure on the health authorities to ensure the ethicality of the database research. Individuals who give their informed consent become the ones who are mainly responsible for contributing to research (regardless of how informed that decision is). If there are any negative effects of that research, be it for groups or individuals, for example through stigmatization, it seems as if the responsibility lies predominantly with the individuals, and to a lesser extent with the health authorities. The individuals that contribute their data will, therefore, become the primary risk-takers. But risk-taking should not only calculate a possible damage, but, just as importantly, should also calculate the opportunity for benefit. For that reason, there were voices during the database debate that demanded that individuals should profit from the database as a business venture. Informed consent should thus be given in turn for a possible profit. The health authorities turned such a proposal down, deeming it unfit (for it is unclear who "owns" the health data). The fact that there was also lack of public interest in this option ensured that it was not developed beyond the stage of an idea. This one minor aspect of the controversy nevertheless sheds light on how the individualistic nature of informed consent can get tied in with neo-liberal conceptions of the individual as a risk-taker, and as a possible commercial entity selling his or her biodata.

Such a possible commercialized version of individualized consent is precisely one main reason for Finnsson's criticism of the idea of autonomy that underlies it. The option of "selling" biodata to the database-company reveals how people's health and genetic data can become a commodity. This form of individualized contribution indicates how the social and political implications of the venture in question can be completely put aside. In other words, developments in this direction would lead to far less scrutiny of the ethicality of the science in question, and the ideology and long-term implications of it. There were voices in the Icelandic debate that criticized the goals of pharmacogenetic database research, pointing at the possible negative social effects of personalized medicine it leads to. Such individualized medicine could in the worst case result in far more expensive drugs, and thereby be a factor in widening the gap between the better and the worse off and hence increase existing power inequities (Erlingsson 2002).

---

11 Non-profit population databases do not necessarily require informed consent, but it is possibly overstated to ban individual consent out of their realm altogether, like O'Neill proposes. There may be types of research undertaken for public health purposes that may be controversial and thus call for consent. Or at least the possibility of withdrawing consent (where it does not endanger the benefits for the public good) should be open.

If the institutions that are supposed to oversee the ethicality of research are strong, independent, and function well, a blindness to the long-term implications of a particular research direction should not be feared. In Iceland the National Bioethics Committee has this function with regard to the database research. The composition of this committee is not prescribed by law but spelled out in a regulation. It was therefore possible with a simple change of regulation to alter the character of the committee at the peak of the database controversy. In contrast to the older regulation, the nomination of members after this change resides exclusively within the executive branch of government. According to the previous regulation, members were nominated by independent institutions such as the Center for Ethics, the Institute of Biology at the University of Iceland, and so on. This arbitrary act, that was a sign of the strong governmental support for the database project, did, however, serve to diminish trust in the independence of the committee.

That fact should be reason enough not to put the bulk of the ethical responsibility in the Icelandic case on such a committee, but instead to try to formulate means to make the principle of informed consent work, and to think of other supportive means for informing the public that contributes its health data to the database. This should be done especially in light of the fact that the old bioethics committee was dissolved in the summer of 1999, and perhaps also because some of its members were seen to be too critical not only of the database legislation, but also of the project itself, in terms of some of the science and health politics it implies. Given this record of the strengthening of a direct executive control of the committee, Árnason's proposal has to be seen as an overestimation of the role of the ethics committee, and an underestimation of the need for a more robust form of consent than a written authorization. In addition to that, Árnason's idea of the substitute for consent is in my view too technical. The authorization entails that the patient be informed about the basics of the use of the data, i.e., about how the data will be processed and the goal of the research. Árnason considers understanding of the research necessary, since the written authorization is internal to the research procedure and includes information on how data will be combined with other databases; who will have access to data; foreseeable risks and benefits; how research will be regulated; privacy secured; and so on (Árnason 2004, 45). Such a bioethical idea of consent promotes conservatism, or in the words of Christine Overall, "the moral question in bioethics is almost always taken to be: what should be done within this given set of circumstances … [and is] chiefly occupied with establishing an ethical rationale for existing practices" (Overall 1996, 169).

The bioethical focus here is exclusively individualistic because it is only concerned with protecting the individual, in the sense that there will be no harm, such as "disrespectful treatment," done to the person (Árnason 2004, 45). This individualistic focus is necessary, but it is one-sided, since the database is a collective resource phenomenon. The individualistic perspective has to be incorporated into a broader perspective that takes the forces that shape people's lives collectively into consideration. Pharmacogenetic database research is one of the biosciences shaping our understanding of health and the future direction of health services. Seen from that angle the written authorization is a unilateral consent. It effectively says, "take me and do what you want with me!" It is like an alibi for a consent. Such a written authorization would be fine for a public health database. However, it lacks the more participatory dimension that an informed consent procedure should have in a privately run database based on public involvement.

If and how informed consent can be made effective

It is worthwhile taking a look at how another commercial biobank has recently dealt with the issue of informed consent. First Genetic Trust is a privately run biobank in Illinois, USA. According to the bank's homepage, the company is a provider of bioinformatic data to support the development and adoption of genetically-based medicines and diagnostics. The company uses information technology to support management of genotypic and phenotypic data in drug discovery. First Genetic Trust obviously sees informed consent as a trump to secure the ethicality of collecting biodata for research, but it also acknowledges the difficulty in implementing this principle effectively. For that reason it announces an "innovative solution to the growing expectations of research study participants and patients ... that their medical and genetic information be safeguarded against unauthorized use" (First Genetic Trust). This solution is called "dynamic informed consent." Dynamic informed consent allows study participants to extend and/or restrict permission regarding the use of their previously collected biological samples and medical and genetic data for follow-up and ongoing studies, as well as newly-initiated research. The participants are therefore kept informed online, and that requires that they give themselves time to keep themselves up to date and to make informed decisions about their data. The goal is not to drown people with more and more detail information, but rather to continually inform them about different research projects for which their data is being used.

This all sounds very bioethically progressive. One can, however, question whether the information given to the research subjects will be unbiased. It would be surprising if a commercial biotech company like this one were to provide their research participants online with any critical, broader discussions about possible negative social implications of genetic research. This example shows that if consent is to be well informed, it requires sources of information independent of the company, and in the Icelandic case, also independent of the governmental health authorities.[12]

---

12 Public participation, like the database is built on, requires public debate. The role of NGOs like Mannvernd, the Association of Icelanders for Ethics in Science and Medicine, in the database debate is therefore important. It is to have function similar to websites like Gene Watch. Such organizations are necessary for public debates on the goals of science. Mannvernd has been very one-sided in its critical position on the database, but it has nevertheless been instrumental in putting the debate into a broader social and political context, as manifested by the collection of articles on their homepage. The scholarly content of many articles of the website is a necessary counterpart to the more laymen oriented debate in the articles people sent to the newspapers during the height of the debate. Mannvernd has been criticized for taking a too critical position on the database, but it also has to be kept in mind that deCODE has also been relentless in promoting its own cause, at times with very high flown promises. When the financial stakes are high and opinions differ on the goals of research of this sort positions will often be extreme in both directions.

Since the database is to be based on public involvement, programs to increase genetic and biotechnological literacy should have been part of the database law. There is no requirement about ways and means to keep the public informed about the research.[13] It is vital that contributors of data are able to continually inform themselves, for the means and goals of a particular research project may conflict with personal values. People may for example want to abstain from research on alcoholism, perhaps out of fear that it could lead to a stigmatization of certain families. So they should have the opportunity of withdrawing from research projects they are in disagreement with. From the perspective of the database operations, it indeed seems more beneficial that people withdraw from specific research projects, than that they only have the opportunity of withdrawing altogether. But such an option requires the kind of interactive online dynamic consent exemplified by First Genetic Trust, although this biobank is apparently not a role model for other biobanks in any other way.[14]

The main advantage of such a consent model is that it can have an empowering effect on the individuals contributing their data to the database. The difference between the First Genetic Trust database and the Icelandic database is, however, that participation in the American database is voluntary, with the goal of the participants earning money from their data. The Icelandic database, on the other hand, is supposed to contain the health data of the whole population (minus the 7% that have so far opted out), and not all citizens can communicate online with the company. But perhaps there are other avenues than the internet to inform and involve the public whose data are being used, such as regular information brochures.

From narrow to extended bioethics

The model proposed here may seem to overburden agents. It is true that many citizens are not interested in informing themselves. So be it, but the ones who are will at least have the opportunity to do so. This model offers a regrounding of informed consent in the context of database research. It is premature to discard this venerable principle of bioethics in the Icelandic database case just because new data gathering technologies have arrived on the scene. These same technologies also give us new avenues for conveying information in order to consent or dissent. And they do not put unreasonably high hurdles in the way of research either. They can thus add to increasing biotechnological literacy. The idea of authorization, Árnason proposes, is, in light of these possibilities, too resigned, too accommodating, and does not make use of the information technologies that can foster informed consent in the form proposed here.

---

13 The strong governmental support of the company should also have been linked to a more general plan supporting biotechnology and non-commercial research. The makers of science policies should bear in mind that longstanding academic research and education in biosciences have laid the foundation for industrialized biosciences and biotechnologies, and continue to do so.

14 On the contrary, the purely commercial nature of this bank is worrisome. Winickoff and Winickoff (2003) have argued against biobanks assuming the role of broker to private-sector biobanks. Instead they advocate a charitable trust as an alternative. They argue that it has clear ethical, legal, and scientific advantages. It can accommodate and foster the altruism, good governance, and benefit to the public that are necessary for the success of such a project over the long term.

Informed consent should nevertheless not be seen as the sole guarantor of ensuring ethicality of research. Its role should not be overestimated. It is merely the private, individual side of regulating research participation. It is necessary if we are to respect the dignity, vulnerability, and self-determination of the people who contribute their data to research, regardless of how well they inform themselves about the research in question. The idea of informed consent sketched here ensures that the firm also has the duty to interact with and inform the participants. Obviously all the other mechanisms, such as an independent bioethics committees, science criticism, and NGOs concerned with the ethics of science, are indispensable contributions to an understanding of the workings, goals, and implications of science. Bioethics itself should take part in this kind of analysis of issues of the workings of science in society, and question its authority. A bioethical approach that merely restricts itself to the implementation of principles that are meant to safeguard individual rights may in the end not be able to protect individuals. Such bioethics is moral plumbing that does not aim to see the whole house. An alternative to such a narrow scope of bioethics is a more extended bioethical approach. Bioethics should, for example, take part in a debate on the developments of "medicalization" inherent in the database research, and on the identity questions posed by a database with the "genes of a nation" (see Thorgeirsdóttir 2004). Bioethics has been too narrow and primarily technical with regard to the implementation of principles, as the debate on informed consent in the Icelandic database shows. It has to incorporate debates on bioethical principles into larger questions of justice and power, and thus become more "political" in the noble sense of the term, in order to live up to its role of providing ethical guidance for protecting subjects of research.

Acknowledgements

For inspiration and help my best thanks to the staff at the Kennedy Institute of Ethics at Georgetown University and Troy Duster.

References

Árnason, V. 2004. Coding and consent: Moral challenges of the database project in Iceland. *Bioethics* 18:27–49.

Beauchamp, T. L., and J. F. Childress. 1994. *Principles of biomedical ethics*. New York: Oxford University Press.

Erlingsson, S. 2002. *Genin okkar. Líftæknin og íslenskt samfélag*. Reykjavík: Forlagid.

Finnsson, H. 2003. Af nýju lífvaldi. Líftækni, nýfrjálshyggja og lífsidfrædi. *Hugur* 15:174–196.

First Genetic Trust. Dynamic informed consent. http://www.firstgenetic.net/products_icf.html

Goss, R. M. 2000. Presumed consent further undermines medical ethics [letter]. *BMJ* 321:1023.

Holm, S. 2002. The role of informed consent in genetic experimentation. In *A companion to genethics*, ed. J. Burley and J. Harris. Malden, MA: Blackwell.

Icelandic Parliament. 1998. Act on a Health Sector Database no. 139/1998. Passed on 17 December. Official English translation available online at: http://eng.heilbrigdisraduneyti.is/laws-and-regulations/nr/659

Icelandic Supreme Court. No. 151/2003. *Ragnhildur Gudmundsdóttir v. the State of Iceland*. Available in Icelandic online at: http://www.haestirettur.is/domar?nr=2566 Available in English translation by Mannvernd online at: http://www.mannvernd.is/english/lawsuits/Icelandic_Supreme_Court_Verdict_151_2003.pdf

Krimsky, S. 2003. *Science in the private interest: Has the lure of profits corrupted biomedical research?* London: Rowman & Littlefield.

Merz, J. F., G. E. McGee, and P. Sankar. 2004. "Iceland Inc."?: On the ethics of commercial population genomics. *Social Science & Medicine* 58:1201–1209.

O'Neill, O. 2001. Informed consent and genetic information. *Studies in History and Philosophy of Biological and Biomedical Sciences* 32:689–704.

O'Neill, O. 2004. Informed consent and public health. *Philosophical Transactions of the Royal Society: Biological Sciences* 1481. Available online at http://www.journals.royalsoc.ac.uk

Overall, C. 1996. Reflections of a sceptical bioethicist. In *Philosophical perspectives on bioethics*, ed. L. W. Summer and J. Boyle, Toronto: Toronto University Press.

Thorgeirsdóttir, S. 2004. Genes of a nation. The promotion of Iceland's genetic information. *Trames* 8:178–191.

Veatch, R. M. and J. B. Pitt. 1995. The myth of presumed consent: Ethical problems in new organ procurement strategies. *Transplantation Proceedings* 27:1888–1892.

Winickoff, D., and R. N. Winickoff. 2003. The charitable trust as a model for genomic biobanks. *The New England Journal of Medicine* 349:1181–1184.

8

Toward a tiered approach to consent in biomedical research

by Peter Lucas

Introduction

BIOMEDICAL RESEARCH FACES a crisis around the idea of consent. International codes of research ethics appeal to an idea of consent originally designed to reflect the ethical exigencies of relatively small-scale non-therapeutic research, performed directly on the bodies of healthy volunteers (see for example Vollmann and Winau 1996, 1448). There is, however, a strong disparity between these sorts of studies and the forms of research associated with the new genetics, which may utilize medical, genetic, and genealogical data and/or tissue samples on a population-wide basis. How, we might reasonably ask, is the principle of consent to be honored in these and other cases?

Some flexibility in our interpretations of consent requirements is clearly called for. There are associated dangers however. Too much flexibility could lead to inadequate protection, double standards, and an increased perception that research ethics is too heavily beholden to the research community. Too little flexibility could lend support to the view that the principle of consent has become a shibboleth, which the pragmatically-minded researcher will treat as an impediment to, rather than a condition of, quality research.

I will not attempt to give a complete solution to these difficulties here. But I do want to indicate the direction in which an adequate solution seems to lie. I want to propose the abandonment of a "one size fits all" approach to consent, and its replacement with a concept of "tiered consent." "Tiered consent" implies the adaptation of consent protocols to the specifics of different types of research. At the same time, the concept of tiered consent is intended to suggest that, for all the potential variety in its application, we need not think that there must be more than one *concept* of consent involved in different cases.

The key to understanding the role of consent in research ethics is to appreciate that the research subject's motives must be, where rational (and in the absence of inappropriate incentives), altruistic.[1] It is by attending to the type and degree of altruism involved, as well as the duty not to take unfair advantage of this altruism, that we can best understand how consent requirements should operate in particular cases.

---

1   I have argued elsewhere that this applies even to so-called "therapeutic research." See Lucas forthcoming.

In this initial sketch of a tiered approach to consent I will identify three different tiers of consent, and I will relate each to a particular type of research. By envisaging more stringent forms of consent as involving the superimposition of successive tiers, it will be apparent how consent protocols can reflect the demands of disparate forms of research in a staged and transparent manner.

Consent as permission

Many types of research affect the person of the research "subject" only indirectly—research on tissue samples, and research utilizing existing medical and genetic data, are prime examples. In these types of cases, we seem entitled, if not obliged, to place the term "research subject" in quotation marks. The subjects of such studies do not serve as subjects in the way that the subjects of, say, a clinical drug trial serve as subjects. They are exposed to risks, they may accrue benefits. But these costs and benefits accrue only indirectly.

Established codes of research ethics already recognize the significance of the distinction between direct and indirect costs and benefits. The distinction is, for example, built into the familiar distinction between therapeutic and non-therapeutic research (therapeutic and non-therapeutic research may both involve direct costs to research subjects, but therapeutic research promises direct benefit, while non-therapeutic research does not). And where the nature of a research project rules out both direct costs and direct benefits to subjects, it seems reasonable to suggest that consent requirements might be modified to reflect this.

At the same time as the distinction between direct and indirect costs/benefits marks a morally significant boundary, the indirect relationship between the research subject and the potential cost/benefits of work with biomedical data and tissue samples compounds the uncertainty that is an ever-present feature of all ethically-conducted research. The consequences of involvement for the research subject may be unpredictable in practice. They may even be unpredictable in principle. In research with data and tissue samples we may well be dealing with the latter type of unpredictability. If this is so, the goal of obtaining fully informed consent from research "subjects" has to be abandoned. The barrier we encounter here is not so much practical as conceptual.

In cases in which only alienable items are going to be directly affected by the research, and where the costs/benefits to the subject may be in principle unpredictable, and are exclusively indirect, it seems sensible to construe consent primarily in terms of *permission*. To consent to serve as a "subject" in this type of research involves little more than permitting the researcher to go ahead using the samples or data in question, in much the same way as we might permit a historical researcher access to a sensitive archive. If this suggestion seems excessively liberal, I should note that it is also necessary in such cases to ensure that researchers do not take unfair advantage of (exploit) research "subjects." Meeting this latter requirement would involve writing in requirements for benefit sharing, and indemnity against possible costs.[2]

---

[2] Establishing how this could be made workable in (e.g.) the Icelandic case is evidently one for the lawyers.

It is an interesting question whether consent-as-permission would have to be unanimous among a research population, or whether majority "permission" would suffice. I don't want to pre-empt debate on this important point. However, it seems worth pointing out that in this case majority consent is not so obviously inappropriate as it would be in the case of, say, non-therapeutic research on competent adults.

Consent as assent

The next tier of consent I want to consider is probably a little more familiar. The term "assent" is currently used in medical research contexts when speaking of the agreement or cooperation of those not competent in law to give their consent (see for example Montgomery 2001, 177). Under English law, for example, children may serve as subjects in ether "therapeutic" or "non-therapeutic" research, subject to the consent of their parents or legal guardians. Legally it is permissible to involve children in research with their parents consent, even without the cooperation or agreement of the children themselves (Montgomery 2001, 178). However, researchers are advised to secure the written "assent" of children in such cases, and I take over the term from this sort of usage.

That the term "assent" is standardly used as an alternative to "informed consent" for those not considered competent to consent is a little puzzling, since genuine assent must surely always be to some degree *informed*—it hardly makes sense to talk of my "assenting" to a proposal that I did not understand. Thus, to "assent" to participation in research implies informed agreement or cooperation. This being so, what serves to differentiate "assent" from "informed consent"? So long as our focus is primarily on the information side, I would say: "little, or nothing." The uninformed assent of any potential research subject is worthless. But it is hard to see why the informed assent of an individual, even one not legally competent to consent, should not be considered tantamount to informed consent for many purposes.

The advice when dealing with vulnerable or incompetent individuals is that, assent notwithstanding, they should not be subjected to research procedures carrying more than minimal risk (a constraint which does not rule out a degree of discomfort or pain, with their informed assent; Montgomery 2001, 179). The legal situation is the cause of some obscurity here, since it is not clear how much weight is being attached to the assent of legally incompetent individuals. However, it seems significant that as things currently stand the assent of legally incompetent individuals is being construed as doing some ethical work; I would like to extend the line of thought implicit in this, and suggest that, for ethical purposes, the suitably informed assent of a research subject should be considered a sufficient condition for their inclusion in minimal risk research. (In the case of legally incompetent minors the subject's assent and the proxy consent of the parent or legal guardian should be considered jointly necessary and sufficient). "Assent," thus understood, would make up our second "tier" of consent.

Fully voluntary informed consent

I have suggested that in many minimal risk situations informed assent should be considered a sufficient form of consent, and I have implied that many individuals not legally competent to consent may be competent to give such assent. A salient question at this point is: why bother with more stringent consent requirements at all?

More stringent consent requirements are necessary because what we look for under the heading of "informed consent" is often considerably more than just informed assent. The wealth of attention paid to the question of when consent is adequately informed has unfortunately tended to obscure the fact that informed consent has other important components besides those of understanding and agreement. Ethicists since Plato have observed that though young people may excel in fields like mathematics, they tend not to do well in ethics. This is not because they lack the capacity to understand ethical arguments, but because they have not achieved the level of psychological maturity required to make suitably considered judgments in ethical matters. If we transfer this important insight to the biomedical research context, then the following picture is suggested: If all that ethics requires is that a given research subject understands what the researcher proposes to do with/to him or her, and to agree to its being done, then the subject's informed assent will surely suffice. In minimal risk cases this may indeed be all that is required, and adequate informed assent may be forthcoming even from legally incompetent individuals—there is no reason to think that informed assent should be beyond many thirteen year olds (though it may be beyond the average eight year old), for example. But when a healthy adult volunteer is asked to participate in non-therapeutic research, or when a seriously ill patient is asked to participate in a clinical trial carrying considerable risks, much more than their informed agreement is required. Fully adequate informed consent in such cases involves not just informed agreement but a distinct act of will. Participation in such cases may be very significantly altruistic. Adequate consent thus involves the subject's appreciation of and endorsement of the goals of the research (DeCastro 1998, 181). When forthcoming such consent is the product of a mature and integrated personality, and an individual who has come to a considered and responsible decision.

It betrays a very one-sided view of the research situation to think that the reason why we do not ask thirteen year olds to engage in research on a more than minimally altruistic basis is because their assent is unlikely to be forthcoming. Many thirteen year olds would be only too ready to engage in significantly altruistic acts, even with full understanding of the risks. The reason why we do not typically ask them to do so is because we do not believe that they have reached a level of psychological maturity sufficient for us to permit them to act in a significantly altruistic way. Permission to act with significant altruism is a privilege we quite properly reserve for mature adults.

Thus, while informed assent will take us through many research situations in which subjects may be exposed to discomfort and brief mild pain, where there is more than a minimal risk of serious harm we must look to subjects who in addition to having the ability to understand the procedure, and are apt to agree to participate, are able to act with sufficient maturity to voluntarily expose themselves to risk, for no obvious benefit to themselves. While there is perhaps some justification for offering a thirteen year old the opportunity to endure discomfort or even brief mild pain for altruistic reasons, there is no justification for exposing them to significant risk of serious harm. This is why we need a further tier of fully voluntary informed consent, over and above informed assent.

Conclusion

I have argued for a three-tiered concept of consent in research ethics. The lowest tier, that of permission, is appropriate to cases in which the altruistic commitment of the research subject is made in a context in which the research is practiced upon some alienated item—typically a tissue sample or medical data—and the costs and benefits to the subject accrue only indirectly. The inherent unpredictability of such procedures suggests that it is unrealistic to expect permission to be more than minimally informed, and thus traditional informed consent procedures are not appropriate. By the same token, benefit sharing and indemnity arrangements should be in place to ensure that such unpredictability does not lead to the exploitation of research subjects in any actual case.

The second tier of consent is that of assent (or "informed" assent). Here the subject is required to understand and agree to the research procedure, and assent thus presupposes and includes permission. This tier of consent is appropriate to research with vulnerable subjects, provided the required level of understanding can be achieved. However, it is only appropriate to minimal risk procedures.

The third tier of consent is voluntary informed consent. This presupposes assent, but involves in addition the requirement that the research subject understands and endorses the overall goals of the procedure, and is of sufficient psychological maturity and stability to appreciate the significance of the risks involved. At this tier exposure to significant risk is permissible, within the limits laid down within existing codes, such as the Helsinki Declaration of the WMA.

The arrangement of these tiers is dictated by the demands placed on research subjects in different contexts. In all cases altruism is a factor. In the third of the three cases a mature appreciation of the risks is a prerequisite.

The only tier at which exploitation is mentioned is the first. This is because it seems to me that this is the only sphere at which it is appropriate to combine the basic altruism of the research subject with a somewhat speculative interest on their part in the possible benefits of the research (when I act with unmixed altruism, by contrast, you do not exploit me by taking advantage of my altruism).

This paper aims at opening up discussion, rather than at producing a definitive statement. I fully anticipate that further tiers of consent will be identified.

References

DeCastro, L. 1998. Ethical issues in human experimentation. In *A companion to bioethics*, ed. H. Kuhse and P. Singer. London: Blackwell

Lucas, P. Forthcoming. Is 'therapeutic research' a misnomer? In *Arguments and analysis in bioethics*, ed. M. Hayry, T. Takala, and P. Herissone-Kelly. Amsterdam and New York: Rodopi.

Montgomery, J. 2001. Informed consent and clinical research with children. In *Informed consent in medical research*, ed. L. Doyal and J. Tobias. London: BMJ.

Vollmann, J., and R. Winau. 1996. Informed consent in human experimentation before the Nuremberg code. *BMJ* 313:1445–1447.

The wolf in sheep's clothing: Informed consent forms as commercial contracts

by Gerard Porter

Introduction

SERIOUS CONFLICTS OF interests have emerged around the commercial usage of human biological materials (Kaye and Martin 2000, Martin and Kasper 2000; also see Lo, Wolf, and Berkeley 2000). Genetic research projects are increasingly likely to have some kind of linkage with the private sector, and this may radically alter the nature of the relationship between participating physicians and subjects (Annas 2001). Under the pervasive logic of for-profit research, subjects may be perceived more as sources of potentially valuable biological materials than as fully autonomous human beings. Katz has observed that "research is not therapy," and it is clear that with these underlying tensions, the interests of the subject and the interests of researchers may not necessarily be one and the same thing (Katz 1997, 16–17).

As we move away from the familiar dynamics of traditional research models and into the new paradigm of commercial genetic research, there has been a corresponding shift in the way in which the doctrine of informed consent has come to be used. Rather than its traditional role of safeguarding the subject against potentially harmful invasive surgery or treatment, informed consent is currently being adapted to promote subject autonomy in a relatively new area—the *ex-vivo* commercial usage and propertization of human biological materials. This was perhaps not a matter of choice, but a matter of necessity based largely on the lack of viable alternatives. Given the generally unresponsive attitude of the courts toward aggrieved research subjects, the doctrine of informed consent is now generally perceived as being the sole remaining means of protecting subject autonomy in a commercial research setting.[1] It would seem that almost by default, informed consent has come to be used as the conceptual bridge between the two very different worlds of bioethics and intellectual property.

---

1 See Moore v. Regents of the University of California, 793 P.2d 479 (Cal. 1990); also see Greenberg v. Miami Children's Hospital Research Institute, Inc., No. 02-22244-CIV-MORENO (S.D. Fla., May 29, 2003) (Slip op. at 3). Both rulings have drawn intense criticism for failing to recognise subject's property rights in their own biological samples.

There are three notable points of contention which have continued to strain relations between subjects and researchers. Firstly, the question of public versus private control of intellectual property; secondly, the issue of freedom of access to research findings by other researchers; and finally, the issue of benefit sharing (Marks and Steinberg 2002). There is a general sense of dissatisfaction with current provisions in all three areas (See Potts 2002). However, rather than constructively engaging with these understandably emotive issues, the doctrine of informed consent is increasingly used defensively by researchers to avoid entering into these kinds of discussions. Informed consent forms will typically request that research subjects waive any claims to a share of the profits of commercial products developed using their biological materials, and grant researchers a free hand with regards to how they structure access to research findings and patented products. Research subjects who may object to these terms are left with few remedies. Given the emergence of this disturbing coercive trend, many scholars are re-evaluating the usefulness of informed consent as a mechanism for providing a voice for the interests of the subject within the turbulent currents of modern genetic research.

Informed consent and autonomy in genetic research

An examination of the streams of current academic debate would suggest that to the extent that informed consent provides a channel for the expression of subject autonomy, its continued usage has its proponents. In their discussion of the Icelandic Health Sector Database, Merz, McGee, and Sankar (2004) have focused on informed consent as a key issue within the project's overall ethical framework. As an opt-out scheme, the Icelandic population's data was gathered under the *modus operandi* of informed *dissent* rather than informed *consent* (Laurie 2002). This, Merz, McGee, and Sankar (2004) conclude, undermines the principle of individual autonomy:

> The importance—indeed the moral necessity—of informed consent for participation in a research project as complex and nuanced as the deCODE endeavor has been well established ... Other regional or national projects around the world are proceeding with express intent to secure informed consent from subjects (Beskow et al. 2001), and this appears to be the emerging consensus ethical approach to performing population genomics (Kaye & Martin 2000).

With regards to the observation that individual autonomy must be respected in the context of commercially oriented genetics research, few would take issue with Merz, McGee, and Sankar. But will a move toward the global standardization of the use of informed consent in genetic research projects provide a complete or even partial solution to these problems? An unconvinced Hoeyer and Lynoe (2004) have questioned the suitability of informed consent as a tool for projecting subject autonomy in the framework of commercial biomedical research. Eloquently capturing the nature of the dilemma, they remind us that in practice obtaining informed consent can quickly become a procedural *end* in itself, rather than a *means* for the expression of the autonomy of the subject in any meaningful sense:

> We fear the authors [Merz, McGee, and Sankar] might unwillingly encourage a deflation of the content of informed consent to mere ritualised routine. (Hoeyer and Lynoe 2004)[2]

The problem is not so much with the ideal of subject autonomy *per se*, but with the limitations of informed consent forms as a mechanism to achieve this objective. We lose sight of the principle of autonomy when it is filtered through bureaucratic structures which may serve to conceal the power dynamic that is at play. To put it another way, there may be cases in which the informed consent requirement has been satisfied as a procedural matter, and yet we may still be far from our goal of giving a voice to the interests of the subject within an institutional setting. As a result, the assertion that we can equate informed consent with autonomy is increasingly being called into question (see generally O'Neill 2002).

An examination of the language in which informed consent forms are drafted and the kinds of conditions under which consent is obtained suggests that there is substance to Hoeyer and Lynoe's concerns. This is particularly true with regards to the divisive issues of research access and benefit sharing. As an illustrative example, we can look to Beskow's suggested model of the language to be used on informed consent forms where profits will not be shared with subjects, presented in the following terms:

> The aim of our study is to improve the public health. Sometimes such research may result in findings or such inventions that have value if they are made or sold. We may get a patent on these. We may also license these, which would give a company the sole right to make and sell products or to offer testing based on the discovery. Some of the profits from this may be paid back to the researchers and the organisations doing this study, but you would not receive any financial benefit. (Beskow et al. 2001)

At a stroke, this single paragraph silences the subject with regards to the issues of public/private control of biomedical knowledge, exclusive versus non-exclusive licensing of patented medical technology, and benefit sharing. Many similar examples of these kinds of coercive consent forms can be found in other countries.[3] Should a subject put his or her name to this form, this may signify to the casual observer that the subject

---

2 In response to Hoeyer and Lynoe's comments, Merz, McGee, and Sankar concede the limitations of the doctrine in practice, but nevertheless stress the importance of informed consent as the only real mechanism for projecting the autonomous will of the subject into the equation: "We agree that informed consent cannot reasonably be relied upon to fix ethical problems in research. Informed consent is but the last and quite imperfect check on researcher—and in this case, State-authority and power" (Merz, McGee, and Sankar 2004, 1213).

3 For example, in Japan, Principle 17 of the Fundamental Principles of Research on the Human Genome stipulates that "in the event that an outcome obtained as a consequence of a research project becomes the subject of intellectual property rights or other rights, these property rights are not attributed to the participant." One informed consent template in Canada (see Weir and Horton 1995) suggests that when informing subjects with regards to the issue of commercial exploitation of research samples, a statement/explanation should be made that the subject has given up his/her right to share in potential commercial benefits.

has expressed his or her autonomy by waiving any claim to rights in commercial profits of products developed from his or her biological materials. Yet, despite the appearance of subject-centered decision making, in reality there is an undeniable power dynamic in operation.

These kinds of informed consent provisions embody a pattern of discourse which separates the commercial value of the body from the autonomous will of the subject (See Pottage 2002). This is achieved whilst simultaneously proclaiming informed consent's innocence by portraying the entire process as deriving from the expression of subject autonomy. To Annas, these kinds of "boiler-plate" consent provisions grant ethical legitimation to a fundamentally disempowering process, and raise serious cause for concern:

> The current system of protecting human subjects unfortunately often focuses almost exclusively on the consent form, *treating it more like a legal contract than a useful document for education and explanation* … The resulting consent form may be more likely to protect researchers and their institutions than research subjects … Consent forms can be used to protect or exploit research subjects. Lawyers who draft documents may view the institution or researcher as their client, and the research subject as an adversary. Thus, consent forms may not serve as the shield they were intended to be (Annas 2001; emphasis added).

Informed consent forms are a "wolf in sheep's clothing." The bioethical terminology which is deployed disguises the true commercial nature of the transaction. Although the *Moore* ruling seems to require researchers to disclose the fact that they have a financial interest in a patient's samples, and whereas Recital 26 of the European Biotechnology Directive states that that the subject must have had an opportunity to express his or her "free and informed consent" in accordance with national law, it would appear that in neither case are researchers under a legal obligation to provide detailed information regarding the potential commercial value of research samples, nor to acknowledge that the informed consent agreement could potentially be structured to allow benefit sharing.[4]

In reality then, informed consent forms only empower to a very limited degree. The subject has the ultimate *power not to sign* an informed consent form, i.e. to refuse to take part in a research project. However, as Laurie (2000) points out, "for one who wishes to participate, a right to refuse becomes meaningless, and so she or he is left with no power whatsoever in respect of the relationship that is to be forged with a researcher." This effectively reduces the ideal of subject autonomy to the right to agree or disagree

---

4 Moore v. Regents of the University of California, 793 P.2d 479 (Cal. 1990). Moore's claim to breach of informed consent was upheld because he had never been told of the potential use of his cells, and correspondingly he had never given his full and informed consent to the initial operation. Recital 26 of the European Directive on the Legal Protection of Biotechnological Inventions (Directive 98/44/EC) provides that:

> Whereas if an invention is based on biological material of human origin or if it uses such material, where a patent application is filed, the person from whose body the material is taken must have had an opportunity of expressing free and informed consent thereto, in accordance with national law.

with the terms and conditions dictated by researchers. What we are left with is something which resembles a standard term contract between a private company and an individual, with the two being of vastly unequal bargaining power. Thus, rather than facilitating subject-centered decision making, this particular version of informed consent may actually be seriously undermining autonomy whilst maintaining the illusion of bestowing power upon research subjects.

The persistence of these problems has bred a kind of disenchantment with the doctrine of informed consent. Have we reached a kind of crisis point? How can we revitalize the mechanism of informed consent so that it becomes a true vehicle for individual autonomy? And yet within Annas' particular conceptualization of the problem—informed consent forms as resembling contracts—lies the possible solution (Annas 2001).

Revitalizing informed consent — toward genetic freedom of contract

In the context of obtaining consent for the commercial exploitation of human biological materials, it may be useful to reconceptualize the informed consent form as being akin to a commercial contract between two autonomous parties. In the context of commercial genetic research, we are becoming significantly removed from the historically continent problems that informed consent was originally designed to solve. As subjects' biological materials begin to be viewed as valuable "raw materials" to be extracted, commercially developed, and exploited, informed consent forms start to look less like a process of bioethical documentation than they do commercial contracts.

As stated above, at the present time subjects only have the power to agree or to disagree to the terms which researchers set. The subject therefore has no direct input into the negotiations, or on the terms of benefit sharing, access, and licensing. The only possible way of revitalizing the doctrine of informed consent in this context is by allowing it to become more of a two-way exchange, rather than simply a unilateral process. Subjects and researchers should be allowed to strike a mutually acceptable agreement on the precise terms on which biological materials will be extracted and used.

The application of the traditional doctrine of informed consent means that before samples can be legally extracted from a subject's body, researchers have a legal obligation to inform the subject that they intend to secure intellectual property protection and commercially exploit the materials should they be of commercial value.[5] The subject then has to give consent to this beforehand. Although the subject only has the power to accept or not to accept the other party's contractual terms, the subject still has something the researcher wants and must consent to all terms before the researcher can legally extract and use materials. Ultimately then, the subject still holds the trump card. If we view consent *as* contract, the expression of autonomy within the setting of commercial contractual negotiations will naturally lead us towards the principle of freedom of contract.

However, we may wonder if freedom to bargain and "genetic freedom of contract" would provide an effective solution, especially given the imbalances of power in the relationship between researchers and subjects. There may be concerns that the asymme-

---

5 See note 4 above.

tries in knowledge and power would simply reproduce themselves even in informed consent forms where subjects have had the opportunity to negotiate. Contract law may give us a different approach to the problem, but it does not increase the importance or value of what you start with in terms of *bargaining power*. A shift toward a more contractual approach does, however, at least acknowledge that one has *the power to bargain*. Furthermore, an examination of the handful of cases in which individuals, families, and genetic disease associations have been proactive and consulted with lawyers to negotiate contracts with researchers would suggest that even this relatively small degree of empowerment can been used to significant effect. Anderlik and Rothstein (2003, 453) report that individuals and organizations supporting research on particular genetic disorders (e.g., PXE[6] and autism), disenchanted and disillusioned with the courts after the rulings in *Moore* and *Greenberg*[7] and wary of the dubious ethical principles of some researchers, have "begun to draft detailed contracts setting forth the rights and duties of all of the parties, including rules on access to the samples by other researchers, publication of findings, and benefit sharing of developed intellectual property." Many similar groups will draw encouragement from these successes.

But this approach may also have its limitations, especially given the great differences in kinds of research projects and the size and type of research groups they study. Freedom of contract may provide an effective solution in cases such as *Moore* and *Greenberg*, where an individual's biological characteristics are known to be commercially valuable, or a genetically homogenous group shares a specific inherited condition. How could we integrate freedom of contract in the context of a large-scale genetic database project such as the UK Biobank, which will collect data from 500,000 participants who will not know each other, and display significant heterogeneity in terms of genetic characteristics?

Yet even in cases of large-scale genetic database projects, examples exist of attempts to utilize the principles of freedom of contract to empower research subjects. For example, in Iceland an initiative was made by a group of lawyers on behalf of subjects to negotiate opting out of the Health Sector Database, and then opting back in after renegotiating mutually acceptable terms with regards to benefit sharing.[8] These kinds of negotiated agreements, far more than the continuing disingenuous use of informed consent forms, would more closely embody the ideal of autonomy by allowing donors direct input and negotiating power in the context of commercial genetic research. Free-

---

6 Pseudoxanthoma elasticum, (PXE), is an inherited disorder that affects tissue in some parts of the body. Elastic tissue in the body becomes mineralised: calcium and other minerals are deposited in this tissue. This can result in changes in the skin, eyes, cardiovascular system and gastrointestinal system. PXE International is a non-profit organisation, founded in 1995, whose mission is to initiate, fund, and conduct research; to support affected individuals and their families; and to provide resources to clinicians. Further details of the contractual arrangements made can be found on http://www.pxe.org/inthenews.html.

7 See note 1 above.

8 As reported in DV, (Icelandic daily) on February 11, 2000: Well-prepared plan from a group of lawyers. Offers the public to sell IE [the Icelandic subsidiary of deCODE] their medical records. Will assist with opting-out and negotiate just payment from IE. English translation available online at http://www.mannvernd.is/english/.

dom of contract, with the additional support of knowledgeable lawyers representing both parties, ensures the expression of individual and collective autonomy. However, we may even consider the possibility of state intervention where a dogmatic interpretation of "freedom of contract" may in fact be perpetuating inequality, exploitation, and imbalances of power.[9]

Another particular advantage of this approach is that it would not be necessary for the legislature or the judiciary to go the extent of recognizing a subject's property rights in his or her own biological materials. We would simply be applying the principle of freedom of contract in negotiations for the terms of use and access to patented products derived from subjects' biological materials—the patented products being "both factually and legally distinct" from the samples themselves.[10]

Some observers would likely suggest that a move toward the principle of freedom of contract in this field would lead to a crude form of bargaining between subjects and researchers—representing the intrusion of market principles into the clinical environment, and an unsightly "battle of the forms" between concerned parties. However, the reality of the situation is that human biological samples are already commodified and instrumentalized by current informed consent procedures. What is at stake now is simply the issue of addressing the imbalances of power within an already commercial context. Furthermore, if subjects were to view financial gain as a crude form of benefit sharing (though it appears to be a particular form of crudeness which many researchers are demonstrably eager stoically to endure), non-financial or more nuanced forms of benefit sharing could also be negotiated. This is the essence of freedom of contract. Benefit sharing and decisions regarding access to patented products are issues which should be decided as a result of negotiations between subjects and researchers, and not by researchers alone.

Conclusion

To the extent that genetic research projects require public cooperation and participation, their success will depend on their ability to earn and maintain *public trust* (Kaye and Martin 2000). This is not just a case of cynically placating dissent in order to ensure a sufficient supply of the requisite raw materials—in this case human genetic data. *Trust flows from justice*.[11] In order to permit the formation of fair and just agreements between researchers and their subjects in the context of commercial biomedical research, in-

---

9 See Laurie (2002: 77–79) discussing criticisms of the ruling in Lochner v. New York 198 US45 (1905), where the Supreme Court overturned a New York Law which sought to protect workers against exploitation by limiting the number of working hours work in any one day or any one week. The Supreme Court viewed the law as violating the principle of freedom of contract, which it viewed as being a fundamental economic right. The ruling was heavily criticised for ignoring problems of inequality of bargaining power between employer and employee and effectively abandoning the protection of individuals to the vagaries of the marketplace.
10 Moore, see note 1 above.
11 "In population studies, benefit to the population has become one of the critical issues in determining the ethical justification for the study itself, and sharing benefits with the population is critical in preventing exploitation" (Annas 2001).

formed consent should be allowed to break free from the constraints of its bioethical foundations and move toward the time-honored principle for the expression of autonomy in commercial transactions—freedom of contract.

References

Anderlik, M. R., and M. A. Rothstein. 2003. Canavan decision favors researchers over families. *Journal of Law, Medicine & Ethics* 31:450–453.

Annas, G. J. 2001. Reforming informed consent to genetic research. *JAMA* 286:2326–2328.

Beskow, L. M., W. Burke, J. F. Merz, P. A. Barr, S. Terry, V. B. Penchaszadeh, L. O. Gostin, M. Gwinn, and M. J. Khoury. 2001. Informed consent for population-based research involving genetics. *JAMA* 286:2315–2321.

Committee on Human Genome Diversity, National Research Council. 1997. *Evaluating human genetic diversity*. Washington, DC: National Academy Press.

Hoeyer, K., and N. Lynoe. 2004. Is informed consent a solution to contractual problems? A comment on the article "'Iceland Inc.'?: On the Ethics of Commercial Population Genomics" by Jon F. Merz, Glenn E. McGee, and Pamela Sankar. *Social Science & Medicine* 58:1211.

Katz, J. 1997. Human sacrifice and human experimentation: Reflections at Nuremberg. *Yale Law School Occasional Papers*, 2nd Series, 2.

Kaye, J., and P. Martin. 2000. Safeguards for research using large scale DNA collections. *BMJ* 321:1146–1149.

Laurie, G. 2002. *Genetic privacy: A challenge to medico-legal norms*. Cambridge: Cambridge University Press.

Laurie, G. 2000. (Intellectual) property? Let's think about staking a claim to our own genetic samples. Arts and Humanities Research Board, Research Centre for Studies in Intellectual Property and Technology Law. Available online at: http://www.law.ed.ac.uk/ahrb/publications/online/GLPaper.htm

Lo, B., L. E. Wolf, and A. Berkeley. 2000. Conflict-of-interest policies for investigators in clinical trials. *NEJM* 343:1616–1620.

Marks, A., and K. K. Steinberg. 2002. The ethics of access to online genetic databases: private or public? *American Journal of Pharmacogenomics* 2:207–212.

Martin, J. B., and D. L. Kasper. 2000. In whose best interest? Breaching the academic-industrial wall. *NEJM* 343:1646–1649.

Martin, P. 2000. *The industrial development of human genetic databases*, House of Lords, Science and Technology Committee Publications, United Kingdom Parliament.

Merz, J. F., G. E. McGee, and P. Sankar. 2004. "Iceland Inc."?: On the ethics of commercial population genomics. *Social Science & Medicine* 58:1201–1209.

O'Neill, O. 2002. *Autonomy and trust in bioethics*. Cambridge: Cambridge University Press.

Pottage, A. 2002. Unitas Personae: On legal and biological self-narration. *Law and Literature* 14 (2): 275–308.

Potts, J. 2002. At least give the natives glass beads: An examination of the bargain made between Iceland and deCode genetics with implications for global bioprospecting, *Virginia Journal of Law & Technology* 7:8.

Thambisetty, S. 2002. *Human genome patents and developing countries*. Study Report 10, UK Government Commission for Intellectual Property Rights.

Weir, R. F., and J. R. Horton. 1995. DNA banking and informed consent, part 2. *IRB, A Review of Human Subjects Research* 17(5–6): 1–8.

10

Gift or duty? A normative discussion for participation in human genetic databases research

by Nadja K. Kanellopoulou

Introduction

GENETIC DATABASE COLLECTIONS and initiatives are subject to considerable current debate. Questions arise as to whether existing frameworks are sufficient to protect the conflicting interests at stake. There are competing interests, raised by a number of different parties. These include individuals and groups as research participants, public health agencies, governmental sponsors, academic scientists, and commercial or industrial partners. There are a number of controversial features inherent in this kind of research, namely features that can be seen as complicating factors. These factors pose a number of regulatory challenges and open issues. A few of these are addressed in this paper, since a sense of direction and clarification is urgently needed.

For the purposes of this paper, examples of current challenges highlight the need to re-think the regulatory focus when contemplating the design, setting up, and management of genetic databases. In assessing these challenges, suggestions will arise that hopefully can offer guidance when resolving regulatory dilemmas in human genetic databases research.

Current complexities

Human genetic database projects aim to promote and facilitate health research. A number of initiatives have been proposed since 1998.[1] These projects aim to locate genes for common diseases, and usually construct large-scale population databases to include medical and health information, linked with genetic data.

The Icelandic database was designed to link health records with genealogical and genotypical information to facilitate research on genetic factors of common diseases.

---

1 At least eight "genebanks" were proposed between 1998 and 2002, starting with Iceland and then Estonia, Sweden, UK, Singapore, Canada. There is a number of regional projects such Generation Scotland and the list is likely to grow. See Austin, Harding, and McElroy 2003.

The UK Biobank aims to provide a resource that will help establish genetic and environmental factors in common diseases (e.g. cancer, cardiovascular). Each of these projects is subject to different social and legal settings that compel different approaches to resolving matters of controversy.

One approach to the issues is to strengthen mechanisms for managing informed consent and maintain confidentiality. This approach draws strength from the traditional understanding of human subjects' participation in research. Starting with the debate around the adequacy of consent as a tool and a safeguard for protection, discussions contemplate the need to review the process, or at least to re-think some of the attitudes that surround it. Provisions under debate include the right to refuse to participate in the database; the right to revoke consent or to withdraw from the study; and the possibility of ensuring that one can be re-contacted, or the option against it if further contact is not desired by the participant. The complexities of the process are rendered more significant and difficult to regulate, given that the necessary distinctions of data in anonymized, coded, and identifiable data have consequences for the level of protection of rights of individuals. It is true that in the context of research ethics, informed consent is regarded as central to the protection of human subjects.[2] Consent is based on the respect for free, fully informed choice, and promotes a highly individualistic view of respect for autonomy.

A number of discussions revolve around the inadequacy of consent to provide continuous control over information and samples given to research (Laurie 2001). There is the potential for other rights to provide additional protections. It has also been maintained, more recently, that a complex and lengthy informed consent process can be burdensome, and that complex consent provisions may well be a hindrance to research, and be irreconcilable with the nature of the research project at hand (see Beskow et al. 2001, 2317). In current contexts of genetic database initiatives, there is an increasing interest in exploring alternative frameworks that avoid complicating the already complex consent process.

More complexities revolve around adequate protections against the potential for discrimination or stigmatization. Most of these concerns have already been addressed but there exists little literature—and very little legal literature—on risks involving adverse discrimination against groups or communities. Perhaps this is the case because there exists little evidence of such discriminatory attitudes, especially since database projects are new. Concerns over securing confidentiality and privacy are addressed through clauses underpinned by data protection legislation. Instruments are in place for effective implementation, as long as the data and records in question are covered by data protection legislation. Concerns over potential commercial involvement have been expressed, but will not be addressed in this paper. Different notions of public ownership could be explored, and it would be worthwhile to examine these in a different paper altogether.[3]

---

2 Compare, for example, MRC 2001, and NBAC 1999. See also, Chadwick 2001.
3 See the proposals in the Ethics and Governance Framework of the UK Biobank (Department of Health, MRC, and the Wellcome Trust 2003).

As indicated earlier, the social contexts for database projects differ, as they differ in their settings and expectations. Comparisons can be seen in the long-standing interest in genealogy in Iceland combined with the homogeneity of the population, in contrast to the heterogeneity of the UK Biobank. It is essential to understand that these contextual differences will affect the ethics, governance, and regulation of each project accordingly.

Proposed direction

The purpose of the current outline is to offer a normative direction as to what approach may best represent the values at stake. Seen on a benefit versus risk scale, fundamental values surround the need to recognize respect for the individual participant, but equally important and noble values indicate the need to acknowledge the significance of benefits from research. A critical understanding of the risks at stake has to be put forward. In addition, it is essential to address potential risks to the benefits of all interested parties involved. Moreover, it is also essential to secure public trust and to acknowledge what the public may or may not be comfortable with. Any governance needs to closely follow and understand public views. In the case of consent, for example, studies on public attitudes highlight the importance of maintaining consent provisions, and indicate that the public may not be comfortable with a lowering of the legal standard to facilitate research (see Caulfield 2002, 577). In managing this, it seems to me essential to tailor and to contextualize risk communication in order to ensure understanding of risk for the potential research participants.

What is also critical in establishing an alternative approach, is that this approach must satisfy ends towards both directions—namely, individual interests and the general public good. This means it must allow for understanding of the individual and the social interests at stake. It can be argued that social interests could be seen as having an indirect effect on individual interests, and this applies both in the case of assessing risks and with benefits from research.

I suggest working toward an approach that adopts the not necessarily conflicting notions of a conditional gift rationale, combined with considerations of social responsibility and solidarity. According to the conditional gift idea, a gift is seen as a voluntary transfer without expectation of any return. A similar notion can be found in existing proposals in the UK, where the MRC suggested that donation of human tissue samples or material is to be construed as a gift.[4] I note that for this notion of conditional gift to operate in a way that takes account of volunteers' interests in research, it is essential that it allows for the acknowledgment that participants volunteer for a reason—perhaps with an expectation of improving the health of their family, or of their community, or of the population. By accepting this, protection needs to accommodate the conditions, stipulations, reservations, or special obligations (in a quasi-contractual rationale) that may ac-

---

4 MRC Guidelines (2001: article 2.2): "We recommend that tissue samples donated for research be treated as gifts or donations ... This is preferable from a moral and ethical point of view, as it promotes the 'gift relationship' between research participants and scientists, and underlines the altruistic motivation for participation in research."

cumulate, and thus has to be supplemented by elements that take account of these expectations, no matter how remote or general they are.

This approach is better served when combined with the values of solidarity and equity. Solidarity promotes participation in research for the benefit of others. Here, one should understand solidarity in one of two ways: either a) as a benevolent act or b) as an altruistic act, the distinction depending on whether one cares to include oneself amongst those who will receive any benefits that may accrue. This could lead to clarifications in current regulatory proposals that favor altruism. They could adopt the former so that there is room for potential participants to be inspired by the latter.

Equity encourages sharing the benefits of research. The idea of equity is based on concepts of distributive justice, namely an ideal of equal opportunity, seen perhaps as a moral duty. It draws on the suggestion that "those who are well off share with the poor," as a duty to help the less fortunate.[5]

Conclusions

The above thoughts do not consider the different current approaches as mutually exclusive. They attempt to extract what is best from them and try to build on their strengths. It has been indicated that different contexts result in different approaches. I suggest that we learn from these different approaches, and that we select the elements that best suit the regulatory circumstances relevant to our research. It was indicated that public consultation findings have displayed that adherence to informed consent is of central importance to potential research participants (see Caulfield 2002; Chadwick 2001). It would not then be wise to underestimate its value. Even if there are significant complexities in maintaining informed consent processes, the value of informed consent is more than symbolic. It is true that it is no longer a response to totalitarian state crimes against humanity. It is also true that it needs to be re-evaluated as a process to address current research problems, but this does not mean it ought to be abandoned altogether or be replaced by an entirely solidary or equalitarian counterpart. Such a swing in focus would overlook current preferences, fail to conform to international human rights standards, and give rise to new problems. What is being suggested here is that a) its role should be re-interpreted as in a conditional gift position; b) strong ethical and legal protections should be set to safeguard against misuse, abuse, and exploitation; and c) that room should be allowed for broad provisions (and/or waivers) that facilitate research progress. It is arguably not defensible to expect potential research participants to show solidarity in the absence of adequate protections against anticipated risks.

---

5 Reflections of this echo in notions of benefit-sharing, as in the HUGO proposals that pharmaceutical industries set aside a proportion of their net income to provide for humanitarian assistance and to promote healthcare infra-structure in developing countries, see the HUGO Ethics Committee's *Statement on Benefit-Sharing* (2000).

References

Austin, M. A., S. Harding, and C. McElroy. 2003. Genebanks: A comparison of eight proposed international genetic databases. *Community Genetics* 6:37–45.

Beskow, L. M., W. Burke, J. F. Merz, P. A. Barr, S. Terry; V. B. Penchaszadeh, L. O. Gostin, M. Gwinn, and M. J. Khoury. 2001. Informed consent for population-based genetic research. *JAMA* 286:2315–2321.

Caulfield, T. 2002. Gene banks and blanket consent. *Nature Reviews Genetics* 3:577.

Chadwick, R. 2001. Informed consent in genetic research. In *Informed consent in medical research*, ed. L. Doyal and J. S. Tobias, 203–210. London: BMJ Books.

Department of Health, Medical Research Council (MRC), and The Wellcome Trust. 2003. *UK Biobank ethics and governance framework*. Version 1, 24 September.

Human Genome Organization (HUGO). 2000. *Statement on benefit-sharing*. Prepared by the Ethics Committee of the Human Genome Organization. Available online at: http://www.hugo-international.org/hugo/benefit.html.

Laurie, G. T. 2001. *Genetic privacy: A challenge to medico-legal norms*. Cambridge: CUP.

Medical Research Council (MRC). 2001. *Human tissue and biological samples for use in research: Operational and ethical guidelines*. London: MRC Ethics Series. Available at: www.mrc.ac.uk.

National Bioethics Advisory Commission (NBAC). 1999. *Research involving human biological materials: Ethical issues and policy guidance*. Rockville, MD: NBAC.

Consent, biobanks, and genetic databases

11

Broad consent — the only option for population genetic databases?

by Jane Kaye

THERE IS GROWING support for broad consent (also described as open, generic, or blanket consent) to be used for population genetic databases. The World Health Organization (1998) considers that a blanket informed consent is "the most efficient and economical approach." This approach has been endorsed or recognized by a number of organizations such as the German Nationaler Ethikrat (2004), the European Commission (2004), the French Comité Consultatif National d'Ethique (2003), and the HUGO Ethics Committee (2002). Two main arguments have been used to justify the use of broad consent rather than informed consent in population genetic databases. The first argument is that it is impossible to obtain informed consent for practical reasons and therefore broad consent should be used. The second argument states there is minimal risk to participants in a population genetic database as information will be anonymous. This paper will analyze these arguments. It will argue that broad consent is only acceptable when it is combined with "opt-out" consent for secondary research use, and there are accountable, transparent security and oversight mechanisms in place. This paper lays out what such a model should look like and in doing so illustrates the limitations of relying on consent as the sole means of protecting participant interests in a population genetic database.

Concerns about broad consent

To seek broad consent runs counter to the commonly accepted principle of medical research that individuals should give their informed consent prior to participation in a research project. The principle of informed consent finds its origins in the condemnation of medical experiments conducted by Nazi physicians during the Second World War and the later Tuskegee case (Corrigan 2004). The principle is meant to protect individual autonomy and self-determination, by allowing individuals to choose whether they will be involved in medical research on the basis of sufficient information. Informed consent is most clearly articulated in the Declaration of Helsinki which requires that individuals be given specific information on "the aims, methods, anticipated benefits and potential hazards of the study and the discomfort that it may entail" (World Medical

Association 2002, article 22). The historical origin of the principle is reflected in the emphasis on providing information about a specific research project and the physical harm that may result. By not seeking informed consent, researchers run the risk of undermining the values that have been seen as a bulwark against further atrocities.

The main concern about broad consent is that researchers are not tied to one specific project, but are given a *"carte blanche"* for research which could be subject to abuse without appropriate safeguards. The rationale behind informed consent is that by providing participants with specific information on the research planned and its risks and benefits, individual autonomy is respected and both researchers and participants are aware of the boundaries of the consent. With broad consent the boundaries are expanded beyond the limits that are seen as appropriate in the Declaration of Helsinki. Another concern is that the consent should be carefully crafted so that it is not so broad that it fails to respect autonomy and self-determination. For instance, to have the "aim of increasing knowledge in order to improve health and health services" (Icelandic Parliament 1998, article 1) runs the risk of being so broad that it could be argued to make the consent meaningless, compared to the specificity required by the Declaration of Helsinki. Despite these concerns, the following arguments have been made to support the use of broad consent in population genetic databases.

Practical arguments for using broad consent

The first argument in support of broad consent rests on the fact that it is impossible to inform individuals of all of the researchers or the possible secondary research uses of the population genetic database at the time that information or DNA samples are first collected. Informed consent requires that individuals are given very detailed, specific information prior to the commencement of the research. In the case of population genetic databases this is impossible to do at the time of collection as these databases are designed to be a resource that can be used by many researchers, and for many research projects well into the future. Therefore it is argued that broad consent can be the only possible option as the requirements of informed consent cannot be met. A possible solution is to seek consent when all of the research uses become known. However this is impractical due to the difficulties of contacting all the people who have information in a population genetic database. In Iceland, the Health Sector Database would have contained information on 285,000 people, and the proposed Estonian Genome Project will have information on 1.3 million people. Some of those people may have become incapacitated, died, or have moved out of the country. If the principles of informed consent were applied strictly it would mean that at the beginning of every new research project all the people in the population genetic database would have to be contacted. This would make a population genetic database unworkable, but may also unduly inconvenience participants if they had to consent to every enquiry that was made of the database. Therefore in terms of practicalities the argument for broad consent is appropriate, even though it has the potential to undermine the reasons for seeking consent.

Minimal risk as a basis for broad consent

The second justification for using broad consent in population genetic databases is that the risk to participants is minimal, as the data will be anonymous and there will be safeguards in place to protect individual privacy (Arnardóttir, Björgvinsson, and Matthíasson 1999). While it is true that there will be safeguards in place to protect privacy, the data in a population genetic database are not anonymous. This is because there needs to be a personal identifier in order to link different types of information from different sources together. The aim is to accumulate a comprehensive body of sensitive personal information over time that can be used for multiple research projects, which will provide a profile on individuals and families in the whole population. It will also contain information on the deceased and people who become incapacitated. This profile may reveal information about an individual's health that the individual is unaware of and may have implications for other family members. The data will be directly, as well as indirectly, identifiable, simply because of the comprehensiveness of the information. The information that may be drawn from a DNA sample and the implications of this for treatment will increase as technology and knowledge improve over the life of the population genetic database. This information could have implications for health, employment, and insurance, as well as being able to be used for crime prevention. There will also be access by a number of third parties which may not be known to the individual. Many population genetic databases are being established by private companies or financed by investors which will also influence management decision making and will have implications on the way the resource is used. It is hard to sustain an argument that there will be minimal risk to participants when sensitive data is identifiable; can be linked, manipulated, and interrogated in a way that has not been possible before; accumulates over time; will be used for unforeseen research purposes; and is accessible by a number of third parties.

What should we have for population genetic databases?

Broad consent should be used at the time of collection as long as it is made clear that it is not consent for research but is consent for the inclusion of personal information and DNA samples in a population genetic database. At this time it is only possible to inform participants of the managers of the database, the kinds of information collected, and to provide a broad description of the research use and possible researchers, as well as the system for withdrawal, the security measures in place, and systems of oversight. This is not consent for research use which is central to the formulation of informed consent. The use of such sensitive and comprehensive information would usually require a very clear description of the research uses and a high degree of control by participants over its use. As discussed earlier, information on the research and the researchers planned for the population genetic database could be provided when it is known. Information on the research planned could be made available six months in advance through the internet or by a newsletter. By establishing a "cascade of information," system participants could access as much information as they wanted on research projects and then decide whether they were involved in specific research projects. If individuals failed to "opt-out" of a research project within the required time, then information would be included in a study. Although this is still not a positive consent, it goes some way to stemming some of the concerns about broad consent, by providing participants with some control over the use of information in a population genetic database. This is also in conformity with the basic tenets of the Declaration of Helsinki.

a)   *The limitations of consent*

The use of a broad consent combined with "opt-out" consent is an acknowledgement of the practical constraints of research on populations, but it only addresses the issue of respect for individual autonomy. A focus on individual consent does not acknowledge that genetic information in the population genetic database also has implications for people other than the individual. It reduces the issues around the use of the genetic database to an individual concern which does not give voice to the concerns of other family members, a group, or those of a nation, who will also be effected by the use of such information. It is also limited in its scope, as it only allows individuals to say "yes" or "no" to research projects—it does not determine the research that makes it onto the agenda. Therefore, in population genetic databases there must be other mechanisms in place in order to deal with the wider issues of information gathering at the family, group, and population level.

b)   *Involvement of participants*

Any population genetic database proposal must be accompanied by a community debate to engage the population "as partners in planning and carrying out the research and not just as research subjects" (North American Regional Committee of the Human Diversity Project 1997). Community consultation enables people to debate these issues in a meaningful and inclusive way, before a collection is established, "either through the democratic process or through the media" (Knopper 2001, 3). Being informed about the wider implications of the research allows participants to be more informed before deciding whether to give individual consent. This approach treats the population as participants in the research, and acknowledges their contribution. It also allows the public to be involved in the planning of the population genetic database and whether it should be allowed to proceed. Once the population genetic database is established, participants must be represented on the committees that approve research projects and oversee the management of the population genetic database. Such representatives would provide the litmus test for understanding when the "people in the subject population take the view that certain aspects of their lives are particularly private matters, onto which the researchers should be especially reluctant to intrude" (Capron 1991, S86).

Genetic information is at the collective level, the human genome that is shared by us all and is "considered the common heritage of humanity" (UNESCO 1997, article 1). This interest should be properly respected and protected. At the personal level the genome is unique and belongs to the individual, but is also shared by other family members, groups, and populations. Therefore, such information must be properly cared for and protected for future generations. In the case of population genetic databases there are concerns that involvement of private companies and market forces may mean that the databases can be sold as part of a company's assets. If data are accumulating over many decades, many of the donors of information will die or cease to have the capacity to consent in a meaningful way. There needs to be a structure that can protect the interests of these people, as well as the population as a whole, which can span generations and exist in perpetuity.

One such legal mechanism is that of a trust, which could hold the legal title to information and be run by trustees. All information could be held in a trust for perpetuity and the trustees overseeing the information would have obligations to individuals. The trustees would act on behalf of the people who had altruistically provided information to the population genetic database. They would be accountable to individuals but could also act as representatives for the community as a whole. The benefits of a trust are that it would acknowledge the fact that genetic information has a number of dimensions to it. Placing the control of information in the hands of the trustees who have a legal obligation to individuals also means that information will stay within the community. Such mechanisms should ensure the security of data and mean that participants will have confidence in entrusting information to a population genetic database. A trust keeps this information as a resource for the country rather than being able to be bought and sold subject to the market. It also means that ownership is retained by those who have donated information, which would at the same time respect and acknowledge their contribution.

c)   *Procedural safeguards*

There also need to be technical, procedural, and supervisory mechanisms to maintain the security of the data and public confidence in population genetic databases. If an individual is giving up intimate data for medical research they need to be certain of the trustworthiness of the institution to which they give information. There need to be mechanisms in place to ensure the security of data through encryption and security measures, and supervisory bodies with appropriate powers of enforcement to ensure that these procedures are followed. There need to be separately constituted bodies to oversee the establishment and running of the database, with clear separation of duties and responsibilities and powers of enforcement and supervision. Such a structure ensures that the handlers of the raw data, the encrypters of the data, the researchers, the enforcement bodies, and the management of the population genetic database belong to different organizations. This transparency combined with very tight technical computer procedures and requirements builds public confidence in the population genetic database. The managers of the population genetic database need to "offer individuals simple and realistic ways of checking what they consent to is indeed what happens, and what they do not consent to does not happen" (O'Neill 2001, 702). Such procedures ensure that privacy can be maintained, and while they are not a substitute for allowing individuals choice about their private matters, they make it possible for information to be used in the public interest without infringing privacy. These mechanisms must be able to engender trust and confidence as well as protect the privacy interests of the participants involved in the population genetic database.

In conclusion

The debate over the appropriate consent for population genetic databases has become polarized between the extremes of informed consent and broad consent without considering other alternatives. This paper proposes an alternative model that attempts to respect individual autonomy, but also recognizes that there are other interests that need to be considered in genetic research involving populations. In the world of genomics, advances in technology can mean rapid changes in the information that can be derived from a DNA sample and what it can tell us about disease susceptibility. This research can have implications for participants in terms of insurance, employment, and family relations. Therefore it is not possible to hand a DNA sample or access to personal information over to researchers in perpetuity, without allowing participants some control over what happens to the sample or the accumulated personal information in an unpredictable future. This paper has presented an alternative way to recognize individual as well as family and community rights over information in a population genetic database, and offers an alternative to the currently polarized debate.

Acknowledgements

This paper was produced as a part of the ELSAGEN project (Ethical, Legal and Social Aspects of Human Genetic Databases: A European Comparison), financed between 2002 and 2004 by the European Commission's 5th Framework Programme, Quality of Life (contract number QLG6-CT-2001-00062). I gratefully acknowledge the support of the European Community. The information provided is the sole responsibility of the author; the Community is not responsible for any use that might be made of data appearing in this publication.

References

Arnardóttir O. M., D. Th. Björgvinsson, and V. M. Matthíasson. 1999. The Icelandic Health Sector Database. *European Journal of Health Law* 6:307–362.

Capron, A. M. 1991. Protection of research subjects: Do special rules apply in epidemiology? *Journal of Clinical Epidemiology* 44:S81–S89.

Corrigan, O. 2004. Informed consent: The contradictory ethical safeguards in pharmacogenetics. In *Genetic databases: Socio-ethical issues in the collection and use of DNA*, ed. R. Tutton and O. Corrigan. London: Routledge.

European Commission. 2004. Ethical, legal and social aspects of genetic testing: Research, development and clinical applications. Brussels: European Commission.

French Comité Consultatif National d'Ethique. 2003. Ethical issues raised by collections of biological and associated information data: "biobanks," "biolibraries." Opinion No.77. Paris: French Comité Consultatif National d'Ethique.

Human Genome Organization (HUGO). 2002. Statement on Human Genomic Databases. Prepared by the Ethics Committee of the Human Genome Organization.

Icelandic Parliament. 1998. Act on a Health Sector Database no. 139/1998. Passed on 17 December. Official English translation available online at: http://eng.heilbrigdisraduneyti.is/laws-and-regulations/nr/659

Knopper, B. M. 2001. Of population, genetics and banks. *Genetics Law Monitor* Jan/Feb: 3–6.

Nationaler Ethikrat. 2004. Biobanks for research: Opinion. Berlin: Nationaler Ethikrat.

North American Regional Committee of the Human Diversity Project. 1997. Proposed model ethical protocol for collecting DNA samples. *Houston Law Review* 33:1431–1473.

O'Neill, O. 2001. Informed Consent and Genetic Information. *Studies in the History and Philosophy of Biological and Biomedical Sciences* 32:689–704.

UNESCO. 1997. Universal Declaration on the Human Genome and Human Rights. Adopted on 11 November. Available online at: http://www.unesco.org/shs/human_rights/hrbc.htm

World Health Organization (WHO). 1998. *Proposed International Guidelines on Ethical Issues in Medical Genetics and Genetics Services*. Geneva: World Health Organization, Human Genetics Programme.

World Medical Association. 2002. *The Declaration of Helsinki*. Available online at: http://www.wma.net/e/policy/b3.htm

12

Databases and informed consent: Can broad consent legitimate research?

by Sigurdur Kristinsson

1. Introduction

THE ADVENT OF research databases with genetic and health information confronts us with new questions about informed consent. Informed consent to participation in research is conventionally construed as a potential subject's explicit agreement to participate in a specific research project, based on information about its goals, procedures, foreseeable risks, possible benefits, the confidentiality of data, and the right to withdraw (see, for example, WMA 2002a, article 22; and CIOMS 2002, guideline 5). These requirements do not square well with database research. As data are entered into an enduring database, their future research uses cannot be disclosed insofar as they have not yet been determined. Apparently, then, informed consent in the conventional sense requires that contributors of personal data be contacted again for every new research project that makes use of the data. This is impractical for many reasons. Apart from the expense involved, donors may be dead or difficult to locate (see O'Neill 2003, 5; and Greely 1999, 740). Even if they are not, they may neglect to respond for irrelevant reasons. In fact, being re-contacted for the sake of every new project might create a nuisance that makes people want not to have their data stored in research databases. The practice of informed consent could thus deter people from participating by creating a new "risk" of future nuisance. The UK Human Genetics Commission states that "many patients very explicitly do not wish to be re-contacted for such consent."[1] In the same report, the commission claims that practical difficulties in securing re-consent for every new project "would seriously limit the usefulness of large-scale population databases" (HGC 2002, paragraph 5, 17).

One way to avoid these problems is to make the data anonymous by unlinking them permanently from any identifying information. Even if contributors are still owed some consideration as the original sources of the data, that consideration need not include

---

[1] Human Genetics Commission 2002, paragraph 5,20. Caulfield, Upshur, and Daar (2003) similarly state that research involving DNA data banks "may require multiple requests for consent, thus burdening both researcher and participant, which was noted on a recent public consultation to be a potential disincentive to participation."

informed consent if the data are irretrievably anonymized.[2] However, this way out of the problem has its costs. It may handicap the database by making it useless for all study designs where original contributors are contacted again for further information. It also limits possibilities for linking the data with databases that are non-anonymous, or linking two or more databases that are anonymous if isolated, but non-anonymous if combined.[3]

Another way out of the problem, and the one I will be concerned with here, is to broaden the scope of the initial consent. In addition to being informed about the aims and methods of the particular study at hand, the potential subject is asked to consent to her information being used in future studies falling under some broad characterization, such as studies of the same disease or studies of related diseases.[4] This practice tests the limits of the concept of informed consent, however, because part of the information that is conventionally required is not provided for these future studies. Should this type of broad, as opposed to narrow, consent nevertheless count as informed consent?

Ultimately, what matters is not this conceptual question as such, but rather the normative question of whether broad consent legitimates future research without re-contact. To answer that question, we need to consider the role informed consent plays in legitimating research, and try to determine whether this role is seriously compromised by the practice of broad consent. In what follows, I will take three small steps toward this goal. First, I argue that the mere fact that broad consent is at odds with conventional requirements does not rule out that it might serve the legitimating purpose of informed consent. Second, I argue that what has been referred to as the "referential opacity" of consent (see O'Neill 2001, 692; and 2002, 42–44) does not count against broadening its scope from what convention dictates. Finally, I argue that whether broad consent

---

2 Informed consent is not normally sought for anonymous research, but some argue that it ought to be. For example, Greely (2001, 224–25) argues that anonymization is not a sufficient reason for omitting informed consent because even if data is anonymous, it is "a harm if people who did not agree to be research subjects are unhappy to learn that their medical records—and their genomes—have been used for research, research of which they might, or might not, approve." However, this argument begs the question of whether use of anonymous health data turns patients into research subjects. On that issue, the WMA Declaration of Helsinki (2002a, article 1), states that medical research includes research on identifiable data. While this wording does not logically rule out that research on anonymous data might also be included, it suggests that it should not (see also WMA 2002b, article 24 and 25). Similarly, the UNESCO International Declaration on Human Genetic Data (2003, article 16b) states that consent is not necessary when data is "irretrievably unlinked to an identifiable person." One reason to resist Greely's claim is that much epidemiological research would be impracticable if individual informed consent had to be sought. Another reason is that a written consent might constitute the only record linking a particular person to the data, thus creating an independent privacy risk. For an overview of the issue and some of the literature, see Nomper (ch. 13 in this volume).
3 For a discussion of further problems with viewing anonymization as a "panacea," see Greely 1999, 759–762.
4 See Árnason 2004, 34–35, for a description of how deCODE has utilized this option of broad consent in the collection of biosamples and disease data in Iceland. Proposals for adjusting regulatory frameworks so as to make room for broad consent (or "general permission") have been made by Greely 1999, 752–756, and Caulfield et al. 2003.

legitimates depends in each case on the relevance and foreseeability of the specific facts that are left out of the broad description, to which the potential subjects consent. When the omitted facts are either irrelevant or reasonably foreseeable, the mere fact that consent is broad rather than narrow does not detract from its ability to legitimate research.

2. Rejecting the argument from convention

Most commentators see five elements as essential to a definition of informed consent. These are (1) competence, (2) disclosure, (3) understanding, (4) voluntariness, and (5) consent.[5] A person who satisfies all five conditions can be said to have given his or her *genuine* informed consent to participation in research. By this I mean that the potential subject's informed consent in fact confers moral legitimacy on the subsequent use of her contribution, in the way that informed consent should. In other words, saying that someone has given genuine informed consent means that whatever it is that makes informed consent morally relevant to legitimating research has in fact occurred.

It might be argued that since broad consent violates some conventional requirements for informed consent, it does not fall under the concept and therefore cannot play its legitimating role. However, this argument would mistakenly assume that conventional requirements for informed consent determine the meaning of genuine informed consent. At most, conventional requirements define what counts as legally or institutionally effective consent.[6] In this latter sense, informed consent is relative to prevailing rules, laws, and regulations, and these are variable across time and place. Effective consent is no guarantee of genuine informed consent, however. A potential subject may give all the required signatures, be deemed competent by the appropriate parties, and be of legal age, without for example having in fact adequately understood the necessary information. Neither is genuine informed consent guaranteed by the fact that researchers have performed all the duties that codes and guidelines impose on them. Effective consent is thus possible without genuine consent, even though the most important purpose of the rules that define effective consent, and of informed consent clauses in ethical codes and guidelines, is presumably that of promoting genuine informed consent, or perhaps maximizing the chances of it occurring (see Faden and Beauchamp 1986, 285).

Conversely, it is possible to imagine someone giving genuine informed consent without satisfying the formal conditions for effective consent, and also without aid from researchers' dutifully performing their role. All we need to imagine is someone who, for whatever reason, already understands all the relevant information and is not coerced or manipulated into consenting.

---

5 Beauchamp and Childress 2001, 79. Arguably, the element of disclosure should be omitted if the idea is to list logically necessary conditions for informed consent. As a practical matter, however, disclosure is normally necessary in order for understanding to occur.

6 Faden and Beauchamp 1986, 277–283, similarly distinguish between informed consent as autonomous authorization and informed consent as effective consent. While I accept their characterization of effective consent, I leave it open whether genuine consent is best analyzed as autonomous authorization.

Clearly, then, the concept of genuine informed consent cannot be read directly from codes or guidelines, no matter how well crafted. What they provide is evidence of the conception of genuine informed consent that has in fact guided efforts to institutionalize it. In particular, they tell us how people have interpreted the scope of the disclosure element, i.e. what should be regarded as necessary information for the subject to understand. In the CIOMS 2002 guidelines, for example, this takes the form of a list of no fewer than 26 specific items. Codes similarly tell us something about how the notions of voluntariness, competence, and understanding have been interpreted. Still, even the best of recommendations do not determine the meaning of genuine informed consent.

Two implications emerge from these considerations. First, we must distinguish between inquiry into what genuine informed consent is and inquiry into the procedures and duties described under this heading in codes and other guidance documents. Second, when inquiring into what genuine informed consent is, we must regard prescribed duties merely as tentative evidence of the conditions that contribute to genuine informed consent. Keeping these two points in mind, let us turn to the question of whether genuine informed consent needs to be narrow, as opposed to broad.

3. *The argument from the opacity of propositional attitudes*

Consent is a particular kind of propositional attitude, distinct from e.g. believing, wanting, and intending (see O'Neill 2003, 5–6; 2002, 43; and 2001, 692). It may be construed generally as agreeing that what a proposition describes should happen. As Onora O'Neill has pointed out (2001, 692), however, "consent will not automatically transfer ... to the logical implications, the causal consequences or the more specific aspects of a proposal, situation or action towards which consent has been accurately directed." It might be argued that this referential opacity prevents consent to a general proposition from legitimating action described by a more specific one.[7]

The referential opacity of propositional attitudes may be explained as follows. Propositions can describe the same state of affairs at different levels of generality and specificity. For example, "X is driving a motor vehicle" is more general than "X is driving a BMW," while both are true of the same state of affairs. The more specific proposition entails the more generic one. Nevertheless, a particular attitude to the specific proposition does not automatically carry over to the generic one. I may believe that X is driving a BMW but not know that this is a form of driving a motor vehicle. Or, to take a relevant real-world example, I may believe that an organ is being removed but not know that this is a form of tissue being removed, and consequently not believe the more generic proposition, "tissue is being removed" (see O'Neill 2003, 6). Conversely, an attitude to the generic proposition does not necessarily carry over to the specific one. I may

---

7 In fairness to O'Neill (2001, 2002, 2003), it should be noted that her purpose in introducing the notion of referential opacity here is not that of arguing against broad consent. Instead, she uses it to warn against excessive reliance on—and blind faith in—informed consent as a safeguard for the protection of patients and research subjects.

believe X is driving a motor vehicle but not that X is driving a BMW. I may believe tissue is being removed but not that a whole organ is being removed.

If we substitute consenting for believing we see that it is possible to consent to a specific proposition without thereby consenting to the generic one that it entails, and vice versa. I might consent to "x will drive a BMW" and not to "x will drive a motor vehicle," since I might not know that a BMW is a motor vehicle. I might also consent to "x will drive a motor vehicle" without thereby consenting to "x will drive a BMW," since I might falsely believe that no motor vehicle is a BMW, or I might approve of x's driving motor vehicles as long as they are not BMWs. Similarly, consenting to the generic proposition that my data will be used in future cancer research does not mean that I have thereby consented to having it used in the specific cancer research project that is eventually launched, because the full description of that research project might be incompatible with what I had understood to be included under the general phrase "future cancer research." I might approve of (and consent to) the fact that cancer is being researched, but disapprove of some other aspect of the research project, and this disapproval might bring about an overall judgment of not wishing to contribute to it. So, the argument would go, since consenting to the generic proposition does not include consenting to the specific one, it does not legitimate the research.

4. *Rejecting the argument from opacity*

If successful, the argument from opacity would mean that broad, prior consent could never be used to justify proceeding without renewed consent in research using stored data. Fortunately for the parties involved, the argument is ultimately not convincing. First, even when propositions describing research are made specific enough to satisfy conventional requirements, they could easily be specified even further, perhaps endlessly. The project of providing every description that could possibly be relevant to the subject's decision to participate is always incomplete.[8] The demand that every relevant description be explicitly stated would therefore not only condemn generic consent, but conventional consent as well. If followed, the demand could also undermine genuine consent by overwhelming potential subjects with enormously detailed descriptions (O'Neill 2003, 5). Conventional informed consent procedures do not follow this demand; if they did, they would hurt their chances of bringing about genuine informed consent.

Second, the reason why conventional consent in fact manages to legitimate research, despite its necessary incompleteness, is that the minimally competent consenter can be reasonably expected to know certain things and be able to make certain inferences. Consent to a generic proposition is normally accompanied by awareness that the property it describes at a relatively abstract level could realistically take various concrete forms. When the concrete instantiation that actually emerges falls within the range of specific forms that could be reasonably expected, consent to the generic proposition can be

---

[8] As O'Neill (2003, 5) points out, "the quest for perfect specificity is doomed to fail since descriptions can be expanded endlessly, and there is no limit to a process of seeking more specific consent."

taken to *imply* consent to the specific one, even though it doesn't *constitute* such consent. For example, consent to participation in future projects aiming to study cancer may be taken to imply consent to participation in a study of liver cancer, since it can be reasonably expected that a study of cancer might focus on liver cancer. The issue is ultimately no different from that of determining which causal outcomes are "covered" by informed consent, even if they are not mentioned specifically to the consenter. In both cases, the answer depends on whether the consenter ought to have known that the actual outcome could reasonably be expected.

Reasonable expectations depend on circumstances and determine how far implied consent extends. While it is true that consenting to "x will drive a motor vehicle" does not include consenting to the specific fact that the motor vehicle turns out to be a BMW —or that the drive turns out to occur on a Sunday, for that matter—the minimally competent consenter can reasonably be expected to know that what she is consenting to might take this particular form. By contrast, it would not have been reasonable to expect her to assume that it might take the form of x driving a stolen BMW at twice the speed limit! The original consent did therefore not extend by implication to all possible instantiations of x driving a motor vehicle, but only to those that were consistent with reasonable assumptions circumscribing the consent. Similarly, broad consent to participate in future studies on cancer may imply consent to the specific focus on liver cancer, while not implying consent to, say, a specific policy on whether individual results will be returned. Instead of arguing that broad consent never legitimates, we should consider the assumptions embedded in each particular context of consent. These assumptions determine the extent of the potential subject's implied consent.

5. *Foreseeability and relevance*

Despite the logical point that consent is opaque, then, broad consent can in principle legitimate unspecified research projects. Whether it does in fact depends in each case on two things. First, it depends on whether the potential subjects could reasonably have been expected to know, at the time of consent, that the later studies might take the specific forms they ended up taking. Second, the legitimating power of broad consent depends on whether specific aspects of the later studies, that the consenting subject could *not* be expected to foresee, are reasonably regarded as relevant to the subject's decision as to whether or not to participate. If all the specific information missing from the original consent is either irrelevant or foreseeable, the original consent seems to legitimate the research. Conversely, if the actual research has aspects that are both relevant and unforeseeable, the original consent seems inadequate. The legitimating power of the original consent thus crucially depends on the relevance and foreseeability of facts not described at the original consent.

It might be objected that even if all this is accepted, the problem remains that judgments of relevance and foreseeability cannot be made without re-contacting the potential subjects. For example, whether any given description is relevant to a potential subject's decision of whether or not to participate seems relative to her particular aims and concerns. Knowing what these are seems hard without actually getting in touch with her and perhaps engaging in extended conversations with her. So even if her original, broad consent might possibly legitimate the use of her data in a new, concrete research project, we cannot know whether this is so in fact without asking her whether she has a problem with any of the new details. As a result, the practical advantages of broad consent disappear.

This objection can be answered by pointing out that conventionally, the practice of informed consent is not taken to require that the researchers' disclosure of information be tailored to each potential subject's aims and concerns. It is normally accepted that researchers, instructed by codes and guided by ethics committees, will make general judgments concerning which descriptions can reasonably be considered relevant and should therefore be disclosed to every potential subject. In the conventional setting, this judgment is made before the act of disclosure, whether it is made by the researcher who is putting together an informed consent form, by the framers of codes and guidelines deciding what type of information should always be included on such forms, or by ethics committees reviewing research protocols. By contrast, the approval of a new research project, using data that have already been contributed based on broad consent, requires a retrospective judgment of whether the new research project has non-foreseeable features that are reasonably relevant to whether the individuals who originally contributed the data would agree to participate. The mere timing of the judgment does not seem crucial, as long as it is made before the actual research takes place. In both cases, the judgment concerns the morally relevant issue of whether the subjects in research that is about to take place will have been exposed to all reasonably relevant information before consenting to participation.

Time can matter in at least two ways, though. First, the foreseeability of a particular specification of a general description may depend on background knowledge that was not common at the time of original consent, but has become common knowledge when the research project is eventually launched. For example, suppose people generally did not assume, at the time of the original consent, that the research they were consenting to might be carried out by a commercial company rather than a university or research institution. Suppose also that in the time that has elapsed since, it has become common knowledge that some of the researchers using the data work for commercial companies. Assuming that this is a relevant fact, should it be considered foreseeable? It seems not. We cannot assume that the original consent implied consent to research done by commercial companies, because this was not a possibility that people could reasonably be expected to be aware of at the time.

This does not necessarily mean that the subjects need to be re-contacted, however, because their current consent, to participation in research done by commercial companies, might be *inferred* from their current behavior. For example, if reliable measures were taken to make sure every subject now has easy access to the relevant, new information, together with a quick, convenient, and confidential way of opting out, their not opting out could be interpreted as a sign of current consent. Of course, this does not apply to those subjects who are no longer competent to consent, including those who have died. Many of the problems discussed in connection with presumed consent thus raise their ugly head here as well.

The second way in which time can matter is that individuals may change over time as regards their attitudes toward participation. If a significant amount of time lapses from the original consent to the commencement of the research, it seems legitimate to ask whether the original consent is still binding, even if we assume that no unforeseeable, relevant facts have emerged. The answer to this question is not clear, and for that reason, it would seem respectful to offer the subjects an easy way to opt out based on regularly provided, unbiased information about upcoming projects and the results of past projects. This could enable participants to protect themselves against the decisions of their past selves.

6. Concluding remarks

Informed consent is a normative concept, essentially informed by assumptions about what makes it morally relevant. We are in fact making a normative judgment when we say that genuine consent need not be explicit consent to all relevant descriptions, but can also include implied consent to relevant descriptions that could reasonably be expected to become true. We are also making a normative judgment when we say that original consenters do not need to be consulted individually before a judgment is made, concerning whether specific facts about a new research project are relevant to their decision to participate. And we are of course making a normative judgment when we say that non-anonymous data from individuals should not be used unless it is reasonably likely that they currently approve of the way in which it is being used. In general, the contours of the concept of genuine informed consent are shaped by the normative framework within which the concept is being used, and which define its moral purpose. A critical examination of the relevant frameworks would strengthen whatever conclusions we come to concerning the moral status of broad consent, but that is a topic for another occasion. For now, let us simply say that if we believe in the moral significance of informed consent, we should be prepared to accept broad consent as morally significant as well, in cases where it fulfills requirements of relevance and foreseeability.

Acknowledgements

This paper is a product of the ELSAGEN project (Ethical, Legal and Social Aspects of Human Genetic Databases: A European Comparison), financed 2002–2004 by the European Commission's 5th Framework Programme, Quality of Life (contract number QLG6-CT-2001-00062). Research for this paper was also aided by participation in workshops funded by the Nordic Academy for Advanced Study (NorFA), through the NorFA Network "The Ethics of Genetic and Medical Information." I gratefully acknowledge the support of the European Community and NorFA. I also wish to thank fellow-participants in the ELSAGEN project for useful comments on an earlier version of this paper.

References

Árnason, V. 2004. Coding and consent: Moral challenges of the database project in Iceland. *Bioethics* 18:27–49.

Beauchamp, T., and J. Childress. 2001. *Principles of biomedical ethics*, 5th ed. New York: Oxford University Press.

Caulfield, T., R. Upshur, and A. Daar. 2003. DNA databanks and consent: A suggested policy option involving an authorization model. *BMC Medical Ethics* 4:1. Available online at: http://www.biomedcentral.com/1472-6939/4/1

Council for International Organizations of Medical Sciences (CIOMS). 2002. *International ethical guidelines for biomedical research involving human subjects*. Geneva: CIOMS and the World Health Organization (WHO). Available online at: http://www.cioms.ch/frame_guidelines_nov_2002.htm

Faden, R., and T. Beauchamp. 1986. *A history and theory of informed consent*. New York: Oxford University Press.

Greely, H. 1999. Breaking the stalemate: A prospective regulatory framework for unforeseen research uses of human tissue samples and health information. *Wake Forest Law Review* 34:737–766. Available online at: http://www.law.wfu.edu/prebuilt/LR_v34n3_Greely.pdf

Greely, H. 2001. Human genetic research: New challenges for research ethics. *Perspectives in Biology and Medicine* 44:221–29.

Human Genetics Commission (HGC). 2002. *Inside information: Balancing interests in the use of personal genetic data*. Report from the Human Genetics Commission. Available online at: http://www.hgc.gov.uk/insideinformation/index.htm

O'Neill, O. 2001. Informed consent and genetic information. *Studies in History and Philosophy of Biological and Biomedical Sciences* 32:689–704.

O'Neill, O. 2002. *Autonomy and trust in bioethics*. Cambridge: Cambridge University Press.

O'Neill, O. 2003. Some limits of informed consent. *Journal of Medical Ethics* 29:4–7.

UNESCO. 2003. *International declaration on human genetic data*. Adopted on 16 October. Available online at: http://portal.unesco.org/en/ev.php-URL_ID=17720&URL_DO=DO_TOPIC&URL_SECTION=201.html

World Medical Association (WMA). 2002a. *The Declaration of Helsinki*. Available online at: http://www.wma.net/e/policy/b3.htm

World Medical Association (WMA). 2002b. *Declaration on ethical considerations regarding health databases*. Available online at: http://www.wma.net/e/policy/d1.htm

13

What is wrong with using anonymized data and tissue for research purposes?

by Ants Nomper

ONE OF THE tasks of social sciences is to critically assess applicable norms and practices, and, if necessary, take steps to introduce new norms that would offer a better solution to the altered situation. During the last couple of years, several proposals have been made aiming to increase research subjects' control over their data and tissue, culminating with suggestions that persons should control even the use of their anonymous data and tissue. This article focuses on proposals that concern use of anonymous data and tissue for research purposes.

I   Current concepts — no protection

The data protection "bibles" on the European level are the Data Protection Convention of the Council of Europe (1981a) and the European Union (1995) Data Protection Directive 95/46/EC. In both documents, Article 2.a defines personal data as any information relating to an identified or identifiable, whether directly or indirectly, natural person. Basically, there are two ways of excluding the applicability of personal data protection requirements—one can collect anonymous data (data without personal identifiers) or render the data anonymous after collection using various methods of anonymization. This conclusion goes *mutatis mutandis* also for the tissue, at least for the purposes of this article. According to the current practice, research on anonymous tissue (for example epidemiological research) is "just research on cases, not persons," (Lowrance 2002, 27) and therefore has fallen outside the scope of major ethical guidelines and national laws. For instance, in Article A.1 of World Medical Association (2002) Declaration of Helsinki, research on non-identifiable human material or data is not deemed to be research involving human subjects. At the national level, the majority of the European states that have adopted biobanking laws expressly exclude any rights to persons after the tissue has been anonymized in the biobank. One example of this is Article 10.1 of the Estonian Human Gene Research Act (Estonian Parliament 2000), which allows a person to require the anonymization of the data and tissue contained in the national genetic databank, but not to have further control over the data and tissue.

Although data protection regulations, research ethics, and biobanking laws fail to protect anonymous data and tissue, another legal avenue is available. The concept of privacy dominating in Europe is developed by the European Court of Human Rights (ECHR) under Article 8 of the Human Rights Convention (Council of Europe 1950). Based on the case law of the ECHR, Bygrave (1998) has distilled three criteria that the ECHR takes into account when assessing whether privacy has been interfered with:

* whether the data reveal information about personality;
* whether the data are processed against the person's reasonable expectations; and
* whether such processing has negative implications for the person.

Considering these criteria, we can conclude that the ECHR is unlikely to find interference with privacy under Article 8 of the Convention in cases where anonymous data and tissue have been used in research. It is hard to imagine that anonymous tissue and data reveal something about the personality of one particular person and that the knowledge is going to be used in a way detrimental to this person. Moreover, given the long tradition of extensive epidemiological research, no one could reasonably claim that he/she honestly believed that his/her tissue and data will not be used for research purposes.

II   Some arguments for introducing protection

There are several arguments proposed in favor of amending the current situation and giving the research participants some control over their anonymous data and tissue. The underlying concept of the first kind of argument is freedom of belief, whether religious or not. For instance, a true Catholic does not want his/her data or tissue to be used in research pertaining to contraceptives (see Beyleveld and Histed 1999, 73). We may also assume that Jehovah's Witnesses do not want to be involved in research promoting use of blood in medical care. And pacifists may strongly disagree with studies on the impact of chemical and biological weapons.

Another group of arguments deals with the risk of discrimination, stigmatization, and stereotyping. Clayton (1997, 133) illustrates her concerns with the example of Afro-Americans, who might not want to be involved in IQ studies and Native-Americans who hesitate in participating in alcohol addiction studies. With similar reasons, several homosexual men refuse to take part in studies assessing the prevalence of HIV in the gay community. Thirdly, there are arguments on a purely egoistic level. Some people just do not want to be in a database or biobank or, if they still are forced to be, they might want to get some money for the use of their data and tissue (Beyleveld and Histed 2000, 296).

But let us take a second look and consider the broader implications of some of these arguments. If our aim is to introduce regulation that permits people to control the fate of their tissue, we have to, for instance, tolerate the cost of allowing choices such as donated blood "not to be transferred to suicide terrorists"; *post mortem* donated organs under the condition "not to be transferred to women undergone an abortion;" egg and sperm cells that are "to be used only for fertilization of heterosexual couples." Thus, before accepting any of these arguments one has to think carefully of the conclusions that may logically derive from the argument.

III   Some empirical data

Before exploring the proposals that have been made to meet these concerns, let us see what people think. Usually, papers arguing for the need to increase the level of protection cite studies showing that the general public is very concerned when medical records are accessed for research without the patient's consent (Robling et al. 2004). However, the numbers change drastically, if the data have been anonymized before conducting the research. According to one American study, only 12.1% of participants required their consent if the samples are to be anonymized before the research is conducted, and 91.9% did not impose greater safeguards on future research on a different disease (Wendler 2002). Some European studies have ended up with similar results —about 80% of the participants were happy to give their general consent to genetic research as long as the research was conducted on anonymized data (Stegmayr 2002). Additionally, when it comes to real life, people sometimes "eat their words." An example is the Icelandic Health Sector Database that also contains allegedly anonymous data, and if a person does not want to be included in the database, he/she has to submit a formal notice—only 7.3% of the population bothered to send such notice (Mannvernd 2004). And finally, polls evincing the high concern of people do not usually mention to the people the economic impacts of increasing their control over the data and tissue. For instance, in respect of the UK it has been argued that adopting the requirement of asking consent for the use of tissue in assessing the results of autopsies conducted in the UK yearly, would require some 150,000,000 minutes of working time. That equates to 1,399 full time jobs which, in turn, equates to the entire staff of one medium-sized NHS hospital (Furness 2004). Are we willing to pay such a high price to satisfy more or less theoretical concerns of some people against the background where the overwhelming majority does not require additional protection of anonymous data and tissue? And is it even possible to trace back each sample to each donor in order to ensure that all, no matter how egoistic and unreasonable, are taken into account?

IV   Proposed concepts and their critical assessment

There are two well reasoned proposals that support protection of anonymous tissue and data. One of them advocates recognition of a special relationship between a person and his/her tissue sample. The second concept broadens the meaning of confidentiality so as also to provide protection after personal data have been anonymized.

The main proponent of the first approach is Trouet, who maintains that current regulation does not protect anonymous tissue (Trouet and Sprumont 2002, 11–15), but that protection is needed since "even anonymized they [tissues] are not neutral to the person from whom they derive" (Trouet 2004, 100). She holds that people should have the right to decide upon the use of their tissue even if the tissue is anonymized, "because the tissues may be sensitive (for example fetal tissue after abortion) or because they do not want their tissue to be used for commercial purposes" (Trouet 2003, 413).

Although Trouet's proposal is a legitimate one, it should not be followed. First of all, there are differences between different kind of biological samples. It is one thing to protect the heart of a dead child and another thing to protect surgical waste or a urine sample. Secondly, it is unlikely that the European Convention on Human Rights (ECHR) recognizes a positive obligation of states to introduce legislation that considerably widens the control over biological samples. If there is a common understanding among European countries in respect of this question at all, it most likely favored rejecting Trouet's proposal given the burdens it imposes on medical research. Thirdly, Trouet confuses interests and valid interests. Not all the interests people might have in controlling their data and tissue can be recognized as legally valid interests. For instance, the ECHR has expressly refused to recognize a general right to access personal information under Article 8 of the Convention,[1] although people certainly have an interest in knowing their medical records.

Another new approach is supported by Beyleveld and Histed (1999, 73). They aim not to introduce a new special right under Article 8, but rather to extend the data privacy approach recognized by the ECHR. They hold that the duty of confidentiality is only one part of the general duty not to use information obtained in confidence in any unconsented for manner (a so called broad duty of confidence), and even if anonymized data are not personal data, they are still private and require extra protection. Lack of such protection might result in violation of privacy and religious freedom, as well as a decrease in public trust and health.

These are serious concerns and should be properly addressed only if they are actual risks, which I doubt. I doubt whether the law should protect interests, violation of which can, in fact, cause emotional distress only in rare cases.[2] I doubt whether the proposal would have any real impact on public health given that medical data are already used in several non-medical purposes even in non-anonymized form.[3] I doubt whether religious freedom outweighs society's interest in research. The consensus achieved in Europe[4] is that the interests of patients are not violated where data have been disclosed in accordance with the professional secrecy rules either in identifiable or non-identifiable form, and the above mentioned theoretical arguments do not justify alteration of this consensus.

V   Conclusion

Research on anonymous tissue and data has raised little debate until very recently. Probably this lack of debate has contributed to the current situation, where neither the data protection requirements, biobanking laws, research ethics, nor the privacy rules furnish research participants with extensive control over their anonymous data and tissue. This

---

1   Gaskin v. the United Kingdom; 07.07.1989, application No 10454/83. Section 37. Confirmed by M.G. v. the United Kingdom; 24.09.2002; application No 39393/98, Section 27.
2   Beyleveld and Histed acknowledge the fact, that a woman will only in very rare cases learn that her data was used in research aims of which are unacceptable for her. Therefore they do not impose the test of likeliness of actual harm but are satisfied if harm and distress would occur if the women knew about such use (Beyleveld and Histed 2000, 307).
3   Public health argument as underpin of the confidentiality has been acknowledged by the ECHR mainly in respect of stigmatizing inflection diseases such as HIV. See Z. v. Finland; 25.02.1997, application No 22009/93. Section 96.
4   See Article 5.4. of Recommendation No. R (81) 1 (Council of Europe 1981b).

might change, as evinced by the arguments and new concepts referred to in this article. However, the author does not share the opinion that anonymous tissue and data deserve additional protection. These arguments and new approaches are certainly thought-provoking but they fail to see the consequences of their implementation. Implementation of these concepts would unduly hamper research that is done for public good and constitute a next step toward individualism in the field of genomics, which should be our common "heritage of humanity."[5] Let me conclude with the strong words of Professor Buchanan (2001, B-17): "It is one thing to argue that the prevention of the certain and uncontroversially serious harms of nonconsensual bodily invasion and disrespectful treatment justifies a serious restriction on research, quite another to argue that the mere possibility of various harms, some of which are not so serious and which are very unlikely to occur, provides an equally compelling reason to restrict research."

Acknowledgements

This paper was produced as a part of the ELSAGEN project (Ethical, Legal and Social Aspects of Human Genetic Databases: A European Comparison), financed between 2002 and 2004 by the European Commission's 5th Framework Programme, Quality of Life (contract number QLG6-CT-2001-00062). I gratefully acknowledge the support of the European Community. The information provided is the sole responsibility of the author; the Community is not responsible for any use that might be made of data appearing in this publication.

References

Beyleveld, D., and E. Histed. 1999. Case commentary: Anonymisation is not exoneration. *Medical Law International* 4:69–80.

Beyleveld, D., and E. Histed. 2000. Betrayal of confidence in the court of appeal. *Medical Law International* 4:277–311.

Buchanan, A. 2000. An ethical framework for biological samples policy. In *Research involving human biological materials: Ethical issues and policy guidance*. Volume II, Commissioned Papers. Rockville, MD: National Bioethics Advisory Commission. Available online at: http://www.georgetown.edu/research/nrcbl/nbac/pubs.html

Bygrave, L. A. 1998. Data protection pursuant to the right to privacy in human rights treaties. *International Journal of Law and Information Technology* 6, 3: 259–270.

Clayton, E. W. 1997. Informed consent and genetic research. In *Genetic secrets: Protecting privacy and confidentiality in the genetic era*, ed. M. A. Rothstein. New Haven, CT: Yale University Press.

---

5 UNESCO (1997) Universal Declaration on Human Genome and Human Rights, Art. 1.

Council of Europe. 1950. Convention for the protection of human rights and fundamental freedoms. ETS No. 005, 4 November.

Council of Europe. 1981a. Convention for the protection of individuals with regard to automatic processing of personal data. ETS No. 108, 28 January.

Council of Europe. 1981b. Recommendation No. R (81) 1 On regulations for automated medical databanks. Adopted on 23 January. Available online at: http://cm.coe.int/ta/rec/1981/81r1.htm

Estonian Parliament. 2000. Estonian human gene research act. Passed on 13. December. Available online at: http://www.geenivaramu.ee/index.php?lang=eng&sub=18&eetika=1

European Union. 1995. Directive 95/46/EC of the European Parliament and of the Council of 24 October 1995 on the protection of individuals with regard to the processing of personal data and on the free movement of such data. *Official Journal of the European Communities* 23/11/1995, L. 28:31–50.

Furness, P., and R. Sullivan. 2004. The human tissue bill. *BMJ* 328:533–534.

Lowrance, W. W. 2002. *Learning from experience: Privacy and the secondary use of data in health research*. London: Nuffield Trust.

Mannvernd. 2004. Opt-outs from the Icelandic Health Sector Database. Online article available at: http://www.mannvernd.is/english/optout.html

Robling, M. R., K. Hood, H. Houston, R. Pill, J. Fay, and H. M. Evans. 2004. Public attitudes towards the use of primary care patient record data in medical research without consent: a qualitative study. *Journal of Medical Ethics* 30:104–109.

Stegmayr, B., and K. Asplund. 2002. Informed consent for genetic research on blood stored for more than a decade: A population based study. *BMJ* 325:634–635.

Trouet, C. 2003. Informed consent for the research use of human biological materials. *Medicine and Law* 22:411–420.

Trouet, C. 2004. New European guidelines for the use of stored human biological materials in biomedical research. *Journal of Medical Ethics* 30:99–103.

Trouet, C., and D. Sprumont. 2002. Biobanks: Investigating in regulation. *Baltic Yearbook of International Law* 2:3–19.

UNESCO. 1997. *Universal declaration on the human genome and human rights*. Adopted on 11 November. Available online at: http://www.unesco.org/shs/human_rights/hrbc.htm

Wendler, D., and E. Emanuel. 2002. The debate over research on stored biological samples: What do sources think? *Archives of Internal Medicine* 162:1457–1462.

World Medical Association. 2002. *The Declaration of Helsinki*. Available online at: http://www.wma.net/e/policy/b3.htm

Informed consent for donating biosamples in medical research — legal requirements in Iceland

by Hördur Helgi Helgason

1. Relevant public authorities

A. OVERVIEW

A general legislation on biobanks was enacted in Iceland in 2000 (Icelandic Parliament 2000a). Separate acts on specific database projects containing genetic and health data were enacted in 1998 and 2001 (Icelandic Parliament 1998; Icelandic Parliament 2001).

By comparison, no central legislation in Iceland deals with medical research. Applicable statutes are dispersed through legislation dealing with patients' rights (Icelandic Parliament 1997) and privacy (Icelandic Parliament 2000b), and regulations (Ministry of Health 2001; Data Protection Authority 2001b; 2001c) that are based on those acts.

According to these statutes, mainly two public entities are entrusted with controlling and monitoring the execution of medical research projects, i.e. the Data Protection Authority (DPA) and the National Bioethics Committee (NBC). In addition, Interdisciplinary Ethics Committees at the country's largest hospitals have a role to play in the evaluation of research proposals originating within those institutions (Ministry of Health 1999, article 2).

B. DATA PROTECTION AUTHORITY

The Icelandic Data Protection Authority was established in 2000 (Icelandic Parliament 2000b), taking over from the Data Protection Commission, a committee that had operated within the Ministry of Justice and Ecclesiastical Affairs since 1981 (Icelandic Parliament 1981). The new Data Protection Authority is an independent governmental agency, entrusted with enforcing national legislation regarding informational privacy. These national statutes are mostly derived from an European Union Directive on the subject (European Union 1995). This Directive aims to balance prevailing declarations of the rights of individuals to privacy[1] and the EU's principle of free movement of infor-

---

[1] Principally, Council of Europe 1981.

mation within the inner market. Consequently, the Data Protection Authority's mandate is to enforce privacy rights, while providing for the free movement of personal data within the European Economic Area, and other countries that the EU considers to ensure an adequate level of protection for the processing of personal data (European Union 1995, article 25, paragraph 1).

According to the general privacy legislation that the DPA is entrusted with enforcing, several different criteria exist for making the processing of personal data legitimate. These include processing that is necessary for compliance with a legal obligation to which the controller is subject, processing that is necessary to protect the vital interests of the data subject, and processing that relates to data which are manifestly made public by the data subject (Icelandic Parliament 2000b, articles 8 and 9). The main legitimizing criteria, however, is the data subject's consent. According to the legislation, that consent is further qualified when the processing relates to data that is considered "sensitive," including medical data (Icelandic Parliament 2000b, article 2, paragraph 1, part 8). For the processing of such data, "informed" consent is required under the act.

C. NATIONAL BIOETHICS COMMITTEE

Any research conducted with the aim to "achieve further knowledge, making it, *inter alia*, possible to improve health and cure diseases" is subject to prior evaluation by the National Bioethics Committee or a separate ethics committee. Such an evaluation must have revealed that scientific and ethical views do not oppose its implementation (Icelandic Parliament 1997, article 2, paragraph 3). The National Bioethics Committee was founded in 1999 and operates within the Ministry of Health and Social Security (Ministry of Health 1999, article 1).

The Minister of Health and Social Security shall issue rules of procedure for the NBC, which are to adhere to the statutes of several international legal instruments.[2]

D. INSTITUTIONAL BIOETHICS COMMITTEES

Institutional Bioethics Committees are to operate within the largest hospitals in Iceland, evaluating research proposals originating from within those institutions. These committees are to report their findings to the NBC (Ministry of Health 1999, article 3).

E. OTHER

In addition to the DPA, NBC, and Institutional Bioethics Committees, a few other government agencies have a role to play with regards to medical research in Iceland, e.g. the Icelandic Medicines Control Agency.

---

[2] These are, according to the Regulation on scientific research in the health sector (Ministry of Health 1999, article 7): Recommendations of the Committee of Ministers of the Council of Europe for the Member States, the International Medical Association's Helsinki Declaration, Recommendations to Guide Doctors in regard to Medical Scientific Research on Humans, and International Ethical Recommendations on Medical Scientific Research on Humans.

2. Approaches to standardizing permit requirements

A. A PROPENSITY TO STREAMLINE

The main governing bodies in the field of medical research in Iceland have demonstrated a strong tendency towards standardizing and streamlining the application process for permits that are required for conducting such research. This propensity is mainly exhibited in rules on what types of processing require permits; in standardized application forms; in general guidelines for applicants; and in the standardizing of consent forms for potential research participants.

Although the aspiration to normalize an otherwise potentially unwieldy application process is shared by the regulatory bodies in the field, noticeable differences exist between the approach taken. It is safe to say that the DPA's efforts have focused on streamlining the formal requirements of the process, while the NBC's main emphasis has been on the material requirements for conducting medical research.

B. FORMAL REQUIREMENTS

As mentioned previously, the DPA has emphasized the formal requirements for conducting medical research, in its efforts to clarify the institution's application process. This becomes apparent when its initiatives to this end are examined.

The DPA can, at its discretion, formulate rules on which processing of personal data requires a permit from the Authority (Icelandic Parliament 2000b, articles 33–35). The DPA exercised this option in 2001, requiring a permit to be issued for various types of processing of personal data, e.g. for processing that involves the joining of sensitive data on individuals with other data on the same subjects; for processing of personally identifiable genotypical data; and for processing of information on medicine and intoxicant consumption by individuals (Data Protection Authority 2001a).

In that same year, the DPA also issued rules regarding the process of obtaining consent for participation in medical research (Data Protection Authority 2001c). The rules delineate how informed consent shall be obtained, but not when that should be done. The principal questions that the rules are intended to answer include: Who may enquire about potential participants' will to take part? What methods may be used to conduct such an inquiry? What categories of information must potential participants be presented with, prior to giving their consent? What methods may be used to seek the participation of a subject's relatives?

In addition, the DPA issued in 2003 internal rules on how applications for access to data in medical records, for use in retrospective medical research, are to be processed (Data Protection Authority 2003). The rules list the requirements that the DPA may set for granting access to said type of data, e.g. regarding deletion or encryption of the subjects' national identification number, and what types of data in medical records is categorically not to be exported to the research data. The rules also contain statutes on monitoring by the DPA, in addition to listing general information security requirements for this type of processing.

The DPA has also issued an open letter to medical researchers, with general information on which types of processing require a permit from the Authority, and which types only require the Authority to be notified.

C. Material requirements

While the DPA has taken a distinctly formal approach to its standardizing of the application process for the Authority's permits for the processing of personal data in connection with medical research, the NBC has focused its efforts to homogenize its application process mainly on material requirements, specifically regarding the acceptable contents and boundaries of types of consent given for participation in medical research. However, the NBC has also composed brief guidelines on several formal aspects of its application process, e.g. general criteria for requiring a permit from the Committee for a research project; application process guidelines for student projects; and views on research participation by children. It has also published a general application form for this purpose, with appropriate instructions.

3. Efforts to categorize types of consent

A. Overview

A substantial proportion of medical research projects conducted in Iceland, which involve the donation of biosamples, are carried out by or in partnership with a small number of organizations. Among the most prominant of these are Íslensk erfdagreining ehf.,[3] UVS-Urdur,Verdandi,Skuld ehf.,[4] and Rannsóknarstöd Hjartaverndar.[5] Due in part to the volume of applications from these relatively few entities, the NBC has sought to pigeonhole a small number of clearly defined types of consent declarations that the Committee accepts for use in these research projects.

These main categories of consent are three in all, and can be loosely described thus:

★ A consent for the donation and usage of health data and specific biosamples, for use in a single, clearly defined research project. For the sake of clarity, this type of consent will be referred to here as a "narrow" consent.

★ A consent for the donation and usage of health data and specific biosamples, for use in a clearly defined group of particular research projects. This type of consent will be referred to here as a "broad" consent.

★ A consent for the donation, usage and permanent storage of health data and specific biosamples, for future use in a group of research projects that will have been permitted by the NBC and the DPA. This type of consent will be referred to here as an "open" consent.

As mentioned in chapter 1.c, the NBC is required to adhere to the statutes of several international legal instruments in its work. Among those is the Helsinki Declaration (World Medical Association 1964), which sets forth, as a basic principle for all medical research,[6] that the subjects must be volunteers and informed participants in the research

---

3 Wholly-owned subsidiary of deCODE genetics Inc. (NASDAQ: DCGN).
4 Wholly-owned subsidiary of Iceland Genomics Corporation Inc. (privately held).
5 Icelandic Heart Association — Heart Preventive Clinic and Research Institute (non-profit organization).
6 The Declaration does not define the concept "medical research," but does stipulate that medical research involving human subjects includes research on identifiable human material or identi-

project, and that each potential subject must be adequately informed of specific facts regarding the project[7], before their "freely-given informed consent" is obtained.

It is safe to say that the concept of informed consent in medical research is generally used with inference to these basic principles of the Helsinki Declaration, e.g. in Icelandic legislation, albeit with varying degrees of specificity, especially when it comes to what information must be presented to prospective participants, and what methods are to be considered adequate for ascertaining that they have understood the information.

B. NARROW CONSENT

Since this type of consent declaration is restricted to a single, clearly defined research project, it is generally considered to be the type of consent that best conforms with the requirements of informed consent, as put forth in the Declaration of Helsinki. The reason is mainly the Declaration's aforementioned requirement that potential participants be presented with information on the possible risks and benefits of the project. This requirement is arguably easiest to fulfill in the case of a single, clearly defined research project.

This category of consent declarations provides for the subject to consent to biosamples from him and health data on him to be used in the research project. This includes blood drawn for the purpose of the research, tissue removed from the subject as part of a medical treatment related to the subject of the research project, and data from the subject's medical records associated with treatment of a condition which is the subject of the project.

An additional requirement made in this category of declarations is that biosamples, gathered for the research project, be destroyed and data deleted when the project has been completed.

C. BROAD CONSENT

Consent declarations in this category are not limited to a single research project. In fact, these types of consent allow for biosamples to be stored in a biosample collection for future use in other research projects that were not described to the potential participants before they were obtained. However, differences in operations at each of the major entities conducting or partnering in medical research in Iceland lead to differences in this category of consent declarations. For example, UVS-Urdur, Verdandi, Skuld ehf.'s main research emphasis is on a large-scale cancer research project.[8] This has lead to the broad consent, that has been approved by the NBC for use in UVS's research,

---

fiable data (see article 2), and that the primary purpose of medical research involving human subjects is to "improve prophylactic, diagnostic and therapeutic procedures and the understanding of the aetiology and pathogenesis of disease" (see article 6).

7   These are the aims, methods, sources of funding for the research project, any possible conflicts of interest, institutional affiliations of the researcher, the anticipated benefits and potential risks of the study and the discomfort it may entail. The subject should be informed of the right to abstain from participation in the study or to withdraw consent to participate at any time without reprisal, see article 22.

8   "The Icelandic Cancer Project," i.e. „Íslenska krabbameinsverkefnid."

to be limited to future studies within that large-scale project. Thus, the materials collected are to be disposed of at the end of that project.

In contrast, the broad consent approved by the NBC for use in Íslensk erfdagreining ehf.'s research does not contain a requirement for materials collected to be disposed of within a certain period. Instead, it provides for collected biosamples to be stored in a biobank, the operations of which is regulated in the Act on Biobanks. The aims and purposes of future research to be conducted on those samples are not specified in the consent declarations, but a requirement is made for such research projects to be approved in advance by the NBC and the DPA. Because of the open nature of this type of consent, used by Íslensk erfdagreining ehf., it is arguably more akin to the category discussed below in chapter 3.d.

D. Open consent

Lastly, the NBC has approved a type of consent[9] for use in medical research conducted by UVS-Urdur, Verdandi, Skuld ehf. that provides for the permanent storage of biosamples, collected in conjunction with studies approved by the NBC and DPA. As pointed out in chapter 3.c, this consent category is in practice similar to a type of consent approved for use in research conducted in collaboration with Íslensk erfdagreining ehf.

This consent class does, like broad consent (see chapter 3.c), arguably not meet the criteria of informed consent as easily as narrow consent, discussed in chapter 3.b. The reason is that since the aims, purposes, methods, etc. of future research projects, which the biosamples in question will be used in conjunction with, are not known at the time when the consent is given, it can become impossible to meet the aforementioned criteria in the Helsinki Declaration for informed consent. However, storage of biosamples in a biobank for later use may, for the same reason, be inherently incompatible with the Declaration's criteria.

4. *Conclusion*

The governmental entities that have the largest role in controlling issues relating to medical research conducted in Iceland have endeavored to standardize the application process for permits that they issue for such projects.

The Data Protection Authority has issued several rules, e.g. to clarify which projects are subject to a permit from the Authority and to streamline the process to obtain such permits. Meanwhile, the National Bioethics Committee has sought to clarify the usage of consent declarations in medical research, e.g. by dividing the most used types of consent into three separate classes. A case can be made for the opinion that some of these classes do not meet the criteria for informed consent, as put forth in the World Medical Association's Helsinki Declaration. However, the same logic may indicate that the storage of biosamples in a biobank for undetermined future use is basically inherently incompatible with the Declaration's requirements.

---

9 „Samþykki fyrir þátttöku í Íslenska krabbameinsverkefninu og einstökum áföngum þess. Valkostur C. Rannsókn á öllum krabbameinum óháð upprunalíffæri. Varanleg vardveisla upplýsinga og lífsýna til annarra krabbameinsrannsókna."

Acknowledgements

This paper was produced as a part of the ELSAGEN project (Ethical, Legal and Social Aspects of Human Genetic Databases: A European Comparison), financed between 2002 and 2004 by the European Commission's 5th Framework Programme, Quality of Life (contract number QLG6-CT-2001-00062). I gratefully acknowledge the support of the European Community. The information provided is the sole responsibility of the author; the Community is not responsible for any use that might be made of data appearing in this publication.

References

Council of Europe. 1981. Convention for the protection of individuals with regard to automatic processing of personal data. ETS No.108, 28 January. Available online at: http://conventions.coe.int/Treaty/EN/Treaties/Html/108.htm

Data Protection Authority. 2001a. Rules on types of personal data processing that require notification or a permit, no. 90/2001. Available, in Icelandic, online at: http://stjornartidindi.is/servlet/stjrtid/B/2001/90.pdf

Data Protection Authority. 2001b. Rules on the security of personal data in biobanks, No. 918/2001. Available, in Icelandic, online at: http://stjornartidindi.is/servlet/stjrtid/B/2001/918.pdf

Data Protection Authority. 2001c. Rules on how informed consent shall be sought for the processing of personal data in connection with medical research, no. 170/2001. Available, in Icelandic, online at: http://stjornartidindi.is/servlet/stjrtid/B/2001/170.pdf

Data Protection Authority. 2003. Procedural rules on the processing of applications for access to medical records in connection with retrospective scientific research, no. 340/2003. Available, in Icelandic, online at: http://stjornartidindi.is/servlet/stjrtid/B/2003/340.pdf

European Union. 1995. Directive 95/46/EC of the European Parliament and of the Council of 24 October 1995 on the protection of individuals with regard to the processing of personal data and on the free movement of such data. *Official Journal of the European Communities* 23/11/1995, L. 28:31-50. Available online at: http://europa.eu.int/comm/internal_market/privacy/law_en.htm

Icelandic Parliament. 1981. Act on the registration of information regarding personal matters, no. 63/1981 (deprecated).

Icelandic Parliament. 1997. Act on the rights of patients, no. 74/1997. Unofficial English translation available online at: http://eng.heilbrigdisraduneyti.is/laws-and-regulations/nr/34

Icelandic parliament. 1998. Act on a Health Sector Database, no. 139/1998. Available online at: http://althingi.is/lagas/nuna/1998193.html

Icelandic Parliament. 2000a. Act on biobanks, no. 110/2000. Unofficial English translation available online at: http://eng.heilbrigdisraduneyti.is/laws-and-regulations/nr/31

Icelandic Parliament. 2000b. Act on protection and processing of personal data, no. 77/2000, as amended by Acts no. 90/2001, 30/2002, 81/2002 and 46/2003. Unofficial English translation available online at: http://www.personuvernd.is/tolvunefnd.nsf/pages/english

Icelandic Parliament. 2001. Act on a law enforcement genetic database, no. 88/2001. Available online at: http://www.althingi.is/lagas/nuna/2001088.html

Ministry of Health. 1999. Regulation on scientific research in the health sector, No. 552/1999. English translation available online at: http://eng.heilbrigdisraduneyti.is/laws-and-regulations/nr/1085

Ministry of Health. 2001. Regulation on the keeping and utilization of biological samples in biobanks, No. 134/2001. English translation available online at: http://eng.heilbrigdisraduneyti.is/laws-and-regulations/nr/684

World Medical Association. 1964. The Declaration of Helsinki. Adopted June, 1964, and amended in 1975, 1983, 1989, 1996, and 2000. Available online at: http://www.wma.net/e/policy/b3.htm

15

Why we should not relax ethical rules in the age of genetics

by Tuija Takala

THE DOUBLE HELIX of DNA was discovered more than half a century ago, but the actual applications of (human) genetics are only now starting to challenge our ethical frameworks. Of the many challenges that genetics poses to our ethical thinking, I will consider here two in more detail. They both have to do with the rights of individuals, and the available and forthcoming technologies to acquire genetic information and make use of it. The first of these relates to the nature of genetic information. The fact that genetic information is always also information about our blood relatives creates various problems for modern individualistic ethics. The second aspect that I am going to consider is the issue of consent. The vast possibilities that genetics is expected to provide in the future have been brought in to challenge the need for informed consent when, for instance, people participate in genetic research projects. This could be seen as a significant step back as far as patients' and research subjects' rights are concerned.

Bye, bye individual rights?

When bioethics was born in the late 1960s, one of the predominant themes was patient rights. It was thought that patients needed to be empowered to fight the paternalism of doctors. This was also a time of liberation in many other fields of life. Sexual minorities began to get recognition, religions lost some of their authoritative power, women gained increasing independence, and the individual became the center of her own life.

This development was always criticized by paternalistic doctors, public health figures, conservatives, many religious authorities, and communitarians, who all, for different reasons, felt threatened by individualism. Although the age of the individual seems to have its shortcomings, as there simply are not enough safety nets to keep everyone afloat, it has still looked as if the return of oppressive measures would be unlikely to gain majority support. Rather unexpectedly, genetics came to unite the various critics of individualism. Now we had scientific proof that we are all connected,[1] and that when

---

1 "The human genome as the common heritage" as the matter is explicated in UNESCO's (1997) Universal Declaration of the Human Genome and Human Rights.

it came to people close to us, to our blood relatives, we seemed to be bound by our shared DNA.

Genetic information is always also information about one's relatives and consequently, it has been argued, this gives rise to special duties and responsibilities among family members (e.g., see Rhodes 1998). Further claims have been made that because our genetic inheritance is to a very large degree shared by all other humans, my genetic information (information about my own genetic makeup) is not actually mine in the sense that I should necessarily be the one to decide what happens to it.

These lines of thought, made possible by the scientific and technological advances in genetics, are contrary to the idea of individual rights. Genetic information can be very sensitive in nature; it can reveal unknown family secrets, like paternity issues, and it is linked to our health status and perhaps even to our traits of character. Those who have access to genetic information about us can, in theory, know more about us than we ourselves do. It should, of course, be remembered that genetics is a new and upcoming field of study, and that its long-term impacts are still far from clear. The degree to which genes determine our health status, character, and personal abilities is not known. The current understanding is that genes alone are responsible for a small minority of our features and that various environmental factors play a significant role. However, this could well change in the future. I, for one, would consider this state of unawareness quite an insufficient basis for relaxing individual rights.

The advances in genetics are challenging the individualistic ethics of our time. If I were to take a genetic test that revealed that I have a genetic mutation which predisposes me to breast cancer, that information would not only be relevant to me but also to many of my female relatives.[2] This is the basis of the argument put forward by those who think that in the genetics era we should weaken some individual rights. They argue that it is in the interests of all to know about adverse genetic predispositions, and that therefore I should share my test results with the relatives in question. This is especially the case if something can be done to prevent the prevalence of the condition. That is, because most gene-based conditions are not caused by genes alone, their occurrence can be hindered by lifestyle changes. Also, for some conditions, such as many cancers, an early diagnosis enhances the prognosis considerably. If I were to tell my female relatives of my test result, they could then have themselves tested, and should they turn out to be carriers of the same mutation, they could have yearly mammograms to secure an early diagnosis and a better prognosis. Some have gone even further to argue that even if the condition is such that nothing can be done, people would still be better off knowing this, as this knowledge would enable them to make their life plans more realistically (e.g., see Harris and Keywood 2001).

Against these apparently reasonable sounding claims it could be noted that perhaps people do not want this kind of information. After all, it is a well-established ethical principle in medicine that people should be asked whether they want tests to be taken. So, one might argue that people should somehow be asked whether they want to be

---

2 Some types of breast cancer affect men also, but it is still predominantly women's disease.

informed of their genetic makeup in the first place.³ Further, information about adverse genetic disposition is not an unambiguous blessing because the lifestyle changes needed might be drastic and costly, and the follow-up studies required may not be available, or may be risky or painful. Then there are the issues of discrimination and other wrongful uses of genetic information. So far in most countries, for instance, insurance companies have chosen not to use genetic information, but this could change (e.g., see Takala and Gylling 2000). We would also need to consider the psychological burden that unexpected bad news imposes on people, especially when the bad news might turn out to be mistaken as genetics develops.

Another, more philosophical, argument for giving individuals detailed information about their own genetic makeup relies on a Kantian-type concept of autonomy, according to which we need to know all the relevant facts of our situation to be able to make autonomous decisions (e.g., see Harris and Keywood 2001). Against this, two main arguments can be presented. First, it is questionable whether and to what degree results of genetic tests count as "facts." Genetics is a field of study which evolves and in which the interpretation of results, as well as the methods of testing, change over time. When the mutations BRCA1 and BRCA2 responsible for some breast cancers were first discovered, they were thought to predict cancer in their bearers with a high certainty, and this led to some very drastic decisions. Only later was it discovered that without a family history, the mutation itself had very little predictive power. The second challenge against this type of argument is to question the need for all relevant information as a prerequisite for autonomous decision-making. For one thing, we simply cannot know all the relevant things (and who defines what is relevant?). But it can also be argued from a Millian viewpoint that while to make autonomous decisions we do need to know that genetic information exists, we can also autonomously decide against seeking it, if we so choose (e.g., see Takala 2001; Häyry and Takala 2001; Takala and Häyry 2000).

A potentially alarming yet interesting feature in this discussion is the return to biology as the basis of ethics. We have long fought the view that biology is all there is to us, and placed special value on our mental abilities and our spiritual side, but with the emphasis on genetic inheritance the clock seems to be turning back.

What is so important about consent anyway?

The fact that technologies to acquire genetic information exist and that the information is familial in nature might, regardless of the problems indicated above, force us to reconsider some of the individualistic tendencies of our ethics. Another instance in medical ethics where gene technologies are given as grounds for changing our ethical thinking is the issue of consent. I have already pointed out how the familial nature of genetic information has led to views which seem to forget the issue of consent altogether. This occurs when disclosing information to family members is discussed without stopping to think whether they would want to know, and how their consent could be

---

3 To arrange this in practice will, of course, be difficult. When people are approached, they will need to be told something.

secured (e.g., see Hallowell et al. 2003; Rhodes 1998). In these discussions the emphasis is usually on the person being tested and whether she is willing to share her test results with her family members. For disclosure, harm-to-others or for-the-good-of-the-others types of arguments are usually used. Many seem to be of the view that in the interest of the greater good, an individual's unwillingness to share her genetic makeup with her relatives should be overruled. This in itself can be seen as controversial, but a more major issue, in my view, is that the rights of the relatives who will get the information out of the blue are dismissed, or they are only considered in terms of the "objective medical good." This is in stark contrast to the requirement of consent, which was introduced partly because the medical good is not the only relevant value to be considered. People are allowed to refuse medical treatments and tests, however medically beneficial, based on their own values and preferences. This is what consent is all about; it was introduced to guarantee that individual choice and autonomy are respected.

Another reason given for relaxing the requirement of informed consent in relation to genetics is "in the interests of science." In most countries there is a legal duty to obtain informed consent from research subjects (Goldworth 1999), but many people are now arguing that we should not demand full informed consent for participating, say, in large-scale genetic database studies. To fulfill the traditional criteria, consent needs to be given freely (without undue influence or coercion), by a person capable of giving her consent (of age and of required mental capacities), and it needs to be based on relevant information (what is being proposed, what are the risks, what are the implications, etc.; e.g., see Holm and Bennett 2001). The problem with genetics is that as science evolves very quickly, it is very difficult to predict all the possible future research purposes for which genetic samples could be used. For these future uses, therefore, informed consent cannot be given, because a person cannot be fully informed about something which is still largely unknown. The current research protocol requires a new consent to be obtained if samples given for a particular purpose are used for another purpose. This, it has been argued, would not work with the national databanks being established in Estonia and Iceland, or with the large-scale gene banks in the United Kingdom and elsewhere. It would be very costly and difficult to contact all the people involved, and as the UK Biobank, for instance, is intended to serve as a research base for any number of studies, acquiring fresh permissions from all those participating could mean that people would be approached daily (see Estonian Genome Foundation; deCODE genetics; UK Biobank).

It is clear that genetic data banks would be easier to utilize if informed consent were not made a requirement for each new study. Perhaps we should, however, consider why informed consent became such an important factor in research ethics to begin with, and whether the expected benefits of genetics are indeed sufficient to overrule it, and whether there are other related factors to be considered.

Since the Nuremberg trials, the idea of informed consent has served as an expression of respect for individual autonomy (Goldworth 1999, 393). To seek an informed consent from a research subject is to respect the research subject's autonomy. The discussions on what type of consent should be required for genetic databases has brought the two types of autonomy, the Kantian and the Millian, back to the centre of the debates.

An interesting feature in these debates is that some of those who say that individuals should know their own genetic constitutions (even if they themselves do not want to), because only this would enable them to make truly autonomous decisions, also suggest that we should settle for less than informed consent when people donate tissue samples or data to biobanks (e.g., see Harris and Keywood 2001; Harris 1999; also Rhodes 1998). In the former case individual choice is overruled by demands of (Kantian) autonomy, and in the latter case (Kantian) autonomy is trumped by utilitarian principles and scientific optimism. Perhaps one way to proceed would be to try to find out which type of autonomy the principle of informed consent is meant to protect.

Leaving, however, issues of autonomy on one side for a moment, there is another angle from which the matter of consent can be approached. We can assume the utilitarian stance, but ask whether the expected benefits of genetics are significant enough to give us conclusive reasons to alter our ethical frameworks. Arguably, all research with human subjects would always have been easier without rules about consent, and more results would have been gained in less time and with fewer resources. Until now, however, it has been held that expected scientific benefits cannot take precedence over the entitlement individuals have to protection. What, if anything, has genetics changed or added to this equation? I would argue that the burden of proof lies upon those who wish to disregard consent. They, I believe, are the ones who should show what undeniable benefits will flow from applying more relaxed rules of consent in genetic research. But even if we viewed the possibilities of future gene technologies with great hopes, we would still also need to show that the benefits would be justly distributed. The latter requirement is, I think, frequently overlooked by science enthusiasts. Yet, what is the justificatory force of benefits to a few, if many are harmed, either directly or indirectly, in the process?

In addition to considerations of autonomy, utility and justice, a number of more practical issues surround a more open consent in genetic research. At least the following questions need to be answered. To what degree would the individual who consents to give a DNA sample be able to dictate the type of research for which her material may be used? How do we make sure that the criteria remain constant over time—as perceived by all those involved? Who should have access to the material and under what conditions? If the material is not completely anonymized, can anyone guarantee its safety? May genetic databases be sold? Under which circumstances can the participants be contacted by the researchers? It seems unlikely that any "scientifically satisfactory" combination of these questions would be able to respect individual autonomy in any of its traditional forms.

In terms of scientific breakthroughs, genetic research holds many promises. In terms of individual well-being, the question is more complex. We do not yet know what genetics will reveal, nor do we know who, if anyone, is going to benefit from it. We do know that genetic information is very sensitive and, even taking into account its familial aspect, very personal. But we do not know who exactly would be interested in using this sensitive and personal information, and for what purposes.

With all these uncertainties, I see very few reasons for relaxing the rules of consent in the context of genetics. It has taken us a long time to get where we are now in terms of ethics, and especially in terms of individual human rights, and if we want to stay on this path, we should not let the ambiguous promises of science lead us astray.

Acknowledgements

This paper was produced as a part of the ELSAGEN project (Ethical, Legal and Social Aspects of Human Genetic Databases: A European Comparison), financed between 2002 and 2004 by the European Commission's 5th Framework Programme, Quality of Life (contract number QLG6-CT-2001-00062). I gratefully acknowledge the support of the European Community. The information provided is the sole responsibility of the author; the Community is not responsible for any use that might be made of data appearing in this publication.

References

deCODE genetics Inc. Website. http://www.decode.com
Estonian Genome Foundation. Website. http://www.genomics.ee
Goldworth, A. 1999. Informed consent in the genetic age. *Cambridge Quarterly of Healthcare Ethics* 8:393–400.
Hallowell, N., C. Foster, R. Eeles, A. Ardern-Jones, V. Murday, and M. Watson. 2003. Balancing autonomy and responsibility: The ethics of generating and disclosing genetic information. *Journal of Medical Ethics* 29:74–83.
Harris, J. 1999. Ethical genetic research on human subjects. *Jurimetrics: The Journal of Law, Science and Policy* 40:77–93.
Harris, J., and K. Keywood. 2001. Ignorance, information and autonomy. *Theoretical Medicine and Bioethics* 22:415–436.
Häyry, M., and T. Takala. 2001. Genetic information, rights, and autonomy. *Theoretical Medicine and Bioethics* 22:403–414.
Holm, S., and R. Bennett. 2001. Genetic research on tissues stored in tissue banks. *ISUMA: Canadian Journal of Policy Research* 2:106–112.
Rhodes, R. 1998. Genetic links, family ties, and social bonds: rights and responsibilities in the face of genetic knowledge. *Journal of Medicine and Philosophy* 23:10–30.
Takala, T. 2001. Genetic ignorance and reasonable paternalism. *Theoretical Medicine and Bioethics* 22:485–491.
Takala, T., and H. Gylling. 2000. Who should know about our genetic makeup and why? *Journal of Medical Ethics* 26:171–174.
Takala, T., and M. Häyry. 2000. Genetic ignorance, moral obligations and social duties. *The Journal of Medicine and Philosophy* 25:107–113.
UK Biobank. Website. http://www.ukbiobank.ac.uk/
UNESCO. 1997. *Universal declaration on the human genome and human rights*. Adopted on 11 November. Available online at:
http://www.unesco.org/shs/human_rights/hrbc.htm

The public discourse on genetics and databases

16

Long-term trends in public sensitivities about genetic identification: 1973–2002

by Martin W. Bauer

Introduction

IN RECENT YEARS, various countries have been trying to set up genetic databases linking the genetic status and private health records of their populations. The pioneering example of Iceland's deCODE genetics Inc. has found imitators in Britain, Estonia, Tonga in the South Pacific, and other places in the world. These developments have been variably anticipated and debated in public.

Biobanks, genetic profiling, genetic testing and screening, and genetic passports are elements of an emergent society in which "genetic information" is widely supplied and demanded. Public imagination abounds as to how these developments will enhance individual and public health, forensic investigations, national security, and economic prosperity. Public concerns are raised over civil liberties in the collection, ownership, storage, validity and verification, and use of such information. The capability of pre-natal and pre-symptomatic diagnosing and treatment (if at all possible) of "abnormalities" point to a potential revival of eugenic ideas, albeit with the liberal face of individual choice. Blockbuster movies such as GATTACA (1997) explore these issues through the medium of science fiction.

The issue of *"self and identity"* attracts considerable attention in the social sciences. Social mobility and the melting away of traditional social structures that buttressed past certainties of personal identity have made social identity fluid, and put it within the remit of agency, design, and choice. The post-modern predicament is one of multiple identities, fluid identities, or hybrid identities, with all its opportunities and downsides. There is an increasing market for identities. Genetics may become one among many dimensions in competition with national, cultural/religious, ethnic, bodily, or lifestyle identity.

Genetic identity is not a novel idea. It is part of traditional notions of family inheritance and pedigree. What may be novel is increased public awareness that identity, and what derives from it, depends also on attribution by others. This mediation of self-identity through the gaze of others may limit choice. Attributed identities may be advantageous, for example by opening access to resources, as well as disadvantageous in the form of a social stigma. Hence, public and private consent to genetic identification becomes an issue.

Over the last 40 years several developments in genetics have given prominence to the issue of genetic identification. Genetic testing and screening became part of public health policy for particular communities and raised fears of a new eugenics (Kevles 1995, 278ff), and its potential for other purposes than health makes for "dangerous diagnostics" (Nelkin and Tancredi 1989; Marteau and Coyle 1998). DNA profiling, invented as a forensic technique in 1984, gained prominence in famous trials such as OJ Simpson's, or through the identification of the remains of the Russian Zarevitch family during the 1990s. The reliability of the technique became controversial (Lander 1992; Lander and Budowle 1994), but overall DNA profiling sits in the public imagination as a great achievement of genetics (Durant, Hansen, and Bauer 1996). The cloning of Dolly the sheep made global headlines and kick-started a debate on the morality of human cloning and stem cell research. The public imagination anticipates human cloning as imminent and, albeit with variable religious or cultural sensitivities, as challenging the moral foundations of modern society (e.g. Habermas 2001). The human genome project was launched in 1990 after much controversy (Cook-Deegan 1994; Greely 1998). Metaphors abound: the "book of life," the "code of codes," or the "blueprint of life" inspire quasi-religious awe and admiration. When the project was closed in 2000, well before schedule and through a mass media event featuring the British Prime Minister and the US President, the double helix model became the definite icon of our time. The British magazine PROSPECT (October 2000) offered the DNA code on a free CD to its readers. With the human genome projects came the *biobanks*, the various initiatives of linking individuals' genetic information and their health records on a large scale in order to "study the role of genetics in the nature and nurture of illness and disease" (UK Biobank). Pharmcogenetics promises to use genetic identity markers to tailor drugs to individual susceptibility to illness and individual drug responses. The Icelandic company deCODE was a model as to the economic, social, and medical significance of such enterprises.

The idea of the present paper is to trace and map overall public sensitivities about *"genetic identification."* Genetic information is one among many controversies that characterize the age of biotechnology and genetic engineering. Other recent issues are the *safety* of genetically modified food, the *environmental impact* of GM crops, or the *ethics* of human stem cell cloning, and the *patenting* of life forms. For our present purposes we consider genetic identity, genetic identification, and genetic information as synonyms.

Public awareness of and sensitivities about genetic identity will be traced with the help of indicators constructed from mass media coverage and from surveys of public perceptions. I have at my disposal a database of elite press coverage from seventeen countries between 1973 and 2002, and a survey of public perception in 2002 (see appendix).

National and international climates of public opinion

Since the 18th century the mass media, in particular the press, have been a key feature of a functioning public sphere variously guaranteed by constitutional rights to a free press and the freedom of speech. Nowadays the mass media are made up of a multitude of channels including newspapers, magazines, radio, television, and more recently the World Wide Web, and these are powerful agents in any modern political system.

The relationship between the mass media and public perceptions of new technology is complex. Mass media often stand for controversy, which is part of a process of reaching sustainable and informed public decisions. The influence is not unidirectional. On the one hand the mass media provide information long before any personal experience with a new technology is possible, raise awareness in the capital markets and in the wider public, set the political agenda, and cultivate a framework to talk publicly about the issues arising. On the other hand, mass media coverage is expressive of existing public sentiments. The editorial process selects and elaborates news stories that pander to existing preoccupations and hopes, which may not always be focal to the science and technology. So for example much European sentiment about biotechnology is both about the technology and about European integration (see Gaskell and Bauer 2001). Nevertheless, we must accept that *public opinion*, the resonance of public perceptions and media coverage, is at any moment a constraint, in the double sense of limitation and opportunity, for what is politically possible. This self-description of the political sphere as public opinion is not an epiphenomenon.

The mass media continue to serve mainly a national public sphere, setting the agenda and reflecting concerns of a national public. A new technology is however a global phenomenon, and many actors transcend the national borders. The emergence of a *transnational public opinion*, a European public in the making, is reflected in the synchronization of coverage and the assimilation of news framing. For biotechnology and genetic engineering such a synchronization of public discourse occurred after 1996/97 in Europe and worldwide. Two key events—the arrival in Europe of genetically modified soya in late 1996 and the presentation of Dolly the cloned sheep in February 1997—put genetic news onto the front pages of newspapers. Coverage significantly increased in the subsequent years. Countries, previously dispersed, started to group according to a particular story frame: the separation of a "green" agri-food from a "red" biomedical biotechnology (Bauer 2002; Bauer et al. 2001).

Issue salience: Genetic identification in the quality press

Our mass media analysis assesses the intensity and the framing of biotechnology articles that refer to *genetic identification*, defined mainly by references such as human nature in general, genetic profiling, and the privacy of genetic information. We also identify whether an article focuses on the topic or makes only a passing reference. Figure 1 shows three indices over time on the issue of genetic identification in the international elite press: Intensity of coverage in focus, total intensity of coverage, and issue salience, i.e. genetic identification in relation to total biotechnology coverage. We interpret "salience" as an indicator of public sensitivity. High issue salience means genetic information is a hot issue and part of a public debate.

- ★ The dark bars show the intensity of coverage. Overall the numbers of references to "genetic identification" are low for a long period, only to increase in the 1990s and to reach their peak in 1997 and 1998. This trajectory is more or less in line with the overall intensity of biotechnology news, which exploded during the 1990s, in particular after the watershed years of 1996/97 when it became a major news item in many of the countries we looked at.

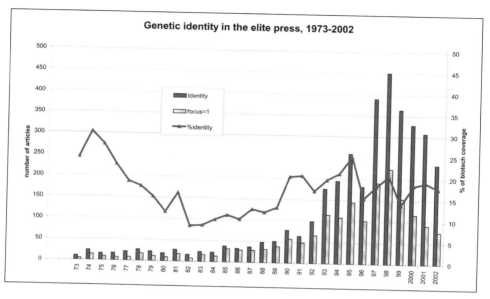

Figure 1: Intensity of genetic identity coverage in the elite press. All articles (dark bar), articles with a focus on these issues (light bar), and salience of identification within total press coverage of biotechnology (line) from 1973 to 2002 in 17 countries: Austria, Britain, Canada, Denmark, Finland, France, Germany, Greece, Italy, Japan, Netherlands, Norway, Portugal, Poland, Sweden, Switzerland, USA. Index of a single elite news source; not all countries cover the entire period, not Canada, Japan, Norway, Portugal and Poland.

* The light bars show that focal reportage on genetic identification follows the overall trends. But the ratio of focal and peripheral reportage declines over time. Peripheral references grow faster than focal ones. This is not surprising if we consider that "genes" have become in themselves a news value that carries other stories.

* The line in figure 1 shows the issue salience of "genetic identification" relative to the total intensity of biotechnology coverage. The issue is hot in public discourse in the early 1970s with about 30% of all coverage. This then declines to below 10% in the early 1980s. Issue salience returns to about 25% in 1995 and recedes since to the current level of just under 20%. Overall we can say that around one fifth of all coverage refers to issues of "genetic identification" over the period 1973–2002. Considering for example that in 1999 a single British quality paper carried 1600 articles, this makes about one article a day on genetic information, or several pieces in an average week in British newspapers. However,

most countries, with the exception of Japan and the USA, carry less biotechnology coverage than Britain, making genetic identification a news item maybe weekly, or less often.

Media sensitivity across countries

If we compare the coverage in the different countries we get a sense of the variable "heat" of this issue. Figure 2 shows the relative frequency of references to genetic identification in various countries for two periods, 1992–1996 and 1997–2002. The intensity ranges from below 5% in Canada and Denmark to above 30% in the UK. There is a step function in the distribution of sensitivities in different countries. We find *low-level* salience in Portugal, Canada, Denmark,[1] Japan, and Greece (<10%); and *mid-level* in Sweden, Norway, Finland, Switzerland, Italy, Germany, and France (10%–20%). However, the Netherlands, Austria, the USA, and particularly Britain show *high* salience of the issue over the entire period (>20%).

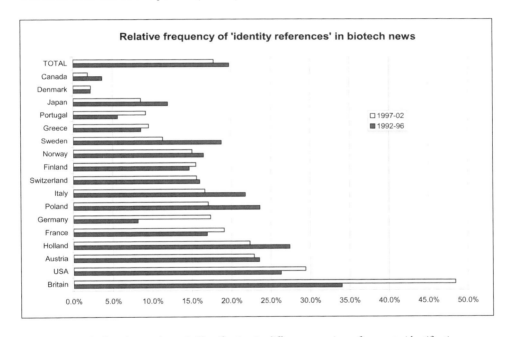

Figure 2: the salience of genetic identification in different countries: references to identification relative to all biotechnology coverage in the opinion leading press. The figures are ordered by salience for the period 1997–02.

---

1 For Poland, Denmark and Canada we have no data from between 1999 and 2002, which is a high intensity period. Therefore, our figures might underestimate the intensity and the issue salience in these countries.

Comparing the countries across the two periods, we observe that in Portugal, Greece, the USA, and Britain the issue is *heating up*, while in Canada, Japan, Sweden, Italy, Poland, and the Netherlands the issue is *cooling down*. There is *little change* in Denmark, Greece, Norway, Finland, and France.

One wonders whether the salience of genetic information is related to the intensity of biotechnology coverage in general. We compare the coverage of biotechnology (counted as numbers of all articles on biotechnology published in a single elite newspaper over the period) with the relative frequency of references to genetic identification. This indicates the relative salience of genetic identification weighted by the attention to biotechnology in general. The results are shown in table 1. There is some association between intensity of biotechnology coverage and issue salience of genetic identification. The more biotechnology coverage in general, the more salient is genetic identification. Portugal and Greece show low intensity of coverage and little concern for genetic identity. Norway and Finland, with low levels of coverage, show mid-level concern for identification. Of the countries with a mid-level of overall coverage, Sweden shows relatively low, Switzerland mid-level, and Austria and the Netherlands high sensitivity about genetic identification. Of the countries that have high levels of coverage of biotechnology, Japan shows low salience, Germany, Italy, and France, mid-level, while the USA and the UK show high salience. This gives a first mapping of the sensitivity of public spheres across the globe and across time about genetic identification.

Genetic identity Issue salience	Biotechnology Low intensity	Biotechnology Middle intensity	Biotechnology High intensity
Low %	Portugal Greece [Canada] [Denmark]	*Sweden*	*Japan*
Middle %	*Norway* *Finland* *[Poland]*	Switzerland	*Germany* *France* *Italy*
High %		*Austria* *Netherlands*	USA UK

Table 1: Intensity of biotechnology elite press coverage (horizontal axis) and percentage of coverage on "genetic identity" issues (vertical axis). (Denmark, Canada, and Poland are excluded from this count because of restricted years in the sample.)

ACTORS OF GENETIC IDENTIFICATION

A striking feature of media coverage is the shift in the actors taking part in the media debate. The debate moves from the scientific to the political arena in the 1990s. Before 1991, 65% of actors referred to in the context of genetic information are scientists, while by 1997–2002 they appear only in 30% of all articles. Before 1991, there are 15% references to political actors, by 1997–2002 these have become 35%. Reference to mass media in the mass media, an indication that the debate itself is debated and becomes reflexive, increases from under 5% to 13%. Increasingly the press begins to comment about its own role in this debate. Contrary to general biotechnology coverage, the issue of genetic information is not staged by corporate actors with any significant frequency.

PROGRESS AND CONCERNS: FRAMING OF IDENTIFICATION

Figure 3 shows the flow of enthusiasm and concerns about genetic identification over the entire period. Our coding distinguishes several ways of framing a gene story. *"Progress"* stands for stories that highlight scientific progress for the good of humanity or the economic prospects of genetic information. *"Concerns"* code for ethical concerns, in the sense that genetics is too important to leave to the scientists themselves; for Pandora's Box as a general warning related to genetic identification; for arguments in the nature/nurture tradition; and for calls to increased or reduced public accountability. *"Fatalism"* codes for arguments such as runaway, according to which the train is rolling and there is nothing to stop it, or globalization, the economic imperative to foster the national science base in a context of international competition.

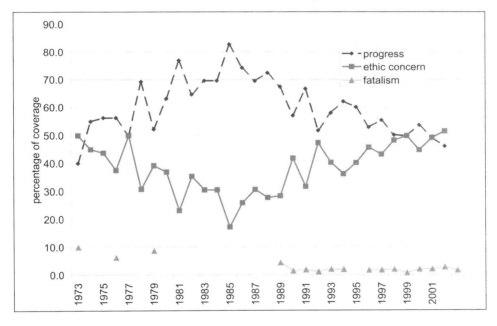

Figure 3 shows the trajectory of public enthusiasm and concern in the press over genetic information and genetic identity from 1973 to 2002 (newspapers in 17 countries).

We observe waves of enthusiasm and concerns over genetic information: The 1970s see controversial discussions on low levels of salience, but a balance of arguments. This reflects the nature/nurture debates of the mid 1970s on IQ and criminality and its heritability (see Kevles 1995). The 1980s sees an opening gap between progress and ethical concern, and rising coverage of biotechnology but lower salience of genetic identification (compare figure 1 above). The second half of the 1980s and the 1990s see increasing coverage of biotechnology, increasing absolute and relative salience of genetic identification news, and increasing concerns in public debates. From the mid 1990s the public arguments are again in balance, as in the 1970s debate, but this time with much higher intensity and issue salience. Fatalisms arguments, either runaway or globalization, have little currency—if anything, they decline. Having carried about 10% of the arguments in the late 1970s, they are nowadays below 5%.

THE STRUCTURE OF SENSITIVITIES IN THE ELITE PRESS

A final observation on the press coverage is the changing structure of topics of genetic identification that our coding is able to detect. Figure 4 shows the distribution of different themes that are co-occurring with basic references to "genetic identification" for two periods, 1973–91 and 1997–2002. Of 6,910 references (in a total of 3,602 articles selected on the basis of references to human nature, genetic profiling, or privacy of information) about 30% refer to DNA profiling for police or other uses. 20% are on human inheritance as human nature. Genetic diagnosis and screening cover 10% of all co-references, while issues of legal regulation, privacy of information, and general ethical issues each catch below 5% of all co-references. These are aggregates over the entire period from 1973 to 2002.

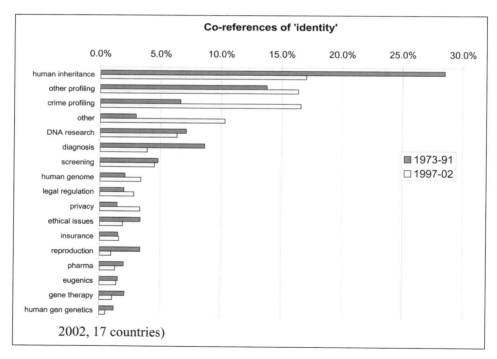

Figure 4: main topics that define and co-occur with "genetic identification" (human nature qua human inheritance, genetic profiling, privacy) in % of all references classified (multiple coding; N=6910 references in 3602 articles, 1973–2002, 17 countries).

Over the two periods, 1973–91 and 1992–2002, the topic of human nature reduces from 28% to 17%. Genetic profiling increases in particular with reference to crime and police investigations. Genetic profiling becomes a stock in trade of crime reportage in the press. Also increasing are cross-references to the human genome project, which became reality during the 1990s. Decreasing are cross-references to diagnosis and ethical issues.

Public perceptions by 2002: Awareness and public consent

Shifting the perspective away from issue salience in the elite press to public perception, we explore two issues, public awareness of genetic testing, and public consent to various uses of genetic testing. The issue of genetic testing serves us as a proxy for the wider issues arising from genetic identification in public perception.

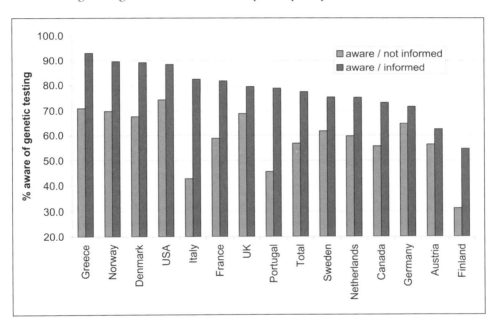

Figure 5 shows the percentage of people aware of genetic testing, for the informed and the uninformed public (Eurobarometer 58.0 2002, EU-15, Canada and USA).

AWARENESS OF GENETIC TESTING

The Eurobarometer 58.0 survey of 2002 asks respondents "have you heard of genetic testing." Figure 5 shows the percentage of respondents who have heard of genetic testing for two groups, the aware and informed, and the aware but uninformed public. The informed respondent has taken notice of news about biotechnology and related issues over the last three months in newspapers, on television, on the radio, or through the internet. The uninformed public has not noticed anything of the kind. Informed respondents are likely to be more knowledgeable of genetics, male rather than female, older rather than younger, more educated, wealthier, and tend to be on the political left rather than on the right.[2] The figures are ordered from left to right by declining levels of informed awareness. In Greece over 90% of the informed public is aware of genetic testing, while in Finland the figure is 55%. Greece, Norway, Denmark and the USA are highly aware, while Austria and Finland are least aware. All the other countries might be said to fall in between. For the uninformed public a different ordering applies. The uninformed seem to be nevertheless aware of genetic testing in Greece, Norway, Denmark, USA, Britain, and Germany, while in Italy, Portugal, and Finland the uninformed are also in the dark about genetic testing.

Conceptually, issue awareness in public perception corresponds to issue salience in the mass media. For some of our countries, we can ask how issue salience in the press and issue awareness in public perception are related. We find very little aggregate correlation. For the informed public there is no correlation at all; and for the uninformed public there is a low correlation (r = 0.23 / n = 14). High issue salience does not go together with high awareness in the public. However, interestingly the awareness gap between the uninformed and the informed public—in figure 5 this corresponds to the difference between the light and the dark bars—is correlated more strongly (r = - 0.38 / n = 14). The relationship is negative, meaning that the higher the issue salience in the press, the smaller is the gap between the informed and uninformed public on awareness of genetic testing. While there is no direct co-variation between salience and awareness in public, there is a relationship between issue salience and the gap in awareness. This finding is consistent with a version of the "knowledge gap hypothesis" on the influence of mass media on public perceptions: increased public controversy over an issue leads to more media coverage and enhanced circulation of information, and to a closure of the information gap between the educated and the uneducated public (see Bauer and Bonfadelli 2002).

Consent to uses of genetic identification

Some questions in Eurobarometer 58.0 explore issues of consent to genetic identification. Figure 6 shows six attitude statements with which respondents were invited to agree or disagree. The figure gives the response distribution for each item. There is a clear hierarchy in people's attitudes to different uses of genetic information. Agreement is highest for doctors to use such information. Pre-natal testing of babies or post-natal testing, provided it is for the diagnosis of serious diseases, also carries a majority of approval. The opposite situation arises for the uses in police and crime investigations, governments having access to it, or private insurance companies making use of genetic information to differentiate their risks. Here the public is clearly divided as in the case of police uses, or clearly skeptical to government and insurance uses.

One may group these six uses into private (private-medical; the first three) and "public" purposes (the latter three), and then create an index to compare the different countries. Figure 7 plots the average number of acceptable private/medical and public uses

---

2 This was ascertained by logistic regression with "having heard of biotechnology before" versus "never heard of" as the dependent variable: pseudo Chi2 = 0.16.

LONG-TERM TRENDS IN PUBLIC SENSITIVITIES ABOUT GENETIC IDENTIFICATION

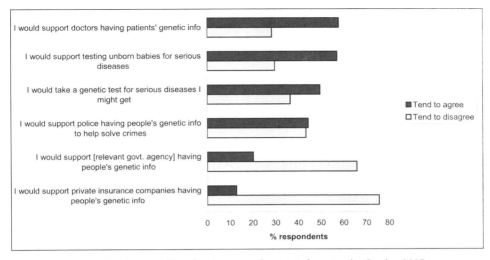

Figure 6: the acceptability of various uses of genetic information by October 2002 across Europe-15; N = 16500 (source: Gaskell, Allum, and Stares 2003).

for each country. There seem to be two groups of countries. Portugal, Spain, Greece, Canada, Italy, France, Britain, and the USA are approving countries, while Sweden, Austria, Finland, Luxembourg, Norway, Germany, Belgium, Denmark, and the Netherlands are clearly more skeptical. The middle ground is held in Ireland, close to the overall average. There is a high correlation between accepting private-medical and pub-

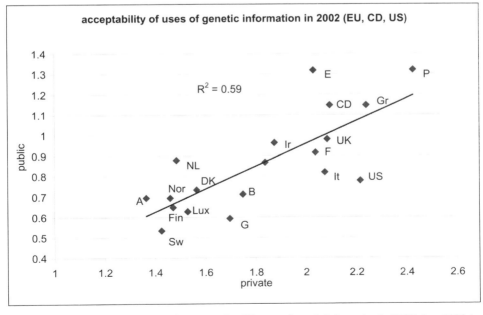

Figure 7: The average acceptability of private and public uses of genetic information in 2002. [n=1000 in each country, Eurobarometer 58.0 2002; plus surveys in Canada and the USA; 19 countries].

153

lic uses (r = 0.77; n = 18). Countries that tend to approve public uses also tend to approve private uses, but overall respondents are more inclined to approve medical than public uses.

Countries above the regression line are more skeptical toward medical uses than expected, in particular Austria, the Netherlands, Spain, and Portugal. They approve less of private-medical uses compared to other countries with similar approval levels for public uses. Countries below the regression line tend to be more skeptical toward public uses than toward private-medical ones, meaning they approve more medical uses than other countries of similar standing on public uses. These countries are Sweden, Germany, Belgium, Italy, and in particular the USA.

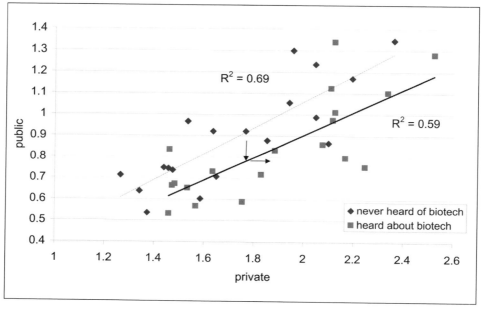

Figure 8 shows the acceptance ratings of private and public uses of genetic information for people who have "heard of biotechnology before" and those who have "never heard of biotechnology before" [Eurobarometer 58.2; plus in CD and US equivalent surveys 2002; n = 19].

Figure 8 shows the average acceptance ratings for private and public uses of genetic information for informed and uninformed members of the public. Generally we observe that with more information, for which the daily newspapers are a likely source, the public is more skeptical of public uses and more favorable toward private-medical uses (see the arrow for the average score of all countries together in the middle of figure 5 moving down on public and right on private uses). Information about genetics accentuates the existing pattern: less support for public, more support for private uses. There are however some exceptions to this general pattern. Informed people in the Netherlands are more skeptical toward both public and private uses of genetic information than uninformed members of the public. Those informed about biotechnology in Ireland,

Finland, Spain, and Belgium are more favorable toward both public and private uses than the uninformed. Overall we note that the correlation decreases with more information, which means there is more variation in average attitudes with increasing information. This is consistent with results on public perceptions of science and technology in general: information affects attitudes *and* increases variance by introducing polarization (Durant et al. 2000).

Summary and conclusion

Genetic identification is not a constant preoccupation in public debates over biotechnology since the 1970s. We detect three phases of public sensitivity. In the 1970s, with general biotechnology press coverage very low, genetic identification has high issue salience. During the 1980s biotechnology coverage is rising, and the issue salience of identification declines. In the late 1990s we see exploding biotechnology coverage and again higher salience of genetic identification within this discourse. The issue salience of the 1970s is matched again in the l990s. The peak of issue salience is reached in 1995, and the peak in intensity is reached in 1998: a year after "Dolly the Sheep" is presented to the world public and raises the issue of identity in relation to human cloning.

The 1990s saw the politicization of the issue. The debates move from an arena of scientific actors to an area of political actors, and there is an increased level of reflexive commentary by the mass media on their own role in the debate. The absence of business actors is characteristic of the genetic identification debate within biotechnology.

Public concern, expressed as high ratios of ethical framing of news coverage of genetic information is high in the 1970s and again in the 1990s. The 1980s is dominated by enthusiasms about scientific progress and economic prospects. Public concerns are a mixture of general ethical concerns, a call for public accountability, nature/nurture arguments, and warnings about tampering with nature. While genetic identity coverage is dominated by stories regarding a putative human nature, in the 1990s the topic of genetic profiling for criminal and other investigations becomes the dominant story. Genetic profiling is a major anchor of achievement in public perception.

Public perception and mass media coverage mirror each other only partially. There is no relationship between issue salience and public awareness of genetic testing. But there is some relationship between issue salience and the awareness gap: higher salience goes with a lower gap between an informed and uninformed public.

Countries clearly vary in their public sensitivity to genetic identification. In the USA and the UK genetic identification is a hot topic with high levels of coverage of general biotechnology. In Portugal, Greece, Canada, Denmark, and Switzerland the issue is of low or medium salience, correlated with the overall biotechnology coverage. In Sweden and Japan issue salience is lower than expected from overall coverage. In Norway, Finland, Poland, Austria, and the Netherlands the salience of genetic identification is higher than expected from overall coverage. In the second half of the 1990s the issue is heating up in Portugal, Germany, Britain, and the USA. It is cooling down in Japan, Sweden, Italy, Poland, and the Netherlands. In the latter countries the issue is more salient in the early than later 1990s.

Country	Media Intensity of Biotechnology	Issue salience of Genetic identification	Consent to genetic testing	Awareness of genetic testing
Greece	Lo	Lo	Hi	Hi
Norway	Lo	Med	Lo	Hi
Denmark	[Lo]	[Lo]	Lo	Hi
USA	Hi	Hi+	Hi	Hi
Japan	Hi	Lo-	--	--
Britain	Hi	Very Hi+	Hi	Mi
Italy	Hi	Mi-	Hi	Mi
France	Hi	Mi	Hi	Mi
Germany	Hi	Mi+	Lo	Mi
Netherlands	Mi	Hi-	Lo	Mi
Sweden	Mi	Lo-	Lo	Mi
Canada	[Lo]	[Lo]	Hi	Mi
Portugal	Lo	Lo+	Hi	Mi
Austria	Mi	Hi	Lo	Lo
Finland	Lo	Mi	Lo	Lo
Poland	[Lo]	[Mi]-	--	--
Switzerland	Mi	Mi	--	--
Spain	--	--	Hi	
Ireland	--	--	Mi	
Luxemburg	--	--	Lo	

Table 2: This summary table compares the countries on overall intensity of biotechnology coverage 1997–2002, issue salience of genetic identification between 1997–2002, consent to uses of genetic testing in 2002, and public awareness of genetic testing in 2002 (Lo = low; Mi = mid; Hi = high level).
Comment: [ ] are countries with incomplete media counts, likely underestimated; A sign (-/+) marks changes in issue salience from 1992–1996 to 1997–2002.

Levels of consent for different uses of genetic testing go together with very different patterns of public awareness and mass media coverage of the issue. Private medical uses gain more consent than public uses of genetic testing. Among the positive countries, Canada, US, Britain, Italy, France, Greece, and Portugal, there seems to be no social need for a public debate in some of them. Greek consent is highly informed, but with low media activity on the topic, as is Canadian and Portuguese consent, albeit with less general awareness in the population. French, Italian, US, and British consent go with high media activity, but less awareness. The skeptical countries are Denmark, Norway,

Sweden, Finland, Germany, Netherlands, and Austria. Of these, Denmark shows high awareness, Sweden and Norway medium, and Austria and Finland low awareness of genetic testing. But Germany, Austria and the Netherlands also show higher issue salience and media activity, while all the Scandinavian countries show relatively low media activity on the issue.

To conclude, genetic identification is increasingly on offer and in demand in a market for post-modern identity. It is not clear whether a privately or publicly discovered and espoused genetic identity is to the advantage or the disadvantage of the carrier. The issue of consent is arising in public debates. Our data allow us to characterize the debate over genetic identification as a climate of opinion in various countries, both in terms of issue salience in the mass media and awareness and consent in public perception. Evidence clearly shows that genetic identification meets very different public spheres in terms of awareness, general level of consent to use genetic testing, intensity of general coverage of biotechnology, and the issue salience of genetic identification. No single explanatory pattern emerges from this complex picture. The search for a model that might explain this pattern of variation remains a matter for future research.

References

Bauer, M. W. 2002. Arenas, platforms and the biotechnology movement. *Science Communication* 24:144–161.

Bauer, M. W., and H. Bonfadelli. 2002. Controversy, media coverage and public knowledge. In Bauer and Gaskell 2002, 149–175.

Bauer, M. W., and G. Gaskell. 2002. *Biotechnology: The making of a global controversy*. Cambridge: CUP.

Bauer, M. W., and S. Howard. 2004. *Biotechnology and the public: Media module Europe, North America and Japan. Integrated codebook: The elite press 1973–2002*. London: LSE Methodology Institute, February.

Bauer, M. W., M. Kohring, J. Gutteling, and A. Allansdottir. 2001. The dramatisation of biotechnology in the elite mass media. In Gaskell and Bauer 2001, 35–52.

Cook-Deegan, R. 1994. *The gene wars*. New York: W W Norton.

Durant, J., M. Bauer, C. Midden, G. Gaskell, and M. Liakopoulos. 2000. Two cultures of public understanding of science. In *Between understanding and trust: The public, science and technology*, ed. M. Dierkes and C. von Grote, 131–156. Reading: Harwood Academics Publisher.

Durant, J., A. Hansen, and M. Bauer. 1996. Public understanding of the new genetics. In *The troubled helix: Social and psychological implications of the new human genetics*, ed. T. Marteau and M. Richards. Cambridge: CUP.

Gaskell, G., and M. W. Bauer, eds. 2001. *Biotechnology 1996–2000: The years of controversy*. London: Science Museum.

Gaskell, G., N. Allum, and S. Stares. 2003. *Europeans and biotechnology in 2002, Eurobarometer 58.0*. A report to the EC DG for Research from the project "Life Sciences in European Society." London: LSE.

Greely, H. T. 1998. Legal, ethical and social issues human genome research. *Annual Review of Anthropology* 27:473–502.

Habermas, J. 2001. *Die Zukunft der menschlichen Natur. Auf dem Weg zu einer liberalen Eugenik?* Frankfurt: Suhrkamp.

Kevles, D. J. 1995. *In the name of eugenics.* Cambridge: HUP.

Lander, E. 1992. DNA fingerprinting: Science, law and the ultimate identifier. In *The code of codes: Scientific and social issues in the human genome project*, ed. D. J. Kevles and L. Hood. Cambridge: HUP.

Lander, E. S., and B. Budowle. 1994. DNA fingerprinting dispute laid to rest. *Nature* 371:735–38.

Marteau, T. M., and R. T. Coyle. 1998. Psychological responses to genetic testing. *British Medical Journal* 316:693–96.

Nelkin, D., and L. Tancredi. 1989. *Dangerous diagnostics: The social power of biological information.* Chicago: CUP.

WAN. 2002. *World press trends.* Paris: World Newspaper Association

Appendix

Biotechnology in the Mass Media, 1973–2002

IT IS WELL known that television has a wider reach than print media. However, this may not be true for any particular television program. On the other hand, it is the case that elite newspapers, with a relatively small audience, lead the development of a public issue such as a new technology. Newspapers raise the issue, and radio and television will pick it up at a later stage, not least because journalists in different mass media take their cues from the leading newspapers. For monitoring purposes it is therefore efficient, convenient, and effective to rely on the analysis of newspaper coverage.

The project "Biotechnology and the Public" conducted continuous international press monitoring of biotechnology news in the national elite press between 1973 and 2002 (for details, see Bauer and Howard 2004; Bauer and Gaskell 2002; Gaskell and Bauer 2001). Figures are for a single source using keywords such as "genes," "biotechnology," "cloning," or "DNA" for on-line searching. The coding follows the rationale of classical content analysis, manually performed by a team in each country, and is based on a random sample of articles stratified by year and country.

Definition of the variable "genetic identification": articles are included which make reference to genetic profiling for crime or other purposes, human nature *qua* inheritance, and privacy of genetic information. References that co-occur frequently with these are unspecific DNA research, genetic diagnosis, human genome project, privacy of information, legal regulations, and ethical issues. The database is held at the LSE, Institute of Social Psychology.

Public perceptions 2002

Eurobarometer is an instrument of the European Commission to monitor attitudes in all European member states. Each survey is a sample representative of the population and with a sample size of n=1000 in each country. Fieldwork of Eurobarometer 58.0 was conducted in October and November 2002. Comparable surveys were conducted in Canada and the USA at about the same time. Our index of consent to various uses of genetic information reports percentage with a margin of error of 3% -/+ with 95% significance for figures around 50%.

THE INFORMED PUBLIC: ISSUE DISCRIMINATION

To assess whether respondents take notice of biotechnology, the following question was asked: "Before this interview, over the last three months, have you heard anything about issues involving modern biotechnology?" Respondents say "no" or "yes," and identify the news source of their information. 39% had *not* taken any notice of biotechnology and related issues before the survey interview. 61% of respondents had heard of biotechnology in newspapers, radio, television, magazines, or on the internet.

The newspaper readership in the various countries 2002

The countries in our study have very different newspaper cultures. Statistics published by the World Association of Newspapers show the following circulation figures of newspapers among the adult population. These figures refer to the reading of any daily newspapers, quality or popular.

	% readers
Norway	70
Japan	66
Sweden	54
Finland	54
Switzerland	44
Britain	38
Denmark	37
Germany	37
Netherlands	36
Austria	36
USA	27
Canada	19
France	18
Italy	13
Portugal	9
Poland	9
Greece	8

Source: WAN 2002 (newspaper circulation per 100 adults)

17

Making genes commonly meaningful: Implications of national self-images on human genetic databases

by Piia Tammpuu

Introduction

THE ESTABLISHMENT OF population-based genetic databases in different national contexts has given rise to extensive debates regarding the principles and implications underlying the foundation of such databanks. While raising a number of similar concerns, the extent and nature of these debates, as well as the responses given by the general public and experts, have varied from one context to another. By focusing on the case of the Estonian Genome Project, this paper examines particular discursive framings applied in the domestic debate on the genetic databank issue, and discusses their relevance in terms of public acceptance of the national genome project, particularly in its initial phase of implementation.

As with the Icelandic human genetic database, the Estonian Genome Project (EGP) has been designed as a population-based database intended to collect genetic samples from the entire adult population. According to initial estimations, this means that the database will collect samples from approximately 1 million gene donors. The idea of establishing a national population-based genetic database was first publicly introduced at the beginning of 1999, and was followed by the adoption of a special legislative act—the Human Genes Research Act (HGRA)—providing the legal regulations for the establishment of the Estonian Genome Project at the end of 2000. The data collection was started in the autumn of 2002 in the form of a pilot project. However, by the beginning of 2004, there were only 10,000 genetic samples included in the database.

Since the EGP is founded on the principle of voluntary participation, where individuals are left with the choice as to whether they are willing to donate their blood sample to the national gene bank or not, public images of the EGP and gene technology are of particular relevance, giving rise to questions about the credibility of the people and institutions involved in the database project.

As the findings of public opinion surveys indicate, the majority of the Estonian public has been generally supportive with respect to the establishment of the EGP (for more detail, see Korts 2004). Also, the political decision to establish the national database by the adoption of the HGRA was passed in the Parliament only three months after the introduction of the initial bill, and without any particular contestation in public.

Drawing mainly on the sources of the domestic media coverage of the Estonian Genome Project, this paper suggests that the positive reception of the project by the public in its initial phase of implementation can at least partly be explained by the ways in which the issue of the project was initially introduced to Estonian society and framed in public discussions (for a more comprehensive study on the Estonian public discourse on the EGP, see Tammpuu 2004). In addition, the prominence and authority of geneticists and biomedical experts initiating and promoting the idea of the genome project appears to be a crucial factor in the achievement of public acceptance of the project.

National self-images and promotion of the database

Debates about new technologies, as Hamper and Renn suggest, are usually not restricted to a single technological method, but include the social and environmental embedding of particular technologies, and can therefore be described as social projects led by certain values and interests (Hampel and Renn 2000). Having long research traditions overall, biotechnology and biomedicine have also appeared as fields of research that have received remarkable attention and recognition both on the domestic and the international level during the years of Estonia's re-independence.

The emergence of the public debate about the establishment of the EGP dates back to 1999 when the idea of the national genome bank was publicly introduced. Here the foundation of the Estonian Genome Foundation (*Geenikeskus*) in January 1999 by a number of recognized Estonian geneticists, biologists, and medical scientists, as well as by politicians, and journalists, serves as one of the first landmarks indicating the beginning of the domestic "gene debate." As explained in the press, the aim of the Genome Foundation was to unify Estonian gene technologists working in different laboratories, in order that "Estonia would stay in the first rank of this rapidly developing field," as well as to "help the society to understand where geneticists have arrived and where they will arrive" (Eesti Päevaleht 27/01/99).

As a systematic analysis of the Estonian media coverage of the databank issue from the period 1999–2000 reveals, the media discourse appeared to be predominantly positive on the issues of gene technology and genetic databases in 1999, mostly emphasizing benefits arising from the genetic research, and paying relatively little attention to possible risks. Besides medical and scientific benefits related to the EGP and genetic research, expectations and promises regarding further economic development and international competitiveness were frequently raised (Tammpuu 2004).

Since the very beginning of the public "gene debate," geneticists and biomedical experts related to the Genome Foundation and the EGP claimed Estonia to have a number of advantages that could allow it to become a "leading country" in the field of gene technology. Strong traditions in molecular biology, on the one hand, and technological innovativeness reflected mainly in the rapid growth of the IT sector and telecommunications on the other hand, were used as premises or evidence to support this assumption. Gene technology was envisaged as one of the few fields in which a country as small as

Estonia could compete with (big) Western countries on an equal footing, or even achieve an advance:

> The idea of the genome project is a strategically well-timed project that would give Estonia the possibility to rise among the forerunners at least in Europe. (*Postimees* 20 November 1999)

> It is important for Estonia that we get to the world arena in gene technology as an increasing and developing economic branch, and compete with other countries. (*Postimees* 13 March 2000)

In this context, the Icelandic Genome Project served both as an example and as a comparison:

> Mini-societies like Iceland and Estonia that are genetically homogeneous and have a good health care system and scientific base can accomplish the leap to the new medicine much faster than big countries that are still standing at the starting line. … an Estonian Nokia may be hidden in our genes and in the Icelandic example. (*Postimees* 18 September 1999)

Even before the detailed plans of the EGP were publicly introduced, the initiators and proponents of the project declared it to be the "Estonian Nokia," drawing a parallel with the Finnish Nokia, a world-leading telecommunications company, and an internationally recognized brand of Finnish origin:

> Estonia's chance is in information and gene technology. … If these two will co-operate, there may emerge the desired Estonian Nokia. (Andres Metspalu, Professor of Biotechnology, *Äripäev* 27 May 1999)

"Branding" the EGP as the Estonian Nokia was thus turned into a powerful metaphor, symbolizing innovativeness and technological advancement as the key factors allegedly determining high technology-based development and success in the modern world. In the context of the Estonian post-communist transition and symbolic. "Return to the West," in which the public debate and the particular discursive framing was embedded, the completion of the EGP appeared as more evidence of Estonia's post-communist "Success Story," and as a "Big Chance" for the country. Emphasizing Estonia's potential in genetics and biotechnology, it was assumed that the genome project would put Estonia on the "world map," and shape Estonia's international image and reputation as an innovative and competitive small country:

> Estonia has a big advantage in biotechnology and gene technology compared to the other Central and Eastern European countries. Gene technology is probably the big chance for Estonia's future. The only problem is that we have not yet founded the gene database that would be an ideal base for the development of technology and enterprising. (Agu Remmelg, Director of Estonian Foreign Investments Agency, *Äripäev* 28 January 2000)

Such expectations were expressed in several headlines, particularly in 1999: "Gene sale will make Estonia well-known" (*Postimees* 24 May 1999), "Estonia's chance is in gene technology" (*Eesti Päevaleht* 31 May 1999), "EGP — The gas deposit of the Estonian state" (*Eesti Ekspress* 04 November 1999), "Gene technology and transit are Estonian trumps for the coming years" (*Postimees* 1 December 1999).

Since the images used in various discursive practices are not only descriptive in their nature, but also interpretive and evaluative, the selection of particular metaphors can be seen as strategic, rather than as an accidental matter of choice (see van Dijk 1998). Repeated and recurring metaphors in turn come to frame and affect the perception and understanding of scientific issues and events (Nelkin 2001, 556). Hence, as Fletcher (2004, 12) has similarly argued, the EGP appears "not only as a scientific project but a symbol of the post-Soviet reconstruction of an Estonia that is advanced and innovative."

However, the influence of framing the databank issue as determining the future technological development of the country does not lie only in its rhetorical weight, but also in the agents behind these discursive constructions. The question of an "Estonian Nokia" was initially proposed by Lennart Meri, the former President of Estonia, in 1999, and was immediately included in the popular discourses on national identity and future scenarios of economic and social development. The symbol was transferred to the context of biotechnology and the EGP by one of the key ideologues and public spokespersons of the project, Andres Metspalu, Professor of Biotechnology at the University of Tartu, and quickly spread through the general journalistic discourse. It is noteworthy that according to the findings of a public opinion survey of 2002, geneticists and the staff of the EGP Foundation enjoy the highest credibility rating in the eyes of the Estonian public among other possible information and opinion-sources regarding the issue of the EGP (Korts 2004).

Conclusions

As Rose has pointed out, the changing relationships between science, the state, and the market, illustrated by biotechnological developments, have required that the new genetics be "sold" to diverse audiences, including investors, lay people, and government representatives (Rose 2000, 67). The idea of building a population-based genome project has perfectly corresponded to the nationally and internationally promoted image of Estonia as a small and technologically rapidly developing post-communist country. Labeling the EGP as Estonian Nokia has provided a commonly meaningful reference point when talking about the significance of the project from the national as well as international point of view.

Popular representations of the Estonian Genome Project as a techno-scientific enterprise embedded in the broader discourses of Estonia's post-Soviet development and "Return to the West" have thus contextualized the construction of the national genebank in the modernist ethos of scientific and technological progress. However, while the public representation and reception of the Estonian Genome Project has obviously been influenced by popular self-images and national aspirations, the domestic public discourse on the database issue has been only modestly informed about possible ethical and moral implications accompanying the expansion of human genetic research and genetic knowledge in society. In this respect, there has appeared insufficient consideration and recognition of the extent to which the establishment of human genetic databanks represents "biopolitical experiments" with individuals and societies at large (see for example Pálsson and Hardardóttir 2002). In addition, as a sociological survey from 2002 reveals, while the Estonian public perceives the genome project *generally* in a positive light, associating it with various benefits in medical, scientific, and economic terms, people tend to be remarkably more hesitant when asked about their *personal* participation in the project (see Korts 2004). This ambivalence may thus reflect a "gap" in the public promotion and perception of the EGP as a joint national venture embodying the technological and scientific potential of the country, and as an endeavor to carry out a search into the very intimate sphere of individual privacy.

Acknowledgements

This paper was produced as a part of the ELSAGEN project (Ethical, Legal and Social Aspects of Human Genetic Databases: A European Comparison), financed between 2002 and 2004 by the European Commission's 5th Framework Programme, Quality of Life (contract number QLG6-CT-2001-00062). I gratefully acknowledge the support of the European Community. The information provided is the sole responsibility of the author; the Community is not responsible for any use that might be made of data appearing in this publication.

References

Fletcher, A. L. 2004. Field of genes: The politics of science and identity in the Estonian genome project. *New Genetics and Society* 23:3–14.

Hampel, J., and O. Renn. 2000. Introduction: Public understanding of genetic engineering. *New Genetics and Society* 19:221–231.

Korts, K. 2004. Introducing gene technology to the society: Social implications of the Estonian Genome Project. *Trames* 8:241–253.

Nelkin, D. 2001. Molecular metaphors: The gene in popular discourse. *Nature Reviews Genetics* 2:555–559.

Pálsson, G., and K. E. Hardardóttir. 2002. For whom the cell tolls: Debates about biomedicine. *Current Anthropology* 43:271–301.

Rose, H. 2000. Risk, trust and scepticism in the age of the new genetics. In *The risk society and beyond*, ed. B. Adam, U. Beck, and J. van Loon, 63–77. London: Sage Publications.

Tammpuu, P. 2004. Constructing public images of new genetics and gene technology: the media discourse on the Estonian Human Genome Project. *Trames* 8:192–216.

van Dijk, J. 1998. *Imagenation: Popular images of genetics*. New York University Press.

Analyzing multiple discourses in the establishment of genetic databases

by Wendy Marsden

Introduction

THE AIM OF this paper is to contribute to the discourse of genetic databases, and to demonstrate how multiple discourses create the social and historical context of this paper, my research project, and the creation of the Generation Scotland genetic database. The text of the paper has two aspects. Firstly, it describes Generation Scotland and the rationale for proposing a discourse study, and secondly, it illustrates particular discourses within the description. The description is followed by a brief example analysis looking at language from those discourses. The ELSAGEN conference also makes a significant contribution to the international discourse of genetic databases, and will be, from the point of view of this project, a fascinating piece of participant-observation fieldwork.

Background

The Scottish Health Survey 1998 reported that: "People living in Scotland experience relatively high rates of mortality from major diseases compared with the populations of England and many other western countries. Deaths and ill-health from coronary heart disease, cancer and stroke are higher than elsewhere" (Shaw 1998). A report published by the Healthy Public Policy Network (Stewart 1998) of the Scottish Council Foundation entitled *The Possible Scot* argued that: "In order to develop policy to improve Scotland's health we first need a better understanding of the underlying causes of Scotland's poor health record." These reports reflect the widely felt concerns about health in Scotland and the need to develop a far-reaching strategy to address the issues of public health policy, planning, and provision.

Against this background two collaborative projects have emerged from the health research community in Scotland: Generation Scotland and a Scottish "spoke" of the UK Biobank. They are both mentioned here because they are occurring simultaneously.

Generation Scotland

The report *A Strategy for Health and Wealth Creation Through Gene Knowledge* produced by the Royal Society of Edinburgh summarizes the project as "a proposal to create an ethically sound, population and family based infrastructure, to identify the genetic basis of common complex diseases. Generation Scotland would focus on the identification, study and follow-up of individuals with a disease diagnosis and their 'at risk' relatives. The aims would be to discover the nature of the genetic contributions to risk, to treatment response and long-term outcome" (Christie 2002). The primary aim of Generation Scotland is to identify genes, initially the genes that predispose an individual to heart disease, cancer, and mental health problems. The identification of these genes will supply the genetic data which can then be related to other factors in health research, and in the development of drug treatments.

UK Biobank

The Medical Research Council and Wellcome Trust have proposed recruiting 500,000 individuals aged 45–69 years within Primary Care settings across the UK to facilitate research into the relationship between genes, lifestyle, and environment (UK Biobank 2004). The data collection is organized into six regional "spokes" with a central organizational "hub." The Scottish "spoke" of the UK Biobank project aims to recruit between 80 and 100,000 people across Scotland. UK Biobank is a prospective project, which will collect data in order that research can be carried out into the relationship between genes, lifestyle, and the onset or outcome of illness in the age group 45–69 years old.

Genetic health in the 21st century

The potential of these genetic databases with associated patient records has powerful implications for health in Scotland. Many of those involved in setting up the Scottish "spoke" of the UK Biobank project are also involved in the Generation Scotland project. Together, the two projects constitute a programme of research into Genetic Health in the 21st Century (21CGH) which is "a multi-institution, cross-disciplinary collaboration led by the University of Edinburgh, embracing all of the Scottish Medical Schools, other important research institutes and the NHS in Scotland to create a novel consortium based approach to addressing important health priorities in Scotland" (Genetic Health in the 21st Century 2004). Their areas of expertise include genetics, epidemiology, clinical specialists, primary care clinicians, law, ethics, social science, statistics, health geography, and informatics.

Generation Scotland genetic database project

The proposal for the Generation Scotland genetic database is led by Professor David Porteous from the Molecular Medicine Centre at the University of Edinburgh. He is the driving force and provides the *charismatic leadership* (Weber 1947) that is necessary to carry forward a collaborative project of this scale and complexity. Professor Porteous has already worked extensively on projects to identify the genes that are causal in the onset of colorectal cancer and schizophrenia. He has gathered together a collaborative team with expertise across the Medical Schools and research institutes in Scotland that are working together on Generation Scotland shaping the genetic database and the way it will be utilized in future research projects. Generation Scotland is not just a data collection project but already has an initial program of research built into it. In this sense it differs from the UK Biobank project.

Aims of Generation Scotland

The aims of Generation Scotland are to identify the multi-factorial genes that predispose individuals to cancer, heart disease, and mental illness.

The identification of these genes will:

- ★ enable a process of testing for the identification and surveillance of "at risk" individuals;
- ★ contribute to the understanding of the underlying disease process;
- ★ create a new and more specific classification of disease;
- ★ contribute to a new approach to treatment using gene therapy;
- ★ contribute to the development of new "targeted" medicines; and
- ★ contribute to a wider understanding of the relationship between predisposed genes and environment.

A family base study

Generation Scotland proposes the collection and comparison of genetic data from families. The project aims to recruit individuals who have already been diagnosed as having a particular illness and their "at risk" family members. In terms of studying genes for identification this is a more effective method than a randomly selected sample from the population. The genes of the diagnosed individual can be compared with genes from unaffected or undiagnosed members of the same family. Because they share the same genetic configuration most of the genetic information is duplicated from one individual to another, so the variants specific to a disease will be easier and more efficiently identified, and those will be the areas of interest for relationships that predispose illness. It will be an exacting process, since the diseases that Generation Scotland aims to research are conditions that have multi-factorial genetic predispositions: cancer, heart disease, and mental health.

Databases and informatics

The development and role of health informatics in research is central to the Generation Scotland project. One of the areas of interest is how the already existing large quantities of data held from disparate studies across Scotland may be linked to create a more detailed and informative picture of health across Scotland. Thus data from a study in one area might be linked to a database containing, for example, genetic data to provide an added dimension to that study. This may, or may not, be possible with already existing databases but has potential for future projects, so that successful health studies and interventions could be rolled out across the country through linking databases. Databases offer the possibility of sharing data already gathered rather than duplicating work that has already been carried out, offering a collaborative and coordinated approach to health research that makes more effective use of skills and resources in Scotland.

Law, ethics and society

For the wider society of Scotland the project has legal, ethical, and social implications. Generation Scotland recognizes this, and has involved people with expertise in these fields from the outset. Dr Graeme Laurie and Johanna Gibson (2003) from the School of Law at the University of Edinburgh have written a detailed report on The Legal and Ethical Aspects of Generation Scotland: "The core concern of this project is the need to strike an appropriate balance between the public interest in ensuring that the value of generation Scotland is realized and brought to the Scottish people and, second, the interest in ensuring that individuals who participate in the project are adequately respected and protected."

Researchers from the INNOGEN Centre at the University of Edinburgh, are currently engaged in a program of Public Consultation. INNOGEN is based in Edinburgh and works in collaboration with the Open University as part of the ESRC Genomics Network carrying out research into innovation in genomics, which includes the public understanding of genomic issues. "INNOGEN's research will provide a sound base for decision-making in science, industry, policy and public arenas and will improve our understanding of each of these groups and their interaction." (Innogen 2004).

Colleagues from the School of Law and the INNOGEN Centre will be delivering papers on their work in this field elsewhere at the conference.

A proposal to study the discourses of Generation Scotland

Discourse analysis is a theory of language and communication, a perspective on social interaction and an approach to knowledge construction across history, society, and culture (Wetherall, Taylor, and Yates 2001, 1–8). The discourse of genetic databases is constructed from multiple discourses, including medicine, science, public health, Primary Care, law, ethics, technology, social science, the media, and a wider public. The discourses have differing frameworks and create different representations. My area of interest is the relationship between the different discourses and how they interact.

Generation Scotland has hosted a public meeting on the creation of the genetic database, and has sent representatives to other meetings that have been organized to discuss genetics and the role of genetics in future research. One of the most noticeable things about these meetings is the multiple discourses that are being deployed. Each discourse has its own historical and social context. It reflects a way of thinking about things which will seem to the speaker to be "common sense," but when that individual speaks they do not create their own language but use terms which are culturally, historically, and ideologically available to them, and when they speak they *act* to promote or contend the issues being raised at these meetings. The most commonly contended issues are the implications of collecting genetic and patient records for research on kinship, "at risk" relatives, screening, disability, discrimination, insurance, employment, and rights. The success of the data collection for the Generation Scotland genetic database is dependent on public support. Public support is contingent on people being able to understand and make informed decisions about participation and consent.

There is a growing body of literature on genetic databases which is incorporated in a wider literature of genetic research and its application. These may be characterized as the "discourse of concern" and the "discourse of great promise" (MacIntyre 1997). Through the paper so far I have included quotations from some of the different discourses that have been involved in the development of Generation Scotland which broadly represent the "discourse of great promise" in technological innovation, medical therapy, and benefit to society. The "discourse of concern" can be exemplified by the following quotations:

From the counter-discourse on family based research:

> Inasmuch as modern medicine increasingly regards diseases as rooted in genetic inheritance, the questions we pose are these: first, how do such explanations influence people's understanding of family and kin? In this context we explore the degree to which medical genetics may place the patient in a double bind between qualitative certainty [it runs in the family] and quantitative uncertainty [risk calculations] of genetic inheritance ... Second, we will propose that family and kinship are being medicalized as a result of the current emphasis on medical genetics and its clinical application (Finkler, Skrzynia, and Evans 2003).

From the counter-discourse on identifying predispositions:

> Observers worry, for example, about the psychological effects of learning one's predisposition to a deadly disease, particularly in the absence of any effective treatments. Not only could this lead to detrimental psychological effects for individuals and their families, but it could also lead to a kind of stigmatization and societal discrimination that goes well beyond the economic concerns of most legislation to date (Everett 2003).

From the counter-discourse on developing new treatments:

> Targeting medicines at those identified as "genetically susceptible" (sometimes known as the "healthy ill") also has serious implications. The potential costs of this approach have never been properly assessed, but it may prove to be both wasteful and unsafe (due to side effects and poor compliance), ineffective (because medication does not tackle the underlying causes of disease), and extremely costly to the NHS (by vastly expanding the market for medicines for healthy people) (GeneWatch UK 2002).

The initial stage of the analysis of the discourses will look at the language used in written documents and published literature and the way in which particular issues are addressed, e.g. consent, privacy, kinship, commercialization, and discrimination. The focus will be on the interaction between the multiple discourses. The analysis of the multiple discourses that are shaping the Generation Scotland genetic database project would contribute to a better understanding of the relationships between and interaction of these discourses.

Brief example analysis

Some of the quotations made through the course of the paper should include familiar language and their genres may be recognizable even without the references. Health policy documents typically include words such as, *better, higher, lower, greater*. Immediately it is possible to relate these words with other words that are associated within this discourse, *better services, higher treatment rates, greater efficiency, lower costs*. The counter-argument uses the same words in opposition: *potential costs, extremely costly, absence of effective treatment*, and invokes *psychological effects* and *deadly disease*.

Legal and ethical discourses typically include words like *ensure, protect, defend*, and we are familiar with the words of reassurance: *ensuring privacy, protecting property, defending rights*. These are the words that particular world views believe will protect from the *serious implications, societal and economic discrimination*, and *exploitation* of the counter discourses.

Conclusion

The study of the multiple discourses involved in creation of a national genetic database offers a way of critically examining the process, contributes to transparency, and engages with the wider discourses of concern. These are the discourses that will inform and engage with the public within an alternative historical and social context.

References

Christie, B. 2002. *A strategy for health and wealth creation through gene knowledge.* Edinburgh: Royal Society of Edinburgh Report.

Everett, M. 2003. The social life of genes: Privacy, property and the new genetics. *Social Science & Medicine* 56:53–65.

Finkler, K., C. Skrzynia, and J. Evans. 2003. The new genetics and its consequences for family, kinship medicine and medical genetics. *Social Science & Medicine* 57:403–412.

Genetic Health in the 21st Century (21CGH). Fact sheet. http://www.erihost.com/gs/21CGH.htm

GeneWatch UK. 2002. Biobank UK: A good research priority? Parliamentary Briefing No. 3. November. Available online at: http://www.genewatch.org/publications/mp_briefs/mpbrief_3.doc

Innogen. 2004. Homepage. http://www.innogen.ac.uk

Laurie, G., and J. Gibson. 2003. *Generation Scotland: Legal and ethical aspects; Full report.* Available online at: http://www.law.ed.ac.uk/ahrb/publications/index.asp?show=op

MacIntyre, S. 1997. Social and psychological issues associated with the new genetics. *Philosophical Transactions of the Royal Society Biological Sciences* 352:1095–1101.

Stewart, S., ed. 1998. *The Possible Scot: Making healthy public policy*. First Report of the Healthy Public Policy Network. Scottish Council Foundation.

Shaw, A., A. McMunn, and J. Field. 2000. *The Scottish health survey 1998*. Scottish Executive. Department of Health. Available online at: http://www.show.scot.nhs.uk/scottishhealthsurvey/sh8-00.html

UK Biobank. 2004. Homepage. http://www.ukbiobank.ac.uk/

Weber, M. 1947. *The theory of social and economic organisation*, transl. A. M. Henderson and T. Parsons. New York: Free Press.

Wetherall, M., S. Taylor, and S. Yates. 2001. *Discourse theory and practice*. London: Sage Publications.

19

Genetic databases and public trust

by Mairi Levitt and Sue Weldon

AS UK BIOBANK is to be run on an opt-in basis it will be dependent on volunteers and their individual consent. However, it will not offer specific or direct benefits to the donor, nor will it offer the possibility of specific and immediate benefit to other people in the way that a donation of blood may do. In the absence of any other incentive it therefore relies on people having the motivation to give. In the absence of direct benefits the HGC report, *Inside Information*, "conclude[s] that there are good reasons in genetics … for stressing the benefits of altruistic conduct" on the part of the individual donor (HGC 2002, 9). As well as requiring motivation, a successful biobank also depends on people having a lack of concern about, or trust in, the way their data will be collected, stored and used. With the experience of the controversy over GM food, all those with an interest in developing genetic research recognize the necessity for public trust. The recent Department of Health report *Our inheritance, our Future* (June 2003) has a chapter entitled "Ensuring public confidence," with the introductory comment that "we want to maximize the benefits of genetic research but this will not be possible without public acceptance and public confidence" (Department of Health 2003, 72). Policy documents recognize that public trust in the regulation of genetic research and applications cannot be assumed; rather the causes must be identified if legitimacy is to be established.

In the UK there has already been extensive research and consultations to determine public perceptions of BioBank, including research commissioned by the MRC/Wellcome (2000; People Science and Policy Ltd. 2002) and the Human Genetics Commission (HGC 2001; 2002). The Eurobarometer survey (Gaskell, Allum, and Stares 2003) also has data relevant to questions of trust in biotechnology and covers all EU member states. One key point from this research, relevant to the issue of trust in relation to genetic databases, is that people are ambivalent about the benefits and dangers of genetic research. This may be seen as a result of a deficit of information. For example the MRC/Wellcome Trust report that "negative associations" are "based sometimes on misinformation and mistaken assumptions," though none of the quotations given support this statement (MRC/Wellcome Trust 2000, 25f). There are concerns about commercial ownership of medical genetic databases and about access by commercial organizations, including insurance companies and employers, to databases. There is a lack of confidence that rules and regulations are keeping pace with new research developments

(HGC 2001). Most reports end with a plea for more openness and more public consultation.

The need for more research on trust and privacy

Surveys are a fairly blunt instrument and are sometimes misleading. For example, they may result in calls for more education which would be seen to be inappropriate once people's views are probed further in ways which enable them to show what they know rather than what they do not know (Levitt 2003). The qualitative research on UK Biobank has been specifically about the barriers to participation. We decided to use focus groups to explore perceptions of privacy and trust more broadly, starting with perceptions of privacy and trust in the participants' own lives, and leading on to views on data collection and storage, medical databases, and finally biobanking. The method enabled us to listen to the way people discuss these perceptions in an interactive group, for example justifying their opinion with stories and experiences. Since participants had not necessarily thought about genetic databases before the group met, we could explore how people used their existing knowledge to make sense of a new issue.

Five focus groups were held in the north, south, and south west of England, and in Wales and Scotland in November to December 2002. A final sixth group was recruited in the north west of England in April 2003 consisting of ethnic minority groups not represented in the first five. Participants were recruited by market researchers, but the groups were conducted by the authors. Each group had 6–8 participants with a wide range of ages and socio-economic groups, and a mix of men and women. The discussions lasted one and a half hours and were tape-recorded and fully transcribed.

Why don't people trust?

The discussion that follows picks out five main themes which recurred when people expressed a lack of trust. As discussed at the end, expressing concern did not necessarily mean people would not donate in the future.

1. "A LOT GOES ON BEHIND THE SCENES THAT THE PUBLIC DON'T KNOW ABOUT" (MALE, FG6).

This concern is not necessarily addressed by more information, because if the source is not trusted then there will still be the suspicion that something is being concealed.

There is a lack of control over how information will be used; it may start as a donation for research into serious diseases but *"there's no lines to be drawn once the information is given … And it will just go from there, it will just snowball. … I think it is a very, very dangerous thing"* (male, FG4).

2. THE NUMBER OF PLAYERS INVOLVED WITH VESTED INTERESTS

There are complex networks of vested interests: commercial and business interests including pharmaceutical companies, but also the interests of government, the military, and the police. The numbers of different types of complex organizations make it difficult to place trust as it is not clear who exactly you are placing your trust in or who can speak for the organization. People were looking for underlying motives like commercial profit. As the Eurobarometer report concludes "there is little perceived 'harmony of interests' between citizens and business or powerful multinationals" (Gaskell, Allum, and Stares 2003). There is a marked perception of a public/private divide. Publicly funded research is for the "public good," whereas privately funded research is for profit. In one instance a participant used the example of blood donation—where it is common practice to give *"freely and gladly"*—to illustrate his concern that publicly owned resources could be *"turned into profit for someone else. That is not the purpose for which I donate the blood and if I thought that was happening I would stop doing so. Emphatically"* (FG1). Extending this to genetic donation another participant explicitly said: *"You cannot have a government institution with the tax payers paying for it and a private institution making profits out of that information which is being supplied voluntarily by the public. The two are not compatible"* (FG1).

3. THE POTENTIAL FOR HARM

From their own experience participants knew that information is valuable to someone and no system is secure so someone will be ready to buy and to sell it. *"It's impossible to secure it because there's always somebody out there to make a buck out of it"* (male, FG5). It would be naïve to expect anyone who obtained information to be interested in benefiting individuals. People already had examples of discrimination by insurance companies and employers which they could extend to genetic information. Ethnic minority groups gave examples of discrimination on grounds of race, others on the grounds of a past health record or of current health issues where individuals could be helped instead of penalized. The husband of an ex-alcoholic explained that his wife applied for insurance when taking out a mortgage, *"she applied for the insurance and they said 'we've got to write to your doctor.' And they said 'she's an alcoholic.' But that was twenty years ago. Why does that count now? .... and they would not insure us"* (FG1). Someone susceptible to migraine argued that the employer says *"I won't be able to do the job properly, don't think to limit time in front of a computer screen"* (female, FG6). Those who mentioned good ways of using genetic information, like finding a rapist through DNA tests, were still ambivalent. *"I don't know how I'd feel if they had mine and ...everyone was looking at it"* (female, FG2).

While people may be inclined to donate for altruistic reasons "freely and willingly," the collectors and especially users of information are not necessarily altruistic. In fact, participants know that some definitely are not; namely pharmaceutical companies, insurance companies, and employers. Yet policy documents address potential donors on the importance of altruism, not the institutions which will be holding and using the information and which are not necessarily under the control of national governments.

4. THE PROBLEM OF ENFORCEMENT

Without trust the existence of regulation does not reassure people. They will not expect it to be enforced effectively. Onora O'Neill discusses how a culture of audit may achieve compliance and, at best, increase trustworthiness of institutions but will not create trust (O'Neill 2002, 133f). Questions of trust shift to those policing the audit and to what exactly is being measured. *"We may have many rules and regulations but how many enforcers do we have? How often are these places inspected? How frequently are prosecutions brought? And how effective are they? I suspect it's like the Health and Safety Executive. Nothing happens until you have a disaster and then they go out and investigate it. It's not proactive*" (male, FG1). Again, reassurance about the enforcement procedures would not necessarily ensure people's trust.

5. CONSENT GIVES CONTROL TO OTHERS

Participants were not particularly impressed with an emphasis on individual consent, because far from giving control it assigns control to others. Information can travel very rapidly and be appropriated by numerous agencies. People talked about genetic information being patented and its use restricted instead of the findings being available to all. Some took a more sophisticated approach to consent as a two-way process, rather than simply a process of giving up rights and control: you have consented to give information freely; they have a responsibility to use it wisely, not to waste it, not to let it be used against you/to harm you, to give back information, to protect you and your information. "*… to a certain extent … there should be a liability from the other side.*"

Who would you trust?

The common response to the question "who do you trust to be in charge of Biobank?" was *I trust no one* and/or cynical laughter! In practice people do of necessity place trust, and having laughed the groups went on to discuss the features of a trustworthy institution. There were two approaches in the discussion, to set limits or to find "safe hands." It was suggested that there should be limits on the type of research that could be done, on the amount of research that could be carried out without the need to seek fresh consent or new samples, on the role of vested interests including pharmaceutical companies and the researchers themselves, and on the power of any one person or interest group. Participants expressed shock at the arrangements in Iceland; "*so much trust in one person, if it fails they're absolutely screwed. Relying far too much on deCODE.*"

The second approach was to argue that a pair of "safe hands" was needed because guidelines can be broken and regulations cannot be enforced all the time. "*All I want to know is that my DNA is in safe hands with a group of people in Europe, with a group of people in America, with a group of people in the Far East and so on. And they're all answerable to a body that's part of the UN…*" (male, FG5).

There were people in the groups who shared the others' concerns but were prepared to trust because they felt that benefits of medical research would outweigh the dangers. They were prepared to accept abuses, possible disasters, and researchers who mislead the public about what they are doing, so long as no one would be coerced into taking part. For them the emphasis on individual informed consent was the right approach. For the rest individual consent did not address wider concerns about the type of society that genetic applications could bring about, about the commercialization and commodification of information, and about the privatization of public goods. Individual consent assigns control to other people whereas donating a sample for research should entail ongoing responsibilities and obligations on *both* sides. This is worrying for those who hope that public consultation and the language of openness and transparency will establish legitimacy for the governance of genetic databases. For UK policy makers the ideal model for DNA sample donation is blood donation: samples are freely given for altruistic purposes and no return is expected. Donors receive only general information on how donations are used. But participants did not necessarily see donation to a Biobank in the same light. They matched their conduct to their perceptions about those who would use the sample, e.g. that a Biobank sample might be used for commercial success and profit, not for the immediate benefit of a sick person. Therefore they expected both some benefit from, and some control over, their donation. In this context simple altruism seemed naïve and even dangerous.

Acknowledgments

This article was produced as a part of the project Ethical, Legal and Social Aspects of Human Genetic Databases: A European Comparison (ELSAGEN), financed between 2002–2004 by the European Community (QLG6-CT-2001-00062).

References

Department of Health. 2003. *Our inheritance, our future. Realising the potential of genetics in the NHS*. Cm 5791-ll. London: Department of Health.

Gaskell G., N. Allum, and S. Stares. 2003. *Europeans and biotechnology in 2002 Eurobarometer 58.0*. A report to the EC Directorate General for research from the project "Life sciences in European Society" QLG7-CT-1999-00286.

Human Genetics Commission (HGC). 2001. *Public attitudes to human genetic information*. Report prepared by MORI. London: HGC.

Human Genetics Commission (HGC). 2002. *Inside information: Balancing interests in the use of personal genetic data*. London: HGC.

Levitt, M. 2003. Public consultation in bioethics. What's the point of asking the public when they have neither scientific nor ethical expertise? *Health Care Analysis* 11:15–25.

MRC/Wellcome Trust. 2000. *Public perceptions of the collection of human biological samples. Qualitative research to explore public perceptions of human biological samples*. Report prepared by Cragg Ross Dawson for the Wellcome Trust and Medical Research Council.

O'Neill, O. 2002. *Autonomy and trust in bioethics*. Cambridge: Cambridge University Press.

People Science and Policy Ltd. 2002. *BioBank UK: A question of trust; a consultation exploring and addressing questions of public trust*. Report prepared for the Medical Research Council & Wellcome Trust. London: People Science and Policy Ltd.

20

"Public databases and privat(ized) property?"
A UK study of public perceptions of privacy in relation to population based human genetic databases

by Sue Weldon and Mairi Levitt

Introduction

PROTECTION OF PRIVACY is a key issue to be addressed in the setting up of population-based databases, and this study arose out of an EU project to investigate the ethical, legal, and social aspects of such databases in four countries: Iceland, Estonia, Sweden, and the UK.[1] The Icelandic Health Sector Database was the first population-based database to be proposed (1998) but many others are now being planned in other countries, including the UK.[2] So far, no uniform general guidelines exist for building up and operating these databases, but it is expected that new guidelines will be required in all these areas of policy, and particularly to address sensitive issues such as confidentiality and privacy. However, since such issues are social and cultural phenomena, these new frameworks and guidelines will need to be informed by a better understanding of public perceptions and attitudes with regard to them.

The privacy problem in biobanks

Genetic privacy is a concept that has gained increasing recognition (Laurie 2002) since the mapping of the human genome. The advent of functional genomics creates a context for, and a need to protect, increasing amounts of personal information that are now being collected and used for research in this area. The significance of genetic privacy, as opposed to any other form of personal privacy, is in the power and potential of information obtained from genetic tests to be used in ways that might affect how others view us (and the way we view ourselves). Based on the expectation that genetic knowledge will both improve our health and increase our ability to solve crimes, there appears to

---

1 Ethical Legal and Social Aspects of Human Genetic Databases: a European Comparison (ELSAGEN).
2 UK Biobank has now been given the necessary funding to collect samples from up to 500,000 volunteers aged between 45 and 69 and purports to be, so far, the largest study in the world of the relationship between genetics, lifestyle and the environment.

be huge potential for public good. The downside of this optimism is the realization that, without proper governance, the knowledge could be misused.

The situation presented by population-based genetic databases is a particularly stark scenario in which high risk of invasion of privacy is to be balanced against tremendous public benefits (e.g. Etzioni 2004). It is felt that problems arise where individuals are required to choose between risks to their own individual autonomy (and a right to privacy) and their responsibility (as citizens) to contribute to public benefits. This is broadly how privacy issues are framed. There is, however, very little current evidence of how people actually perceive these privacy issues in this context. Here, we report on the findings of a qualitative study, carried out in the UK, that suggests that this conventional wisdom is over-simplistic, and could be misleading in its implications.

The UK study: Our approach

The empirical study for the ELSAGEN project was carried out in four countries: UK, Iceland, Estonia, and Sweden to address, among other things, the following aims: to gain a better understanding of public perceptions of privacy in relation to personal medical and genetic data; and to investigate the extent to which standards of privacy might vary between countries.[3]

The method chosen to investigate people's perceptions of privacy in the UK was qualitative research involving focus groups. Our protocol was designed to begin by situating people in their own experience and only then to move on to discuss the particular issue of genetic databases. As the literature suggests, focus groups are useful in allowing people to generate their own questions, frames, and concepts (Barbour and Kitzinger 1999). In fact they are designed to raise issues that might not even have been thought of by the researcher. In contrast with surveys—which impose researchers' meanings of abstract concepts such as "privacy" and "trust"—focus groups are designed to enable researchers to explore people's perceptions as they operate in a social context. They allow people to respond in their own words and to use their own categories and associations. Careful interpretation of the results enables researchers to move one step further toward understanding the complexities and ambivalences of lived experience.

Six focus groups were carried out from November 2003 to April 2004 in locations throughout Britain. They were chosen for variety of location, urban and rural. The intention was to obtain a generalized selection of "the public" and, to this end, the groups comprised 6–8 people, equal men and women, ages from 18 to 70, and with a variety of socio-economic backgrounds (the final group was adjusted to address an ethnic bias that had inadvertently been introduced). Each of the discussions lasted for 1.5 hours and was tape recorded and transcribed.

---

3 It is often suggested that Icelandic people, for instance, have a more open attitude to the widespread use of personal information whereas, in the UK where the introduction of personal ID cards has been resisted so far, there is a greater sensitivity.

The UK study: Concerns about privacy

Our aim was to get a more grounded feeling for any particular perceptions arising in the UK context. The focus group protocol moved from discussion of people's general perceptions about privacy, and about how they felt able to balance individual rights against the "public interest," toward a more focused discussion of medical and then genetic privacy. We began by talking very generally about privacy and what that means for people: in doing so we tapped into some quite spontaneous and strong feelings about personal space and autonomy. It was notable, however, that when we introduced the topic of invasions of privacy, the conversation immediately switched to information—particularly the kinds of information received in junk mail, emails, and text messages. People felt that there was increasingly an invasion into private life by commercial interests. Another issue discussed was that of surveillance and discussion about the emergence, everywhere, of CCTV cameras. The concept of "big brother" was raised spontaneously without any prompting by the moderator. There was genuine ambivalence about the perceived need for greater security (particularly in the light of recent terrorist threats) that needed to be balanced against erosions of personal privacy. There was an explicit acceptance that "rights to privacy cannot be enjoyed by everyone in every situation." This was not just about balancing individual rights against public interest: in some cases it was also recognized that loss of privacy is a necessary feature of the convenience of modern technology (e.g. use of the Internet for shopping).

Medical privacy arose spontaneously as an issue concerning the need for security and confidentiality because of the perceived potential for discrimination based on health problems such as alcoholism or AIDS. Concerns were also expressed about the increasing evidence of data sharing in public services, although there was an awareness of a need to increase the efficiency of data services, particularly those concerned with medical data. Particular concern related to the idea of moving information from public to private companies, "who are going to target you for purchasing."

Overall, during a discussion that was focusing in on medical research databases, people began to reflect on the very real problems of protecting confidentiality of information stored in the public domain and accessed by multiple users. The problems included the extent to which the public/private divide could be policed. It was felt that information can be moved too easily these days–and sold on to commercial interests. When our respondents discussed issues related to participation, as donors to genetic databases, there was a very clear sense that anonymization of samples would not address the full range of concerns in this area. In particular there was a clear sense that people did not wish to surrender all "interest" in what happened to their sample:

> I would want a series of tick boxes. There are certain fields of medicine I wouldn't want [my DNA sample] used for.

Again, the discussion led back to concerns about who would have access to the samples.

> I would bitterly resent for instance my blood which I gave freely and gladly turned into profit for someone else…

In summary, the privacy issues that emerged from this qualitative study reflected and explained more fully concerns that have been raised in calling for appropriate regulation of genetic databases. In the following sections we look at how these concerns might, or might not, be addressed within existing legal and ethical framings.

How can these concerns about privacy be addressed within a legal framework?

We noted, on one hand, an awareness of the trend away from personal privacy rights as more and more data is placed within the public sphere. On the other hand, participants were noticing an increasing privatization of personal information that is stored in the public domain. In the case of genetic databases this is evident in a trend toward the creation of "intellectual property" and patenting of that property. There is much uncertainty in the law surrounding property rights in parts of the human body. In the UK the law has traditionally stated that there is no such legal protection. It is usually suggested, as the MRC does (2001), that donation to medical databases should be seen as a "gift" to the research organization in charge of the data, after a process of informed consent. Alongside this, confidentiality will be protected by "anonymizing" the samples once they have been placed in the "public domain." But this was not how our focus group participants saw the process. For one thing, anonymizing the data will only protect the confidentiality issue provided the technology can be trusted in this respect! In addition to this it was felt that, when people donate samples, the informed consent process assigned control to others. Consequently, this procedure did not adequately address people's feeling that they still wanted to maintain an "interest" in what happens next, including who would have access to the data and for what purpose.

Graham Laurie (2002) believes that a case could be made for a new construction of genetic privacy that does render it amenable to legal protection. He suggests that a balance could be struck between personal property rights and commercial interests. We do not discuss the pros and cons of property rights here, except to say this would pose a significant challenge for unified governance.[4] Our main point is that, although people did reflect on the commodification issue, they never presented it as a violation of their own property. For most people this appeared to be an issue about loss of personal agency and control, rather than ownership. The implication is that if people are being asked to make a donation for the "public good," they want to be confident that someone, or some institution, can be trusted to act responsibly, and in the public interest: and they wanted to be consulted about that.

Can these concerns be addressed within existing ethical framings?

We also examined our evidence to see how it fits within our existing ethical framings. As we have already pointed out, genetic databases raise a whole new set of issues, particularly in relation to individual informed consent. Ethical concerns have been identified at both the individual level (of participation) and at a wider social level. For instance,

---

4   Colin Bennett and Charles Raab (2003) have highlighted an asymmetrical treatment of privacy in the public and private sectors, in that governments are moving towards diminishing privacy protections in the private sector whilst enhancing protection in intellectual property.

it is suggested that there are tensions to be resolved between public and private interests, and notions of individual rights and "the common good" (Chadwick and Berg 2001). In the setting up of the Ethical Governance Framework for UK Biobank, these concerns have received much attention (Department of Health, MRC, and The Wellcome Trust 2003).

A key UK advisory document that deals with private and public interests in genetic information is a report published by the Human Genetics Commission (2002). The report confirms a general acceptance that genetic information is "a private matter." In seeking to address the tension between the "public interest" and individual rights, the report states that genetic knowledge brings people into a special moral relationship with one another based on the fact that "we all share the same basic human genome" and that we should therefore base our ethical approach on a principle of solidarity to promote "the common good." Our findings support an acceptance of this laudable aim, but with a significant qualification: who gets to decide what exactly constitutes the "common good?" What kinds of genetic research, for instance, would be acceptable?

The report also sets out certain principles designed to respect individual rights. These secondary principles—informed consent, privacy, and confidentiality—are considered to be of great importance in the ethical treatment of genetic privacy. However, again we point to a requirement for a more qualified understanding of how people experience these principles in context. For instance, we have indicated that people's perceptions of privacy are complex and contingent. Informed consent was also an important factor, but it does not address the question of "what happens next?" Confidentiality is important too, but people ask, "is it going to be feasible to guarantee that?"

What are the policy implications for the protection of privacy in relation to population-based databases?

Overall, our research indicates a need to understand better how people perceive privacy issues in the context of their personal circumstances. For instance, our research uncovered little evidence of people wanting to claim property rights over their samples, but neither did they want to see commercial interests doing so. The message is that, in the UK, people are willing to make their donation for medical research and to promote public health, but they are much more cautious about this when they lack confidence in the strict maintenance of the boundary between public good and private profit.

It appears then that this stark separation between individual rights and public interest is an over-simplification. Furthermore, if the surrender of rights to personal privacy is to be negotiated against a principle of solidarity for "public interest," the public sphere must be well regulated. The safe and sustainable operation of databases, the linking of datasets, and the arrangements for access by third parties depend on controlling the misuse of personal data. People need to feel that they can *trust* in the governance of the research agenda and the public sphere.

Acknowledgments

This paper was produced as a part of the ELSAGEN project (Ethical, Legal and Social Aspects of Human Genetic Databases: A European Comparison), financed between 2002 and 2004 by the European Commission's 5th Framework Programme, Quality of Life (contract number QLG6-CT-2001-00062). We gratefully acknowledge the support of the European Community. The information provided is the sole responsibility of the authors, the Community is not responsible for any use that might be made of data appearing in this publication.

References

Barbour, R., and J. Kitzinger, eds. 1999 *Developing focus group research*. Sage: London.

Bennett, C., and C. Raab. 2003. *The governance of privacy: Policy instruments in the global perspective*. London: Ashgate.

Chadwick, R., and K. Berg. 2001. Solidarity and equity: New ethical framework for genetic databases. *Nature Reviews Genetics* 2:318–321.

Department of Health, Medical Research Council (MRC) and The Wellcome Trust. 2003. *UK Biobank ethics and governance framework*. Version 1, 24 September.

Etzioni, A. 2004. DNA testing worth the risk? *Financial Times*, May 11.

Human Genetics Commission (HGC). 2002. *Inside information: Balancing interest in the use of personal genetic data*. A report by the Human Genetics Commission, May.

Laurie, G. 2002. *Genetic privacy: A challenge to medico-legal norms*. Cambridge: Cambridge University Press.

Medical Research Council (MRC). 2001. *Personal information in medical research*. London: MRC.

21

Becoming masters of our genes: Public acceptance of the Estonian Genome Project

by Külliki Korts

Introduction

THE ESTONIAN HUMAN Genome Project (EGP) was launched in 1999 as an initiative of a handful of gene researchers and public figures, with the far-reaching ambition of creating a national genetic database comprising one million samples, i.e. those of virtually the whole population. As the participation in the project is voluntary, public attitudes towards the project constitute one crucial factor in the final success of the project. In this respect the early phases of the project have been characterized by a controversy: though the EGP is enjoying remarkable public support and recognition, the actual participation rates among the population have remained rather low. After the first phase of data-collection, instead of the envisioned 100,000 samples, only some 10,000 DNA samples had been gathered.

In the current article, the reasons for this disparity are discussed in more depth, with the focus on the different factors influencing the support (or lack of it) for the project on two related, but still independent levels: general recognition of the project as a national scientific venture, and the decision to make the personal donation.

Background

Though the gene researchers enjoy a long-standing high reputation in Estonian society, the idea of a nation-wide gene bank can be described as the first "gene issue" catching the attention of the wider public, formerly characterized by low-level personal experience of the existing applications of gene technology. Before and during the launch of the EGP, there was a limited debate over the issue, but it was a debate that never reached far beyond a limited scientific community.

Just as there is general low public awareness, there exists very little research into public attitudes toward genetics or genetics-based medicine. Such research is confined to the studies on the awareness and support to the EGP project financed by the EGP itself. Under such circumstances, in the survey design[1] on which the following analysis is

---

1 The survey was carried out in the framework of the international research project ELSAGEN

based, both information on general attitudes toward science and technology, and hopes and fears in respect to the launching of the EGP was gathered.

Attitudes toward genetic research

The results of the survey show that, compared to Western societies, the Estonian population at large shares a rather unconcerned and optimistic view toward the recent developments in science and technology. By a two thirds majority, the benefits provided by the new knowledge are valued higher than the accompanying risks. Furthermore, the Estonian population is characterized by exceptionally high expectations in respect to the use of new discoveries in genetics in the medical field. The survey indicated strong public support for putting these new applications into use both in terms of diagnosing possible illnesses through genetic testing and making respective "corrections": the absolute majority of the respondents were in agreement with the statement that people should be encouraged to be tested in young adulthood for disorders that develop in middle age or later in life. Almost as many also agree that parents have a right to ask for their child to be tested for genetic disorders that develop in adulthood, and that genetic information may be used by parents to decide whether children with certain disabling conditions will be born. A two thirds majority also thinks that couples who are at risk of having a child with a serious genetic disorder should be discouraged from having children of their own. Though these responses might not reflect people's own potential behavior, they indicate a potential for rather strong social pressure for taking "advantage" of such preventive measures, once they become more widely available. This is where the Estonian public diverges considerably from its Western counterparts—whereas in the West, too, hopes are high for the prospects of genetic research (European Commission), the pressure for setting limits to the use of the new possibilities on ethical grounds is significantly stronger (Korts, Weldon, and Gudmundsdóttir 2004).

Compared to the support for the potential uses of new applications of gene technology, the number of people perceiving risks accompanying the wide use of gene tests and similar technologies is much smaller. Approximately half the respondents consider justified the prediction that insurance companies will start to demand gene test results when determining the level of insurance premiums, and that employers will start to demand gene test results from candidates for certain jobs. Smaller numbers of people consider it possible that knowledge of genetic information will start to influence social and interpersonal relations.

However, the high level of optimism and low level of caution can also indicate a lack of profounder acquaintance with, and reflection on, such issues among the general public. As already mentioned, genetic research did not get too much public attention before the idea of the gene bank. Even in the case of the EGP, after two years of intensive propagation of the project, in late 2002, only 62% of the Estonian population claimed

---

(Ethical, Legal and Social Aspects of Human Genetic Databases). The survey was conducted in December 2002 as face-to-face interviews with a nationally representative sample of 917 respondents.

to have heard about the EGP, with 7% considering themselves well informed on the issue.

Acceptance of the EGP

According to the survey, for the majority of respondents knowledgeable of the gene databank project, its benefits, both personal and for society as a whole, seem to outweigh the probable risky consequences. The major advantages of the EGP are considered to be medical, but its contribution to economic development and international recognition are also considered important. The major risks are seen in the possibility of leakage and abuse of the data collected by the EGP, as well as in the possibility that the actual economic benefit will not reach beyond a small circle of investors and pharmaceutical companies. However, as with attitudes to genetic research in general, in the public eye, the negative consequences of the EGP are evaluated as less important than the benefits outlined above.

The voluntary nature of participation in the project is complemented, however, by a unique feature of the EGP: it is the only population-based gene bank that has granted donors the right to access their own data in the bank. According to the survey, the vast majority of respondents believe that people will personally benefit from participating in the project by getting to know their health risks. In fact, this seems to constitute one of the major appeals of the gene bank project based on voluntary participation. According to the survey, of the potential donors, 83% plan to definitely apply also for a personal gene card "containing the genome of each gene donor,"[2] while only 2% reject it decisively.

The generally positive inclination in public attitudes toward the EGP is in correlation with the level of trust toward different persons and institutions as the most reliable sources of information on the project. Genetic scientists and the employees of the Estonian Genome Project are trusted by more than 80% of the population. People's trust toward the persons connected to the project outweighs that toward, e.g., the Ethics Committee supervising the activities of the project, and family doctors—the actual contact persons of the potential gene donors—and other scientists. Especially low trust characterizes the public attitude toward journalists (20%), although for the majority, printed media and television constitute the principal sources of information about the project, leaving other sources far behind (e.g. family doctor, friends, relatives).

The major perceived benefits of the project, both societal and personal, echo to a large extent the features emphasized in the media discourse (see Tammpuu 2004). This gives reasons to believe that the popularity of the project is based on its successful presentation as an impressive national scientific venture. However, support expressed in these terms is rather abstract in nature and might not reflect the actual eagerness of people to personally become donors. Indeed, despite generally positive evaluation of the project, only 24% of the population knowledgeable about it were planning to take part,

---

2   Cited from Krista Kruuv's article Kas hakata geenidoonoriks [Should One Become a Gene Donor]? (*Postimees* October 23, 2002). Krista Kruuv was the Director of the Estonian Genome Project Foundation.

and more had taken a negative stance (40%). Many, however, had not made up their minds (36%). Moreover, the people who intended to participate and those who were "opting out" did not show significant differences in their attitudes toward the project, or toward genetic research in general.

Conditions of participation in the EGP

As in the Estonian Genome project, voluntary donors are recruited by family doctors, who take a blood sample and interview the donor. All participants sign a consent form, by which they accept that they can be approached again for supplementary health information, and that this can be gathered also from other sources, e.g. hospitals. There also exists the opportunity to sign a special form, whereby their data will be inserted into the data bank anonymously. Later, the donor has the right to demand the removal of data that can be decoded (i.e. make them anonymous). As for the requirement of informed consent, there is a near absolute consensus among the respondents that it is necessary to ask for written consent from the donors. At the same time, less than half consider it necessary to allow donors to demand the anonymization of their data after first consenting. Other survey data also reveal considerably high trust toward the working principles of the project: more than half of the potential donors are willing to give the Genome Foundation a free hand with access to other health databases, while only a quarter have decided to forbid it. However, the majority considers it most important to be informed of what kind of research will be done using their gene data. In contradiction to the current regulation that leaves the consent rather open, there is majority support for the idea that fresh consent should be required before new research is conducted on their existing samples.

Discussion

The results of the survey reveal a high level of technological optimism among the Estonian public, and even an eagerness to make maximum use of the new ways to control events formerly governed by nature. The same feeling of optimism can be discerned in general attitudes toward the gene project: though a certain caution is expressed in public opinion, the EGP has gained a high level of acceptance among the Estonian population. However, this acceptance has several specific characteristics that should be kept in mind: mirroring the low level of personal experience with genetic technology, general knowledge of the project has remained rather low. Most information about it is received from mediated sources, rather than personal contact, e.g. family doctors, who will be the actual mediators between the gene bank and the potential donor. It can be concluded that until now, the project has not been able to inspire genuine interest in the majority.

Though the presentation of the project as a national scientific undertaking harmonizes well with the generally high level of technological optimism, as well as with national pride, it has not had sufficient appeal to mobilize people on grounds of solidarity of contribution to the common cause. Rather, for the vast majority of the potential donors, the possibility of being granted a personal gene card constitutes the major attraction.

Besides, those people who intended to participate and those who were "opting out" did not show significant differences in their attitudes toward the project, nor to genetic research in general. This gives reason to assume that for the majority, the decision to participate or not was intuitive, rather than a reflective decision, which can be altered if by chance (e.g., through a visit to the family doctor) they should get more deeply involved. The "latent potential" seems to be rather high, and this gives us grounds to believe that the actual participation of most people is susceptible to rather concrete factors: e.g. first getting their primary interest triggered, the trust the potential donor has in the family doctor, or the satisfaction of the potential donor with the concrete conditions of gathering and restoring the donation.

The survey also revealed that the question of control over the contributed DNA sample and health information is indeed an important issue for potential donors. It seems, however, that given an adequate level of trust in the people and institutions engaged in the process of gathering, restoring, and processing DNA-samples, rather than wanting to keep strict control over their sample, people just want to be kept informed. People are committed to the requirement of informed consent, and would like to be kept informed of the "fate" of their DNA sample in the future. Also, willingness to allow the GP access to other sources of personal health information means that the GP and the donor will enter a relationship involving a remarkably high level of invested trust on the part of the donor, which, however, will make that trust vulnerable to the smallest violation by the GP.

In conclusion, it can be asserted that general acceptance and support of the EGP has been gained against the background of a generally high level of technological optimism, including high expectations for the new developments in gene technology, as well as the high reputation of the main initiators and designers of the project, giving confidence in the "rightful" aims of the project. However, success in turning this "latent potential" into actual participation is dependent on the way in which recruitment is done "on the ground," both in attracting the first interest, but, even more importantly, in the ability of the EGP to create and maintain a rather precarious trust in each phase of gathering, keeping, and processing the personal donation.

Acknowledgements

This paper was produced as a part of the ELSAGEN project (Ethical, Legal and Social Aspects of Human Genetic Databases: A European Comparison), financed between 2002 and 2004 by the European Commission's 5th Framework Programme, Quality of Life (contract number QLG6-CT-2001-00062). I gratefully acknowledge the support of the European Community. The information provided is the sole responsibility of the author; the Community is not responsible for any use that might be made of data appearing in this publication.

References

European Commission. 2001. Europeans, science and technology. Eurobarometer 55.2. Brussels: Directorate-General Press and Communication, European Commission. Available online at:
http://europa.eu.int/comm/public_opinion/archives/eb_special_en.htm

Korts, K., S. Weldon, and M. L. Gudmundsdóttir. 2004. Genetic databases and public attitudes: A comparison of Iceland, Estonia and the UK. *Trames* 8:131–149.

Tammpuu, P. 2004. Constructing public images of new genetics and gene technology: The media discourse on the Estonian Human Genome Project. *Trames* 8:192–216.

"We don't have that many secrets" — the lay perspective on privacy and genetic data

by Anna Birna Almarsdóttir, Janine Morgall Traulsen and Ingunn Björnsdóttir

Introduction

THE DEBATE / CONTROVERSY

BACK IN THE late 1990s researchers spoke of "witnessing the dawn of the sociological study of the new genetics" (Conrad and Gabe 1999). It didn't take long before the academic literature (including new scientific journals) exploded and popular science began reporting on this topic on a regular basis. Some of the issues explored in the sociological research include: commodification of the body, genetic privacy, the intersection of biomedicine and society, the implications of new genetics for society, and how knowledge of genetics is produced and structured. In the media public debates over issues related to new genetics have flourished; however, there is still a paucity of research concerning "lay" contributions to the bioethics debates. This paper will address that issue.

The Icelandic Health Sector Database

Following almost a year of national debate, the Icelandic parliament passed the Health Sector Database Act in December 1998 (Icelandic parliament 1998). This act authorized the Minister of Health to grant an exclusive license to a for-profit corporation, deCODE genetics Inc., to create a database—the Icelandic Health Sector Database (HSD)—of the medical records of all persons receiving health care in Iceland. The intention was for the Icelandic government to use the database for planning and policy purposes. In return for conceiving, creating, and paying for the database, the government granted deCODE genetics Inc. exclusive rights to market it for twelve years.

The Icelandic HSD was conceived of mainly as a tool for epidemiology, health services, and genetics research. The idea was that for genetics research, the HSD would be used in combination with genetic and genealogical data available in other databases. The combination of the three database types (health care, genealogical, and genetic data) was envisioned as a unique tool for medical research that allows users to access the information in order to ask questions about disease modeling, disease management, and genetic linkage of traits and outcomes (deCODE 2001).

No single issue has been as much debated in Iceland as the HSD (Ólafsson 2002). The debate has been critical, including over 700 newspaper articles, more than 100 radio and television programmes and several town meetings all across Iceland. According to an analysis by Pálsson and Hardardóttir (2002) of the public discourse as it appears in the Icelandic daily newspaper "Morgunbladid," news items were typically neutral and opinion pieces, by definition, usually had a rather strong and clear agenda.

Experts, mainly medical doctors and academics, dominated the debate carried out in the media on both sides. Arguments against the database were concerned with confidentiality and personal autonomy (assumed consent) and the fear that sensitive health data could be used indiscreetly in a way that could harm individual patients. The proponents of the database were led largely by highly educated persons—most with expert knowledge from the fields of medicine, molecular biology and epidemiology—who emphasized the positive implications of the database. These included the opportunities the Icelandic Biogenetic Project provides in terms of medical advance, employment, entrepreneurship and private initiative (Pálsson and Hardardóttir 2002; Sigurdsson 2001).

We found that the two sides of the HSD controversy—the proponents and the critics—shared certain similarities. The major participants in the controversy on either pole were similar in that they expressed strong convictions as to why the HSD was either good or bad for Icelandic society; their arguments were informed by the bioethical discourse and they were mainly academics. Despite the raging debate and criticism, public opinion polls in the late 1990s showed support of the database (Gulcher and Stefánsson 2000).

In 2002, deCODE genetics Inc. planned to launch a nation-wide campaign to collect DNA material from blood samples of the general population throughout Iceland. An appeal to the general public to participate was planned by a group of scientists at deCODE who were separate from the database enterprise. The company had already collected blood samples through research on particular diseases known as "the disease projects." This group of scientists/researchers were very interested to learn about the opinion of the lay public, since without popular acceptance the project would fail.

Objectives

The results presented here are from a secondary analysis[1] of data collected in a pre-marketing pilot study, financed by deCODE genetics Inc., to assess public opinion regarding a campaign to collect DNA material from blood samples. This paper discusses the results derived from the secondary analysis concerning the informants' views on biomedical research, including the issues of confidentiality, data protection, and privacy.

---

1 A secondary analysis is defined by Hakim (1982, 1) as "any further analysis of existing data which presents interpretations, conclusions, or knowledge additional to, or different from those presented in the first report."

Methods

Focus groups (FGs) were chosen as a data collection method because they provide a forum in which participants can discuss a wider range of ideas and issues than would arise in individual interviews or surveys. FGs encourage discussion and reflection on issues of public concern, as well as the emergence of common or shared views (Bowie, Richardson, and Sykes 1995; Kitzinger and Barbour 1999; Kitzinger 1994; Waterton and Wynne 1999).

Participants

The aim of the primary study commissioned by deCODE was to study the lay view on the company's planned activities. The planners of the study were aware that there might be differences of opinion within and among the general public, and therefore decided to recruit members of the lay public in three different ways. Eight participants were recruited for each group.

The first group (Group I) was made up of persons who had already participated in one of deCODE's disease projects. A health care worker at the community primary health care center was asked to recruit persons, based on her knowledge of who had participated, as well as on certain selection criteria. The goal was to get equal numbers of men and women from the town itself and from the rural areas surrounding the town, and to obtain as wide a spread of ages as possible. Group I ultimately consisted of 2 men and 3 women aged 40 to 60.

Group II was made up of randomly selected persons who were not participants in any of the deCODE disease projects. Participants were recruited by a local person from the town. The local person was given similar criteria to those given to the health care worker. The ages of the participants ranged from 36 to 65 years old, and the group consisted of 5 men and 3 women.

Group III was made up of students from the local high school (the only one in the town). A teacher was asked to recruit an equal number of male and female students. Group III consisted of 3 men and 2 women from 19 to 20 years.

We purposely did not recruit anyone who had a job or research interests in the issues being explored. The focus groups were conducted at a site that was neutral with respect to health care and that was convenient for each group. While no monetary incentive was offered, a small gift was presented to participants as a surprise at the end of the session.

Interview guide

The participants were presented with a proposed plan as to how the deCODE research project (a campaign to collect DNA material from blood samples requiring voluntary help from the community) was envisioned and asked to comment on it. The moderator of the FG used an interview guide with selected key words as an aid to make sure that all aspects of the plan were covered. The actual proposed plan itself cannot be shown here, as it is proprietary (privately owned and controlled by deCODE).

Data handling and analysis

All FGs were audio taped and transcribed verbatim. Content analysis was conducted based on the transcripts of the FGs. The analysis consisted of finding, labeling, and compiling an overview of common themes discussed in the FGs that related to the studies' objectives. Thereafter emerging themes were identified and extracts from the FGs, which were typical of the views expressed, were identified and translated into English. The moderator was interviewed after each session as a means of assessing the dynamics of the FG and to hear what was said after the tape recorder was turned off. This method of "interviewing the moderator" also served as a form of validity check for the final analysis of the data (Traulsen, Almarsdóttir, and Björnsdóttir 2004).

Origins of the data

The data used in the secondary analysis were originally collected as part of a pilot pre-marketing study. The primary study was designed for deCODE genetics Inc. in order to gauge the views of the public in a small town on how best to proceed with a new project in order to avert negative reactions from the public. One of the authors (ABA) was employed by deCODE at the time of the study, and was involved in the planning and implementation of the pilot study. When she left the company for an academic position she was given permission to reanalyze the raw data for research purposes. Another of the authors (IB) was an independent consultant hired to plan the study and moderate the focus groups. The consultant moderator made it clear to participants at the beginning of each focus group that although deCODE commissioned the study, she was not an employee of that company and she guaranteed them anonymity.

The primary analysis had been carried out by deCODE employees whose interest was in specific research questions pertaining to the planning and execution of the proposed project, whereas the results reported here were based on an entirely new research question in a secondary analysis (see note 1).

Results

Solidarity and the use of medical / genetic research results

The results of the FGs show that deCODE's work in the field of genetics is viewed as beneficial, as it is seen as leading to improved health care and disease prevention. When the issue of genetic research came up in the older groups (I and II) they often referred to solidarity among the Icelandic people—to the idea that they needed to band together in order for this beneficial research to happen.

> If it is possible to find a cure for diseases by helping each other [Icelanders], then [deCODE] is more than welcome [to my blood and health information]. (Woman in Group I)

> deCODE can use whatever they want from me … if it will improve health care in this country and I do not understand those [people] who are against deCODE, it is a beneficial company and will make good things happen. (Man in Group II)

The young people in Group III touched upon issues of solidarity and talked about "miracle drugs" that could become useful for all. They pointed out the necessity of broad participation if medical research is to be successful. Two members in the group helped each other make the point that people should participate because they might benefit in the end:

> … and if they find some miracle drugs [for diseases] then of course … (Man in Group III)

> … then it will be beneficial for all people. (Woman in Group III)

The older groups (I and II) spontaneously started talking about how deCODE had been unfairly criticized. Both groups agreed that the critics would also expect to benefit from any medical advances resulting from this research.

> How would [the people who do not participate] respond if they had to sign a statement saying that they will not receive any medical help because they didn't participate [in the research leading to the cure]? (Man in Group I)

Group II even went so far as to say that it should be a citizen's duty to contribute to medical research. This group named the medical profession as one of the harshest critics, and commented that the critics as a whole are, in the words of one man:

> A loud minority group in society that likes to show off and is negative. (Man in Group II)

Confidentiality, privacy and data protection

The verbatim transcripts show very little concern with issues of confidentiality, privacy, and data protection in the focus groups of older participants. Group II discussed the ethical dilemma of cloning and the dangers of meddling with nature to a great extent.

> What happens if we artificially change genes—will it impact any other genes? We don't know. (Man in Group II)

Group I talked about data security in relation to bank information, conceding that there is no 100% security; but as one woman put it:

> One could just as well go in and shut the door [forever] (stop living in this society) if one is always concerned [about electronic data security]. (Woman in Group I)

Both older groups agreed that people in their town were not that concerned with privacy since they already know everything there is to know about each other.

> [Health care] information in such a small town as this—here we nearly always know if there is something wrong with a person—the whole town knows about it. (Man in Group II)

> We don't have as many secrets [as people in Reykjavík]—it's no use us trying to have them. (Man in Group II)

The young focus group (III) voiced a fear of exploitation of health care data. They thought that information would be less secure when kept in their own town than if it were moved to Reykjavík. They provided anecdotes of sensitive information leaking out of the primary health care station in their town.

> I think it would be much better if my medical records were kept somewhere in Reykjavík where people don't know me at all, rather than having them lying around in the hospital here. Because I think there is more danger of an information leak from here than from a place in Reykjavík. (Woman in Group III)

The data showed that although Group III was concerned about confidentiality, they were more critical of the local health care professionals than of deCODE genetics.

Insurance companies were mentioned by this group and Group II as inappropriate users of health care information. All groups agreed that it was the government's responsibility to make data secure and prevent misuse of information.

Discussion

SOLIDARITY VERSUS PRIVACY

Another focus group study was carried out in Iceland in 2001 (Traulsen, Björnsdóttir, and Almarsdóttir 2002; Traulsen, Almarsdóttir, and Björnsdóttir 2004), which focused on the hopes and fears of the lay public with regard to medicine and drug therapy (including gene therapy) in the present and in the future. The initial assumption of the research team was that the ethical issues of privacy, individual rights, and confidentiality (that had been prominent in the public controversy) would be important issues for the lay public. This did not prove to be the case; in fact issues of privacy and confidentiality were rarely mentioned. On the contrary, the concerns of the participants were situated in a social discourse in which issues of social equality were of central importance. The participants' arguments for remaining in the database were related to the benefits they saw in the future for their children and grandchildren. The lay public did not adopt the dominant bioethics rhetoric of the HSD controversy, but instead expressed their concerns in terms of social problems reflecting collective values and social solidarity (Traulsen, Almarsdóttir, and Björnsdóttir 2004).

Similarly, the results of this study showed a lack of interest on the part of participants in the concepts so heavily debated by the physicians and academics concerning bioethics. On the one hand, the groups of older participants voiced their solidarity with the rest of Icelandic society with regard to medical research. On the other hand, the young participants were more skeptical of deCODE's and the authorities' ability to protect the individual. Although data protection was discussed, the FGs would inevitably turn their attention to discussing the greater benefits of taking the risk of providing genotype and phenotype information. It was clearly considered to be more important that the entire population contributed and participated in medical research in order for Icelanders to reap the benefits.

This study was the first in Iceland to recruit young people in FGs specifically to discuss issues of medical/genetic research, and it was therefore an exciting and unexpected finding to see how they differed from older adults in their views on privacy and data protection. One possible explanation for these differences could be the fact that the young people did not have children, whereas most of the participants in the other groups had children and even grandchildren. In other words, these people could have a more personal interest in what happens beyond their own anticipated life span and what happens as a result of medical research which may be beneficial in the future.

It is also interesting to note that the Icelandic surveys mentioned earlier in this article found much more positive attitudes toward the database project than American attitudes toward genetic research in general. This points to fundamental differences between the two societies, including not only social and cultural factors, but also the difference in health care delivery—Americans have much more reason to be worried about privacy and other individual interests because they fear the loss of health insurance and job security if genetic information about them is revealed.

The dynamics of the focus groups

The focus group format enabled participants to explore a range of views and thereby concentrate more easily on common, rather than individual concerns. The method encouraged discussion and reflection among the participants on extremely complex issues, enabling participants to make the conceptual leap to issues of common concern. The interviews with the moderator, however, revealed some disadvantages of the FGs. In the group of young adults one person was clearly the leader and dominated the conversation. The moderator had to curb this individual several times and make efforts to draw out the other members of the group. Unfortunately, it was almost impossible to hear new ideas and views from the other participants, which may jeopardize the interpretation of the results regarding young adults. Another problem was that although many persons initially agreed to participate (we recruited at least 8 persons each time), 3 failed to show up in two of the groups.

Conclusion

This study presents results from three focus group discussions in which all the participants were from the same small town in Iceland. The main theme that emerged from the FGs was that of a national solidarity among Icelanders regarding biotechnology research. There was surprisingly very little concern with issues of confidentiality, privacy, and data protection. The results show a difference in opinion across age lines (between the younger and the older groups). The opinion of the more mature participants was that information about individual financial matters was more sensitive a topic than information about health related matters. Although this pilot study is not generalizable to the entire population, the results provide an interesting insight into the lay perspective of the Icelandic public with regard to the research being done by deCODE genetics Inc.

Acknowledgements

The authors wish to thank Ingibjörg Þórhallsdóttir for her part in the planning and conduct of the study on behalf of deCODE; and secondly, deCODE genetics Inc. for allowing access to the data that was collected for the company.

The results and the opinions expressed in this paper are the sole responsibility of the authors. deCODE genetics Inc. has had no influence on the secondary analysis and the results, discussion, and conclusion published here.

References

Bowie, C., A. Richardson, and W. Sykes. 1995. Consulting the public about priorities. *BMJ* 311:1155–1158.

Conrad, P., and J. Gabe. 1999. Introduction: Sociological perspectives on the new genetics; an overview. *Sociology of Health and Illness* 21:505–516.

deCODE genetics. 2001. Homepage. www.decode.com (accessed May 1, 2001; the text has now been removed).

Gulcher, J., and K. Stefánsson. 2000. The Icelandic health care database and informed consent. *The New England Journal of Medicine* 342:1827–1830.

Hakim, C. 1982. Secondary analysis in social research. A guide to data sources and methods with examples. London: George Allen & Unwin.

Icelandic parliament. 1998. Act on a Health Sector Database no. 139/1998. Passed on 17 December. Available online at: http://eng.heilbrigdisraduneyti.is/laws-and-regulations/nr/659

Kitzinger, J., and R. S. Barbour. 1999. Introduction: the challenge and promise of focus groups. In *Developing focus group research: Politics, theory and practice*, ed. R. S. Barbour and J. Kitzinger. London: Sage.

Kitzinger, J. 1994. The methodology of focus groups: The importance of interaction between research participants. *Sociology of Health and Illness* 16:103–121.

Ólafsson, S. 2002. Information policy disputes in Iceland. *International Information & Library Review* 34:79–95.

Pálsson, G., and K. Hardardóttir. 2002. For whom the cell tolls — debates about biomedicine. *Current Anthropology* 43:271–301.

Sigurdsson, S. 2001. Yin-yang genetics, or the HSD deCODE Controversy. *New Genetics and Society* 20:103–117.

Traulsen, J. M., A. B. Almarsdóttir, and I. Björnsdóttir. 2004. Interviewing the moderator, *Qualitative Health Research* 14:714–725.

Traulsen, J. M., I. Björnsdóttir, and A. B. Almarsdóttir. 2002. What do the public want? A study of the hopes and fears of the lay public for the future of medicine. *Zoom*, newsletter of the Community Pharmacy Section of Fédération Internationale Pharmaceutique (FIP) November: 4–6 (http://www.fip.org/).

Waterton, C., and B. Wynne. 1999. Can focus groups access community views? In *Developing focus group research: Politics, theory and practice*, ed. R. S. Barbour and J. Kitzinger. London: Sage.

Do regulations address concerns?

by Matti Häyry

PEOPLE CAN HAVE concerns regarding population genetic databases. Such concerns, if they exist, are likely to develop into political issues, which have to be settled by public authorities. But the public authorities do not always know what concerns people can have, or how they could be properly addressed.

In this paper, I aim to provide an outline of people's possible concerns, and some general remarks on how these concerns could be addressed by alternative policies within different political systems.

What, why, and how?

Population genetic databases are a relatively new phenomenon, and views and perceptions on their nature, purpose, and functions vary considerably. The experiences in Iceland and Estonia have clarified the picture slightly, but the ongoing debates show that different groups still rely on different initial descriptions.

As to the nature of population genetic databases, at least three views have been presented. Icelandic authorities have seen their projected database as a welfare agency; Estonian officials have stressed the commercial features of their model; and many critics have claimed that only essentially eugenic institutions can be created by any such exercise.

The main purpose of population genetic databases is, even in the eyes of their advocates, almost as unclear as their nature. Some say that these databases would primarily be expected to turn a profit; others emphasize scientific advances; and yet others highlight the aspect of public health promotion. These are not always incompatible ends, of course. But there are tensions, say, between an ideal of scientific knowledge as everybody's property and scientific knowledge as a privately or nationally owned commodity.

Those who have participated in ELSAGEN and similar endeavors have learned, during the last few years, much about the details of the operation of particular genetic databases. The variation is, however, substantial, and from the ethical viewpoint, it is still the safest course just to identify the three main functions of these institutions. All population genetic databases will *collect* or otherwise acquire tissue samples or genetic information from individuals or existing databases, *store* these, and *disseminate* them to potential users. These are the functions that give rise to most concerns people may have.

Ethical issues in population genetic databases

People's concerns about population genetic databases can be rather neatly captured by five ethico-legal concepts, namely

- privacy;
- consent;
- confidentiality;
- security; and
- public interest.

Empirical studies conducted within the ELSAGEN project and elsewhere seem to indicate that people's anxieties are, as these concepts suggest, mostly related to others intruding in "my own sphere" and using "my information," possibly without "my permission" and in a way that reveals "my important stuff" to others. Citizens can also have doubts about the trustworthiness of the officials processing the data, and they can worry about excessive control and potential discrimination by the state, by employers, and by insurers.

It is not clear how these concerns could, and should, be taken into account in the regulation of population genetic databases. One point is that people's attitudes seem to vary according to the degree of knowledge and awareness they have. Does this mean that only informed opinions should be reckoned? Or does it mean, on the contrary, that only spontaneous, uninformed reactions should be considered? Another difficulty is that some attitudes call for regulations, while others advocate free market developments or the unrestricted pursuit of scientific knowledge. If people have conflicting opinions, how should these be ranked?

I do not have direct answers to these questions. But let me list some of the main meanings of the words "privacy," "consent," "confidentiality," "security," and "public interest," in an attempt to clarify the issues involved.

Dimensions of privacy

Questions of privacy are dealt with in different ways in different legal and political settings. There are, however, three demands that should be addressed in one way or another by any legal system in the context of population genetic databases.

The requirement of *physical privacy* expresses the idea that other people, including public authorities, should not seize, search, or touch us, at least not without very good grounds for doing so. Insofar as samples for genetic databases need to be acquired by touching individuals—for instance, by sticking needles into them—violations of this requirement are always a possibility.

The notion of *informational privacy* draws attention to the fact that some types of knowledge and information are regarded as personal, sensitive, and perhaps inviolable. It has been argued that genetic data fall into this category, and if this is true, the founders and keepers of population genetic databases should be careful not to compromise people's integrity in this sense.

The principle of *decisional privacy* proceeds from the ethical claim that certain decisions in our lives are legitimately only ours to make. Others may disagree, and we may be called foolish or immoral if we hold on to our own views, but the foolishness or immorality does not justify coercion or constraints on our choices. The case can be made that decisions about genetic data belong to this class.

Varieties of consent

One of the main excuses for bypassing further considerations of privacy is that the individuals involved have consented to the intrusion of their physical, informational, or decisional sphere. Consent, however, comes in many packages, and the strength of the excuse can vary according to the option chosen.

Explicit consent means that the individuals in question have actually and knowingly given their permission to the suggested procedures. The current, although increasingly challenged, ethical ideal is that the unquestionably free and comprehensively informed authorization of indisputably competent persons is needed for any medical, genetic, or research-related interventions. The logic is that if authorization like this is secured, cooperation will be easier and the responsibility is shifted from the intruder to the individual intruded upon.

Implicit or *presumed consent* enters the picture when it is thought that explicit permissions would be too difficult or time-consuming to come by. Individuals who have not actually opted out of an arrangement can be seen to have given their implicit agreement to it. And individuals who have not opted out of similar activities in the past can, so the argument goes, be presumed not to have any problems with it now. This line of thought makes the intruders' actions less suspicious, as it aims to shift the moral responsibility to the individuals intruded upon.

Hypothetical or *rational consent* comes into play when two conditions are met. First, people have not actually consented to what others suggest. And second, it would be nonsensical to claim that they could have opted out now or in the past, but have not done so. The idea here is to insist that no rational person would object to the kind of intrusion proposed. The standard line is, "If they were rational, they would not have these concerns." A further embellishment is to say that only immoral individuals would like to benefit from other people's participation without joining the effort themselves. This type of consent seeks to facilitate and justify coercion by appeals to universal reason and justice.

Proxy consent has two principal senses. It can mean that relatives or friends are asked to estimate the wishes of individuals, when the individuals themselves cannot be asked what they think. It can also mean that legal guardians or community leaders are required to make the choice on behalf of the individuals, perhaps on the basis of the dependant's best interest, or on the basis of traditional or agreed values. The first type provides an educated guess at the will of the individual intruded upon; the second gives a paternalistic justification for the intruder's actions. Both models are particularly popular in medical emergencies.

Which confidentiality?

The consent to intrusions into our private sphere can be qualified in many ways. One of these ways is marked by the requirement of confidentiality, which can be interpreted more or less stringently.

Absolute confidentiality would mean that nothing others learn about our condition and affairs should be revealed to third parties without our prior permission. This demand makes a lot of sense in medical matters, as we do not always want other people to know about our sensitive or embarrassing ailments. The situation can, however, be different in the context of population genetic databases. Information is collected and stored in these databases exactly because third parties are expected to make use of it. It would be futile to let one's information be included on the condition that it will not be conveyed to anyone else.

Qualified confidentiality means that while, as a default position, data should not be passed on to others without our permission, the permission can be read into our decision if certain external criteria are met. One option would be to anonymize the data, and to argue that people's confidence is not betrayed by data sharing which does not make it feasible to track the information back to the individuals in question. Another possibility would be to use the data only for purposes which the individual has consented to. Although this could be cumbersome in practice, it would give us some control over the information we have surrendered. The most radical solution would be to claim that considerations of private and public interest should override confidentiality in cases where harm can be prevented or benefits achieved by the disclosure.

Senses of security

People's concerns about security can, in theory, be addressed in two main ways. One is to establish trustworthy institutions and procedures, the other is to promote people's trust, or sense of security. These strategies are not, of course, mutually incompatible.

The *objective security* or *trustworthiness* of population genetic databases could be enhanced by safety mechanisms, data protection, risk assessment, and good professional practices. Legislators and the advocates of the databases have in most countries assumed that adequate safeguards can be installed by these measures. Others have pointed out that no amount of data protection can prevent the identification of individuals in small populations, when descriptions of any sort are attached to the samples or records.

The *subjective* or *inter-subjective sense of security*, or *trust*, can be created in many ways, some of which are more acceptable than others in current political contexts. In liberal democracies, the ideal is that trust is based on the openness and transparency of the system. People are expected to know about the arrangements, because the information about them is available, and citizens have a general political duty to keep in touch with public developments. In social democracies the trust is implied, and it is related to the trust individuals are supposed to have in the political structure. People are entitled to know about specifics, but if they choose not to find out about and challenge existing rules, their trust is assumed. In non-democratic political environments, the sense of security is created by other means. Citizens do not have a duty or a right to know how

Public interests

Public interest arguments are often employed to support population genetic databases, but they can also formulate people's concerns regarding their establishment and uses. These matters can be classified under three headings, namely "benefit," "harm," and "justice."

The advocates of genetic databases argue that these would be extremely *beneficial* for a variety of purposes, including scientific advances, public health promotion, national welfare, and crime prevention. To the degree that people think that these arguments are sound, they can be concerned that unnecessary regulations would hinder progress and leave humanity's problems unaddressed.

According to skeptics, however, the benefits of population genetic databases would be inconsequential in comparison with the ensuing *harms*. Violations of privacy, autonomy, and human dignity, they say, are more than probable, and proper measures to counteract these are hard to find. The availability and use of genetic information can also cause anxieties and distress, and too much information in the hands of public authorities can lead to the development of a threatening police state.

Skeptics have also pointed out that considerations of *justice* speak against the establishment of population genetic databases. The use of genetic information in medical decisions, employment, and insurance can give rise to many kinds of inequality and discrimination. The ownership issues of genetic information are unsettled, and there are no guarantees that the benefits of the enterprise would be shared fairly. It is also unclear who should have the power to make decisions concerning the collection, storage and use of the data.

What concerns do people have?

Although it can be agreed that the list that I have presented airs some of the most important concerns people can have regarding population genetic databases, it is more difficult to know which concerns are the most prevalent, and which are the most significant. Social scientists could have some answers to these questions, but so far the results of their studies have been less than conclusive, for several reasons.

One tangible problem is that people seem to nurture the concerns that they are asked to nurture. If the interviews are designed to emphasize issues of privacy, those interviewed will stress those issues. If they are geared towards the benefits of establishing databases, the answers will reflect this. (On the other hand, I have not seen studies in which the tentative concerns I have listed here would have been rigorously tested. That could be a start).

The data regarding people's attitudes can, moreover, be challenged whatever its attention to detail or its sensitivity to social and cultural variation. Both social scientists and others working in this field tend to think that some opinions are more important than others. Philosophers often argue that irrational or unreasonable fears and hopes can be ignored. And philosophers and social scientists alike sometimes suggest that attitudes created by unrealistic promises or imaginary threats should be overlooked. The difficulty is that we do not always agree on what fears, hopes, promises, and threats *are* rational, reasonable, realistic, and actual.

Are people's concerns addressed by regulations?

Without knowledge of people's concerns, or legitimate concerns if that route is taken, it is difficult to say whether regulations address them. An added hardship is that lawyers do not seem to be sure what the regulations in this field are. Laws and policies are constantly changing, and the changes and interpretations are heavily influenced by incomplete analyses made by bioethicists, pressures applied by interest groups, and political agendas driven by those in power.

Some say that people's concerns can be addressed by involving them directly in the decision-making. The earlier they are allowed to participate in the process, the better for society. But who exactly are the people who should be involved? It would be very difficult to include everybody. All other alternatives, on the other hand, would mean making choices between individuals and groups. Who decides whose voices should or should not be heard?

It seems to me that as long as social and legal studies try to simulate majority democracy by aiming to unearth the abstract "common will" of the people, all attempts to take actual people's concerns properly into account are doomed to failure. A better model could be provided by the protection of minority rights. Let me conclude by sketching what I mean by this.

We do know, for instance, that some people would like to see their privacy and confidentiality respected absolutely; their explicit permission asked for every use made of their genetic constitution; and safety measures to be so tight that no leaks would be realistically possible. But these people are not necessarily listened to. Why? Because it is possible that only a ban on population genetic databases could take their concerns seriously into account. And the decision already seems to be made that this is not the way. Population genetic databases will come, and the only question is how to regulate them. But if this is true, then regulations do not, and will not ever, address the concerns of this particular group of people.

Acknowledgement

This article was produced as a part of the projects *Ethical, Legal and Social Aspects of Human Genetic Databases: A European Comparison* (ELSAGEN), financed between 2002–2004 by the European Community (QLG6-CT-2001-00062); *The Ethics of Genetic and Medical Information Network* (EGMIN), financed between 2002–2004 by the Nordic Academy for Advanced Study; and *Ethical and Social Aspects of Bioinformatics* (ESABI), financed between 2004–2007 by the Academy of Finland (SA 105139). My thanks are due to these institutions for their support, and to Peter Herissone-Kelly for checking my English.

Bibliographical note

Some of the ideas presented in this article have been further developed in Matti Häyry and Tuija Takala. 2001. Genetic information, rights, and autonomy. *Theoretical Medicine and Bioethics* 22:403–414; and Matti Häyry 2005. Can arguments address concerns? In *Arguments and Analysis in Bioethics*, ed. Matti Häyry, Tuija Takala and Peter Herissone-Kelly. Amsterdam and New York: Rodopi, forthcoming. References to the literature can be found in these articles, and in the chapters Tuija Takala and I will contribute to the edited volume to be completed by the ELSAGEN team during Fall 2004.

Values, knowledge, and dignity

24

Interests, values, and genetic databases

by Ann Bruce and Joyce Tait

Introduction

CONFLICT OVER BIOTECHNOLOGICAL developments such as genetically modified (GM) foods has highlighted concern that new developments in genomics may not be automatically accepted but may instead lead to controversy. On the whole the development of genetic databases has not generated the same negative publicity as GM crops, but there is a great deal of interest in ensuring that development of genetic databases is carried out in a way which learns the lessons from the GM crops debate.

Tait (2001) proposed a model for the GM foods debate by distinguishing between value-based and interest-based conflicts. In practice, individuals and groups involved in conflict are likely to be motivated by a mixture of interest-based and value-based concerns (with often complex links between the two). However, the value-based components may need to be addressed in very different ways from the interest-based ones. We argue that values, whilst not immutable, are rather resistant to change, whilst interests are likely to be more flexible. Values often reflect moral positions and common core concerns which tend to be constant across a range of issues, whereas interests are more specific to particular situations and may or may not reflect a core set of background values.

This paper aims to investigate what kind of insights can be derived using Tait's "values and interests" model in the context of genetic databases. It is not intended as an exhaustive exploration of all possible interests and values, but aims to identify some of the main issues raised by stakeholders.

Interest-based issues

Proponents of genetic databases rely on interest-based arguments in citing the promise of new or better targeted drugs, improved diagnostic tools, and better understanding of disease as the main motivations. The UK Biobank, for example, gives its aims as to "improve the prevention, diagnosis, and treatment of illness and the promotion of health throughout society" (Department of Health, MRC, and The Wellcome Trust 2003, 6).

Donations as gift — altruism

A fundamental tenet of many databases is that results may take a long time to come and that participation is unlikely to benefit the individual. Databases and the associated gene searches deal with populations, not individuals. Donating material to a database is therefore increasingly viewed as a "gift" for the common good. Public debate around UK Biobank (People Science and Policy Ltd. 2002) suggests that despite their initial interest in gaining information about themselves, people are able to understand this wider concept and to accept it. Interests here are not the self-interest of individuals, but a wider concept of altruism in the context of community interests. Laurie (2002), however, argues that the gift paradigm is one-sided, such that the focus is on the loss of interest in donated samples rather than the benefits of ownership of the samples that make the "gift" possible.

From the individual's perspective, donations may benefit members of their family, but is the individual also likely to be disadvantaged? Much of the concern about databases focuses around potential loss of privacy. Who else will be able to access information, and what use might they wish to make of it? For example, could insurers or employers discriminate against individuals? Whilst security measures can be put in place, it may be impossible technically or legally to maintain confidentiality all the time, even if there is a genuine desire to do so (HGC 2003). Since a totally "risk free" database is impossible, trust must be key. Individuals will therefore have to make decisions about what level of risk is acceptable to them, under what conditions. Conversely, excessively precautionary anticipation of risks to individuals (which may be unrealized) could prevent wider benefits for society.

Who profits?

A major concern for some stakeholders is the perceived exploitation by commercial companies who may be viewed as using public-good, population derived information for their own profit. The issues of corporate involvement, greed, and perceived abuse of power are of growing interest to the public. If one of the aims of a genetic database is to result in new drug development, then at present, commercial organizations will need to be involved in some way. Focus groups on UK Biobank (People Science and Policy Ltd. 2002) suggest that people are able to comprehend and accept this, assuming that it is explained to them. Although individuals are very unlikely to benefit financially from commercial developments, in the interests of sharing benefits, various arrangements have been arrived at to achieve community benefits.

Balancing interests?

We have considered some of the interests which might underlie the development of genetic databases:

* The scientific drive to use new knowledge to improve health;
* The need to ensure that people making an altruistic donation are protected from harm because of information derived from this donation;
* Individuals being able to protect their interests adequately;
* A fair distribution of benefits.

These interests may be compatible. For example, the Human Genetics Commission carried out an extensive survey of public attitudes and concluded that there appears to be "a strong public attachment to a socially inclusive response to developments in human genetics" (HGC 2002, 8).

Values questions

Some of the arguments made above under "interests" have a potential "value" dimension of varying strength, which may depend on the degree of altruism involved. For example, the drive to use new knowledge to promote health may be more interest-based if the health of the tissue donor is being promoted, or more value-based if benefits will only accrue to future generations. This type of value dimension needs to be clearly differentiated from questions of value for money, i.e. whether genetic databases are the most cost-effective way to promote health benefits in future.

The UK Government White Paper makes a strong case for the value of genetics research in health care, e.g. "Our vision is for the UK to maintain its role at the leading edge of genetics research and development" (Department of Health 2003, 59). While it is clearly important and valuable for the Government to make sure that the NHS is able to cope with this influx of genetic information (the thrust of this White Paper), not everyone agrees that more genetic information is the most effective way of achieving the desired health benefits. There are signs of a strengthening ideological movement against the techno-genetic medical paradigm in favor of a more "sustainable," "holistic" medicine, and such groups often use value for money as an argument to strengthen their case in policy debates.

This argument of holistic versus reductionist approaches to medicine is not new, nor is it restricted to the area of genetics. Callahan (1999), for example, argues that no matter how much money is spent, it will never be enough to meet the demands on the health service made by an ageing population, new technologies, and increasing expectations. He calls instead for a "sustainable" medicine shifting the focus from medical and technological improvements to the social, economic and cultural conditions that contribute to health.

Similarly, a letter from Charlotte Augst of CancerBACUP to the *Financial Times* on August 23–24, 2003, argues:

> People who get heart disease, cancer or diabetes require the NHS to provide good, patient-focused and responsive "traditional" care, which has hardly been touched by the new genetic insights. Their families need support and information, not only if there is a suspected genetic link but also whenever they are affected by their relatives' illnesses.

Another group which could be seen as using such arguments to promote value concerns is GeneWatch, which has the stated aim to "ensure that genetic technologies are developed and used in the public interest and in a way which promotes human health, protects the environment and respects human rights and the interests of animals" (GeneWatch UK 2004). This may not imply that members of the organization are against the technology as such, but the GeneWatch UK (2002) briefing document *Genetics and "predictive medicine": selling pills, ignoring causes* implies that for them, too much stress is placed on the genetic aspects of illnesses rather than looking to alleviate the "real" causes (diet, lack of exercise, and pollution).

Thus, it appears that an important critique of genetic databases will be on the basis of the public values-related direction of health care generally and the charge of excessive focus on genetic aspects. These group values could begin to coalesce into a powerful ideology capable of challenging a range of genomics applications in medicine, including databases.

Genetic databases and GM crops

Genetic databases will not inevitably encounter a crisis like that of GM crops. There are some key differences. There is not currently a strong moral objection to genetic information as such. Genetic databases are not generally viewed as an instance of manipulating nature, but of seeking to understand disease.

There are however ideological tensions and concerns underlying both the development of GM crops and genetic databases. In the GM crops debate these are competing views on the benefits of organic versus conventional agriculture and the need to address the redistribution of food rather than producing more food. The parallel with genetic databases appears to be the benefits of genetics approaches versus a more holistic approach with more stress on addressing social and environmental issues around disease rather than genetic ones. It is not yet clear, however, whether this will develop to exclude genetic approaches to medicine as organic agriculture had defined itself to exclude GM crops.

Because GM crops affect food, everyone is potentially affected and it is necessary to address issues of how individuals can opt out if they wish to do so (e.g. by labeling and segregation). However, provision of labeling GM foods in the EU has not yet had the effect of making GM food acceptable to the groups that oppose the technology on a value basis. With genetic databases, the present consensus in the UK appears to be that individuals should be invited to opt into them. So individuals can decide to refuse to take part. However, it may be impossible to distance oneself from the information coming from databases and this information may cause difficulties, for example the confirmation of a genetic susceptibility to a specific disease could be very unwelcome information for families who are already bearing the burden of caring for someone with that disease. Of course, the information about risk does not *per se* alter risk, but may alter response, which may be positive or negative. Identifying risk early does offer the possibility of non-genetic strategies for risk reduction, such as altered behavior or lifestyle, or avoidance of environmental exposures.

The benefits of both GM crops and genetic databases are in the future, uncertain, and therefore contested (to a greater or lesser degree). The way in which information about the developments is accepted and futures imagined will be affected by the context in which the developments take place and issues such as public trust in the organizations presenting alternative futures. The potential benefits from genetic databases may be more obvious than those of GM crops. The uncertainty surrounding new therapies could be viewed as a source of hope. It is easy to see how genetic information could potentially be lifesaving. GM crops on the other hand can be seen to increase risk and to provide a threat. It may be difficult to credit GM crops with alleviating hunger based on present evidence of applications developed. Both technologies require the involvement of commercial organizations, and linked into commercial involvement is the question of who is driving the technology and for what purpose. There was a strong perception that the introduction of GM crops was being driven by commercial interests rather than public benefit. The perceived drivers of genetic databases may be more complex and potentially more morally defensible.

Conclusion

Much of the debate about genetic databases to date has been about the interest-based aspects and how to protect people participating in a genetic database. It is as important that current discussions also take into account the underlying values questions. If priority is being given to one set of values, the reasons for this should be clearly argued against the claims of alternative value perspectives, and policy makers should be prepared to make decisions that may challenge deeply held values and also the interests of vocal public and private interest groups.

Genetic databases may represent a less contested situation compared with GM crops in that there appear to be fewer non-negotiable beliefs involved. Most of the concerns expressed are interest-based and therefore potentially resolvable by appropriate structures.

Acknowledgements

We would like to express our thanks to the Economic and Social Research Council who funded this work, to colleagues at Innogen who have provided fruitful discussions, and to Donald Bruce, Graeme Laurie, Donald MacKenzie, and David Porteous, for their helpful comments. Views expressed in this paper are, however, our own.

References

Callahan, D. 1999. *False hopes: Overcoming the obstacles to a sustainable, affordable medicine*. New Brunswick, New Jersey and London: Rutgers University Press.

Department of Health. 2003. *Our inheritance, our future: Realising the potential of genetics in the NHS*. Norwich: HMSO.

Department of Health, Medical Research Council (MRC), and The Wellcome Trust. 2003. *UK Biobank ethics and governance framework*. Version 1, 24 September.

GeneWatch UK. 2002. *Genetics and "predictive medicine": Selling pills, ignoring causes*. Briefing Number 18, May.

GeneWatch UK. 2004. About GeneWatch UK. GeneWatch UK website. http://www.genewatch.org/aboutGW.htm

Human Genetics Commission (HGC). 2002. *Inside information: Balancing interest in the use of personal genetic data*. A report by the Human Genetics Commission, May.

Human Genetics Commission (HGC). 2003. *Memorandum to the House of Commons Science and Technology Committee on Medical Research Council/Wellcome Trust/Department of Health UK Biobank study*. Available online at: http:www.hgc.gov.uk/business_publications_memorandum_ukbiobank.htm

Laurie, G. 2002. *Genetic privacy: A challenge to medico-legal norms*. Cambridge: Cambridge University Press.

People Science and Policy Ltd. 2002. *BioBank UK: A question of trust: A consultation exploring and addressing questions of public trust*. Report prepared for the Medical Research Council and The Wellcome Trust, March.

Tait, J. 2001. More Faust than Frankenstein: The European debate about the precautionary principle and risk regulation for genetically modified crops. *Journal of Risk Research* 4:175–189.

Genetic databases and what the rat won't do: What is dignity at law?

by Mark Cutter

"I hear they have started to carry out experiments on lawyers …the scientists don't get so attached to the lawyers… there are certain things that even rats won't do"
— Robin Williams, *Hook*

Introduction

DIGNITY IS A word that is often bandied around in ethics debates, especially regarding the conduct of experiments on humans in general, and in genetics in particular. This ethical reliance on a concept of dignity is codified within a range of legislative and quasi-legislative instruments. This paper will, for the most part, focus on two of these instruments, namely, the Helsinki Declaration (World Medical Association 2002) and the recent EU Convention on Human Rights and BioMedicine (Council of Europe 1997).

This paper will consider the question of dignity as enshrined within these governance documents, in an attempt to understand the nature of dignity, as it exists as a legal concept, and what if any effect this has on the governance of genetics and genomics technologies and research, with a particular focus on database projects. This paper will not give an account of existing discussions on dignity as included in the works of Kant, Foucault, and others, but instead will focus on the governance instruments themselves and legal, social, and philosophical questions raised therein.

Research on humans

In all areas of medical research, and in particular in research involving new genetics and genomics technologies, the intention is usually for the research to have application to human subjects. In the case of pharmaco-technologies this requires a level of testing on humans[1] prior to the products being made broadly available to the market; in the case of biobanks, the human participation element of research is implicit from the start.

---

1 As accepted in the Helsinki Declaration (World Medical Association 2002) at Article A.4, "Medical progress is based on research which ultimately must rest in part on experimentation involving human subjects."

In general the pattern of research involving pharmaco-technologies might be described as a spectrum that has theoretical research and human clinical trials at its poles, and testing on animals at the mid-point. Various products move along this spectrum. Only a relatively small number reach the stage of human clinical trials, and a smaller number reach the market.

Whilst this model does not fit with the nature of genetic bio-bank research, it gives one perspective on the concept of medical research, and as such may be shown to be important in the discussion of dignity in this context.

Defining dignity

When seeking to interpret the meaning of a legislative instrument, it is appropriate to consider the ordinary meaning of the words. Thus, as one might invite a Judge to do, it is instructive to refer to a dictionary for a definition of the word "dignity" and, in applying this definition to the particular wording, to seek to identify what it is that dignity means in law, and in a sense take judicial notice of the word's meaning.

The Oxford Advanced Learners' Dictionary attributes three definitions to the noun "dignity." The entry appears as follows:

> dig•nity /ˈdɪgnəti/ *noun* [U]
> *1* a calm and serious manner that deserves respect: *She accepted the criticism with quiet dignity.*
> *2* the fact of being given honour and respect by people: *the dignity of work. The terminally ill should be allowed to die with dignity.*
> *3* a sense of your own importance and value: *It's difficult to preserve your dignity when you have no job and no home.*

The word dignity appears in the Helsinki Declaration at article B.10, being the first of the basic principles laid down, in which it is stated that "it is the duty of the physician in medical research to protect the life, health, privacy, and dignity of the human subject" (World Medical Association 2002).

The concept of dignity is more deeply enshrined within the Convention on Human Rights and Biomedicine, appearing in its full title as the *Convention for the Protection of Human Rights and Dignity of the Human Being with regard to the Application of Biology and Medicine*. It then appears in the preamble to the convention, underlining the fact that the signatories would be:

> convinced of the need to respect the human being both as an individual and as a member of the human species and recognising the importance of ensuring the dignity of the human being;
> Conscious that the misuse of biology and medicine may lead to acts endangering human dignity;
> … [and as a result are] resolving to take such measures as are necessary to safeguard human dignity and the fundamental rights and freedoms of the individual with regard to the application of biology and medicine (Council of Europe 1997).

It then follows that the first article of the treaty stipulates that:

> Parties to this Convention shall protect the dignity and identity of all human beings and guarantee everyone, without discrimination, respect for their integrity and other rights and fundamental freedoms with regard to the application of biology and medicine.
> Each Party shall take in its internal law the necessary measures to give effect to the provisions of this Convention (Council of Europe 1997, article 1).

Interpreting dignity

It is apparent from these instances of the word dignity that it is something to be protected, and it is not merely the dignity of the individual that is to be considered, but also the dignity of the human race as a whole.[2]

So the question remains of what it is that is meant by this concept of the preservation and protection of dignity, as enshrined in these documents. In considering the definitions provided above, it appears that definitions two and three are the most appropriate for use in these circumstances: that is, the concept of dignity as "the fact of being given honour and respect by people" and as "a sense of your own importance and value."

From these definitions, we can establish that human beings as individuals, and the human race in general:

a) should be afforded honor and respect (by other members of the human race);

and

b) that the human race's sense of its own importance and value should be maintained.

When discussing the meanings of these two expressions, it is interesting to note the recurrence of the concept of dignity as something to be respected and protected, as it appears in the language of the two governance documents. With regard to this, and in discussion of point (b), we should consider what it is that the human race (or human beings as individuals) should be contrasted with. That is to say, if these governance documents seek to afford protection to the human race's sense of its own importance and value, in relation to what is this a sense of importance and value?

It is at this point that the spectrum of research discussed earlier becomes of interest. Where a key principle in these governance documents is to protect human dignity at the stage of human testing and clinical trials, we can extrapolate that a lower standard or sense or value is attached to trial subjects at other points along that spectrum, namely animals.

This in itself is a highly-charged issue, and it is not the purpose of this paper to debate the philosophy or morality of this position. Nevertheless, in examining the very concept

---

2 As highlighted by use of the phrases "member of the human species" in the preamble, and "…all human beings" in Article 1of EU Convention on Human Rights and Biomedicine (Council of Europe 1997).

of human dignity as defined and protected within law, we may show that, doctrinally speaking, human beings should not be reduced to the level of lab rats.

Humans, rats, and databases

It now remains to apply this paradigm to the concept of a human genetics research database, and to construct a relevant interpretation of the term "dignity." It is interesting that in the debates surrounding human genetics databases—such as the Icelandic (deCODE)[3] and Swedish (UmanGenomics)[4] projects—reference has been made to the Helsinki Declaration and the BioMedicine Convention. This suggests the need to properly understand the principles contained within such governance tools.

As a means of further constructing the concept of dignity in this context, it is useful to consider what steps are taken to preserve dignity by these governance tools. One key idea, common to a majority of governance instruments involving the interaction of science with humans, is the concept of consent. This concept first appeared in governance documents relating to medical research in a judgment from the Nuremberg trials in 1947, which takes the form of the Nuremberg Code (1949). It is recognized in the first of article of the code that "the voluntary consent of the human subject is absolutely essential." It is from this document that the principle of (informed) consent has entered into general governance and ethics codes. From the jurisprudence of the Nuremberg trials, and the nature of the experiments that gave rise to the code, it may be argued that the requirement of consent is a method of "affording honour and respect to the subject of the research," and although it is not directly cited, it was the concept of human dignity, as we have defined it, that this document sought to protect. Furthermore, the proliferation of the concept in other governance documents that are overtly formed around the protection of dignity implies that consent may be seen as one of the key factors in determining dignity. This supposition that consent is one of the determinants of dignity is reinforced within governance documents, where strong measures are taken to protect those who cannot for reasons of competency give consent.[5] These same documents re-

---

3 Examples include: "By opting out after the start of operation of the database an individual can stop entry of further data onto the database but previously entered data will still be used. They will not be removed and an individual can thus not stop participation in research for those data already entered. Mannvernd considers this to be in direct opposition to the Helsinki Declaration of the World Medical Association which states that individuals must able to quit participation in research at any time" taken from the Mannvernd website (Mannvernd 2003).

4 Examples include: "…In turn the government made it clear that any such legislation would need to be compatible with the 1997 European Convention on Human Rights and BioMedicine to simplify eventual ratification" (Rose 2003).

5 As per Article 24 of the Helsinki Declaration: "For a research subject who is legally incompetent, physically or mentally incapable of giving consent or is a legally incompetent minor, the investigator must obtain informed consent from the legally authorized representative in accordance with applicable law. These groups should not be included in research unless the research is necessary to promote the health of the population represented and this research cannot instead be performed on legally competent persons" (World Medical Association 2002). Similar provisions appear at: Articles 6 and 7 of the EU Convention on Human Rights and BioMedicine (Council of Europe 1997).

quire that research on humans is not to be commenced until it has been previously trialed on animals. This serves to underline the distinction in the doctrinal sense between animals and humans, in that humans appear to be afforded a sense of honor and respect that is not afforded to the animal test subjects (which de facto are unable to consent). In doing so, it creates the sense of importance and value of the human race discussed earlier.

Although this human/animal distinction is useful for defining dignity in this context, it may not appear useful in relation to database discussions. However, in rebuttal of this point, it is interesting to note the existence of databases of animal genetic material, such as the Rat Genome Database, based at the University of Wisconsin, which "curates and integrates rat genetic and genomic data and provides access to this data to support research using the rat as a genetic model for the study of human disease" (Medical College of Wisconsin 2004). This, and similar projects, might be considered to be the animal testing stage in the development of human genetic database projects.

While consent may not be the sole characteristic of dignity, it appears to be a key element in the construction of the content at law, by virtue of its recurrence in governance documents. If it is accepted that consent is a key concept in the definition of human dignity in this context, then it is interesting that many disputes surrounding human genetics databases have included the issue of consent. As mentioned above, Mannvernd cites issues of consent among its grounds for objecting to the Icelandic project. Equally one of the perceived failures in the UmanGenomics project centered on consent. Consent is not an issue with animal test subjects. This raises the question of whether we can consider dignity, in the terms we have defined it, to include a question of choice, and self-determination.

A further example of the importance of self-determination can be found in the Tonga database proposal put forward by the Australian company then called Autogen. The proposal met with resistance linked with a lack of public consultation, and as such a de facto lack of consent on behalf of the population. A denial or removal of choice may be shown to be a failure to afford honor and respect to the members of that population, and equally might be seen to detract from the population's sense of its own self-importance. This exemplifies the potential affront to dignity that can be caused by a failure to consider these consent based issues when defining the governance of databases. The potential of a biobank to remove all elements of self-determination from the participant, by transforming genetic information into information stored in other forms, over which the test subject no longer has any control, links tightly with our construct of dignity. It may be that biobanks provide a new position on our spectrum of dignity, reducing the human form to the level of a collection of data. Thus, if dignity is relative to animals in other forms of medical research involving animals, is this reduction, and as such participation in biobank projects, an affront to dignity as we have defined it? The church leaders of the South Pacific appeared to make similar suggestions (as shown in the Tongan study): "the conversion of life forms, their molecules or parts into corporate property through patent monopolies is counter-productive to the interests of the (people) of the Pacific" (Burton 2002).

Conclusions

"Dignity" is a word that has become enshrined in a range of governance materials related to biobanks and other genomics technologies. The term is difficult to define in law, and infringements of dignity are similarly hard to identify, but in taking the word at its dictionary definition we are able to create a construct of dignity which, when applied to specific examples, can identify governance issues in biobanks that might be considered to be related to dignity. This is done by considering those concepts, such as consent, that appear to have been tied to dignity within the governance documents, and then expanding from this position to consider issues that fall outside the scope of these documents. It can be shown that dignity is not a solid concept for governance, but rather an interpretive one that should be considered relative to other concepts and states. However, it will become increasingly concrete if it becomes further enshrined in law.

References

Burton, B. 2002. Opposition stalls genetic profiling plan for Tonga. Inter Press Service. Available online at: http://www.hi.is/~elsagen/tonga.html

Council of Europe. 1997. *Convention for the protection of human rights and dignity of the human being with regard to the application of biology and medicine: Convention on human rights and biomedicine*. Available online at:
http://conventions.coe.int/treaty/en/treaties/html/164.htm

Mannvernd. 2003. Opt-outs from the Health Sector Database.
http://www.mannvernd.is/english/optout.html

Medical College of Wisconsin. 2004. Rat genome database. Website at:
http://rgd.mcw.edu/

The Nuremberg Code. [1947] 1949. In: *Trials of war criminals before the Nuremberg military tribunals under control council law*, 10, 2, 181–182. Washington, D.C.: U.S. Government Printing Office. Available online at:
http://www.med.umich.edu/irbmed/ethics/Nuremberg/NurembergCode.html

Rose, H. 2003. An ethical dilemma: The rise and fall of UmanGenomics. *Nature* 425:123–124.

World Medical Association. 2002. *The Declaration of Helsinki*. Available online at: http://www.wma.net/e/policy/b3.htm

26

Human dignity and technology

by Daniel Statman

A. Introduction

THE NOTION OF human dignity is widespread in moral and legal discourse, where it is assumed to play a central role. It is often referred to in discussions about medical technology, though this reference is rather perplexing as some argue that dignity is promoted by such technology, while others argue that it is undermined by it. This paper seeks to shed some light on these issues by revisiting the notion of human dignity and examining its relevance to practical philosophy in general, and to matters concerning ethics and technology in particular.

The thesis presented here will argue (a) that a basic distinction exists between two understandings of (respect for) dignity and (b) that this distinction has significant implications for applied ethics. According to the first understanding, violation of dignity occurs whenever there is any immoral act directed toward human beings (oneself included). I shall term this understanding "(respect for) dignity as morality," or, in short, "DM." According to the second understanding, violation of dignity occurs whenever there is an act which humiliates, insults, or is contemptuous toward other human beings. I shall term this understanding of dignity "dignity as non-humiliation," or, in short, "DNH." As not all immoral behavior is necessarily humiliating, it is clear that the first approach, DM, reflects a much broader understanding of the notion of dignity than the second, DNH. Whereas DM encompasses DNH—since humiliating behavior is normally an instance of immoral behavior—the opposite is not true: DNH does not encompass DM.

One further point of clarification before I elaborate on these two concepts. In practical discussions, dignity is usually referred to negatively, i.e. we tend to speak of violations of dignity, rather than positively, by reference to sorts of behavior that *promote* respect for dignity. It is always much easier to identify evil than to identify good, easier to identify wrongdoing than right doing. I will follow this common usage here and, for the most part, will talk about violation of dignity, though everything said here can also be applied, *mutatis mutandis*, to the positive aspect of dignity.

B. Two concepts of dignity

1. Dignity as morality (DM)

To clarify this concept of dignity, the ethics of Kant, which in contemporary times is a constant source of inspiration for arguments relating to human dignity, provides the best starting point. On Kant's view, *qua* free and rational beings, human beings possess dignity or intrinsic value, and therefore deserve special consideration. To treat others (and ourselves) with dignity means treating others (and ourselves) in accordance with what is required from a moral point of view, and vice versa: Treating others (and ourselves) in accordance with what is required from a moral point of view means treating others (and ourselves) with dignity. This congruence is evident from Kant's statement that "The three modes of presenting the principle of morality that have been adduced [universalizability, humanity as an end, and autonomy] are at bottom formulae of the very same law." They cover the same area and entail the same practical norms. This is also evident from the fact that Kant uses the same four examples to illustrate the practical implications of the universalizability test and of the humanity-as-an-end test.

By the phrase "dignity as morality," I therefore wish to capture the idea of a complete congruence between respect for dignity and moral behavior. Any time I lie to human beings, steal from them, exploit them, restrict their freedom, discriminate against them, and so forth, I thereby fail to express the appropriate respect due to them. I treat them solely as means, as having a merely external and contingent value, thus denying or disregarding the unique value they possess *qua* human beings: their *dignity*.

Yet, in spite of the enormous importance of the concept of human dignity in Kantian ethics—indeed *because* of this importance—this concept does not play any independent role in moral-practical reasoning. It does not constitute a separate moral reason that might guide us in moral deliberation, a reason distinct from those such as freedom, justice, integrity, and so on. Within the concept of dignity as morality, dignity cannot be invoked to solve practical questions in cases where other moral considerations fail to do so. Put differently, to respect the dignity of human beings is to behave toward them in a fitting manner, that is, according to the manner required from a moral point of view. However, in order to know what kind of behavior is so required, we must first turn to criteria other than dignity, otherwise we will find ourselves caught in a vicious circle.

Hence, in the approach I termed "dignity as morality," dignity is necessarily parasitic on other moral concepts or principles. As a practical standard, the concept of DM is empty. Kant understood this very clearly, and, accordingly, stated that though the various formulae of the categorical imperative were of equal weight, "In forming our moral judgment of actions, it is better to proceed always on the strict method and start from the general formula of the categorical imperative: Act according to a maxim which can at the same time make itself a universal law." Kant explicitly instructs us to avoid using the concept of dignity in practical moral thinking.

Not only is the concept of dignity not useful in practical moral thinking, on Kant's view, it may even be harmful. Although Kant recognizes the positive influence of the notion of dignity on moral feeling, he fears that feelings of respect toward human beings would lead to sentimentality. In his view, to be motivated solely by feelings indicates moral weakness, even if the feelings are positive, as there is a danger that moral judgments will be made in accordance with feelings rather than with the objective and fixed laws of reason. Kant recognized the rhetorical power of claims about dignity, and he feared that such claims would reflect and also inspire unbalanced feelings. Hence the recommendation to base our moral judgments on the test of universality and not on subjective feelings concerning violations of dignity. In the light of the overuse of claims about violations of dignity in contemporary legal and public discourse, it seems to me that the fears expressed by Kant in this connection were not unfounded. In the absence of a moral or philosophical criterion to aid us in deciding what a violation of dignity is, the use of this expression is becoming more and more prevalent, while our ability to discuss this use rationally and critically is being correspondingly weakened.

2. Dignity as non-humiliation (DNH)

According to the second notion of dignity, to violate the dignity of human beings means to humiliate them, to insult them, or generally to behave in a way that causes injury to their self-respect. At times, the humiliating behavior is morally wrong for other reasons too, for example, when it causes physical injury, or when it involves discrimination, violation of autonomy, etc., while at other times it is wrong only (or primarily) because it is humiliating, e.g. in typical examples of sexual harassment. As this sense of (violation of) dignity is closer to the usual usage of this term, it can also be referred to as dignity in the ordinary sense.

As indicated earlier, not all immoral behavior, even if extremely serious, may be seen as a violation of dignity in the sense of humiliation. Murder, for example, does not humiliate the victim but destroys him or her. In lesser moral offences, the element of humiliation is clearly not essential; most people would be angered if their car was stolen, but only a few would see it as humiliating and feel that their dignity had been infringed. The same is true of restrictions on freedom; not all restrictions the state imposes, even if unjustified, cause people to feel humiliated.

Let me list now the fundamental differences between the two concepts of dignity and elaborate on them:

A. DNH plays a real role in practical moral thinking. At times, it is the only source to establish the moral wrongness of some act while, in other cases, it supplies extra moral gravity to an act which is anyway prohibited on other grounds. Pointing to the fact that a certain behavior violates dignity in the sense of humiliation, therefore, helps us reach an answer to the question of whether this behavior is (*prima facie*) right or wrong, whereas stating that it violates dignity in the sense of DM fails to guide us in making such a decision and can only be done once an answer has already been given on different grounds.

B. From a structural point of view, DM is the most fundamental concept of the moral system, whereas DNH is just one moral concept among others.

C. Conceptually speaking, DM cannot conflict with other moral concepts, as they do not exist on the same plane, hence, there is never a need to strike a balance between DM and other concepts. For the same reason, there is no need to strike a balance between respect for John Doe and respect for Richard Roe: if the test of universalizability (or any other moral test) teaches that it is prohibited to harm John Doe even at the price of harm to Richard Roe, then this is the necessary act from a moral point of view, hence it is the act that expresses proper respect for humanity. In contrast, DNH may conflict, and, indeed, frequently does conflict, with other moral values, such as the public good, education, and punishment, and conflicts might arise within DNH between the dignity of John Doe and that of Richard Roe.

   D. DM has a unique and exalted value. The dignity of humans, their intrinsic and non-contingent value, is the basis for their special moral status, and for the obligations owed to them by all rational beings. In that sense it may be said that, in DM, human dignity is the most supreme moral value. In contrast, in DNH, dignity does not have such exalted value, and it cannot be assumed that in every case of conflict, it will override competing values. For example, in ordinary circumstances, the value of dignity as non-humiliation defers to the value of life, and, only in extremely rare cases, would we prefer to die rather than be humiliated. Thus, DNH does not have absolute moral value.

C. Consequences for ethics and technology

The distinctions discussed above enable us to understand the source of the ambiguity in the moral and legal use of the concept of human dignity. On the one hand, there is a desire to use this concept in all its normative force and with all its rhetorical and emotional power, as a concept which, by its nature, overrides any other moral consideration (and *a fortiori* any nonmoral consideration). This desire is tied to the understanding of dignity as morality *a la* Kant. But within the framework of this understanding, dignity can play no separate role in our practical thinking, and therefore will inevitably frustrate the expectation that it guides society in general, or the courts in particular, in solving disturbing moral problems. On the other hand, there is a desire to have a concept of dignity that might help in practical moral reasoning, hence dignity is tied to the non-humiliating understanding. However, within the framework of this understanding, dignity will not necessarily override other moral considerations, and its status will be determined in each particular case according to the nature and weight of the conflicting considerations.

   Thus, when interpreting the concept of human dignity we face a fundamental dilemma: Either we interpret it in Kantian terms, which means it has an absolute moral value but no normative-practical role to play; or, we interpret it as non-humiliation, which means it does have normative-practical value, but lacks absolute moral value. As practical thinking needs concepts which are not empty from a normative point of view, the second concept is naturally the more appropriate one to use. But then when it tries to implicitly rely on a kind of Kantian notion of dignity, the result is rather confusing. Let me now turn to this confusion by reference to a number of questions touching upon ethics and technology.

1. THE ALLEGED THREAT OF TECHNOLOGY TO DIGNITY

The Polish philosopher, Zbigniew Szawarski (1989), writes that "Technology has been developed in order to protect and safeguard human dignity; however, technology may also threaten it." Let us see what can be made of this claim in light of the distinction presented above. Has technology developed in order to protect *dignity*? That strikes me as a rather strange claim. First, I doubt whether the inventors and scientists who advanced modern technology did so explicitly *in order* to protect human dignity. Second, and more importantly, it is hard to see in what sense modern technology can be said to aim at the protection of dignity. If by dignity we mean DNH, I very much doubt whether technology has contributed to its protection. I suspect that the vulnerability to humiliation of both individuals and of minorities has not declined as a result of technological developments. Is the raising of the standard of living, prompted by technology, equivalent to a protection of dignity? Again, I think not. If anything, it might have caused the opposite result, by encouraging the materialism of the developed countries, materialism which causes frustration and low self-esteem among those members of society who fare low on the materialistic scale and cannot afford a new DVD and a vacation in the Caribbean, like most of their neighbors. The more "natural" technologically primitive, and extremely poor societies in the developing countries might prove to be less threatening to individual self-esteem than the technologically sophisticated and rich societies, with their unbearable pressure for self-realization and for "making it" and with their lack of clearly defined social roles providing an anchor for psychological self-respect.

Szawarski, however, probably has nothing like DNH in mind when he refers to the danger of technology. He says that technology "may endanger man's existence [I guess he means woman's existence too…]" in several ways, including the imperilment of the ecosphere with its effects on "human health and life," and the use of weapons of mass destruction. But, in this context, I indeed find the notion *existence* far more helpful than that of *dignity*. Weapons of mass destruction, first and foremost, *kill* millions of people, not *violate their dignity*. By killing a human being we thereby eliminate the possibility of humiliation in the meaning of DNH, since to humiliate a person we need him alive while, if we have DM in mind, we cannot determine whether killing is a violation of dignity before we utilize some independent moral test to determine the morality of the killing. If the killing is justified, e.g. on grounds of self-defense, then it cannot be ruled out on the basis of its failing to express the appropriate respect for humanity. This brief comment on killing and dignity has direct implications for the debate over the morality of capital punishment. Many who argue against capital punishment base their case on the assumed "destruction of human dignity" involved in it. I do not think DNH is relevant here, but DM would not be of much help either. Or more precisely: Dignity as DM could enter the argument only after the main work has been done by other, independent arguments, that would show why capital punishment is morally wrong. It might help to remember that Kant, the great defender of human dignity, saw no contradiction between this idea and the permissibility, indeed the obligatory, status of the death penalty for certain crimes.

2. DIGNITY AND MEDICAL TECHNOLOGY

Developments in medical technology are often thought to pose serious challenges and threats to human dignity. Some of these challenges have to do with the creation of human beings by cloning or other methods of genetic engineering. But as David Heyd (1999) makes clear, the notion of dignity is relevant only with regard to living human beings, not with regard to potential ones. Once a person exists, she ought to be the object of respect and her dignity should not be violated. But she cannot be so treated before she is born or even conceived.

Other challenges in this field grow out of the incredible ability of modern technology to sustain life in spite of terminal illnesses and injuries. This ability is at the root of the ongoing debate over euthanasia and assisted suicide, where the notion of dignity is often referred to. What is its exact role in this context? At times, it is used as an argument *against* these ways of ending life, by assuming that to kill a human being is to violate his or her dignity. Yet, I think that what this argument against euthanasia has in mind is much better captured not by the notion of *dignity* but by the notion of the *sanctity* of life. The notion of dignity is, no doubt, relevant to this debate, but not as a consideration against euthanasia or assisted suicide, but as a (*prima facie*) consideration for it. When people in desperate circumstances ask to be allowed to "die with dignity," what they have in mind is dignity as non-humiliation. They perceive their condition as irredeemably humiliating and degrading, and the pain of this perception is so terrible that they prefer to die in order to escape it. Thus, dilemmas concerning the end of life provide us with another example for the preferability of DNH in practical ethics.

To sum up my argument: I suggested we make a clear distinction between two concepts of violating dignity. On the first, violating dignity is equivalent to any immoral behavior, and hence can play no separate role in practical moral thinking. On the other, violating dignity means humiliating some individual human being or some group. Not keeping this distinction clear causes quite a lot of confusion and ambiguity in moral discussions regarding a variety of issues, among them ethics and technology. Respecting the distinction between two notions of respect for dignity, dignity as morality and dignity as non-humiliation, will enable us to avoid such ambiguity and, ultimately, to achieve better protection of dignity.

References

Heyd, D. 1999. Dignity in Gen-ethics. *Annual Review of Law and Ethics* 7:65–78.
Szawarski, Z. 1989. Dignity and technology. *Journal of Medicine & Philosophy* 14:243–49.

Autonomy and privacy

The collection and management of confidential genetic data

An evaluation of deCODE genetics based on the principle of autonomy

by Pascal Schwarz

What is confidential data and what is information?

WHICH QUALITIES MUST someone have in order to be trustworthy? It is difficult to find an answer to this fundamental question. From our everyday-life experience we know that there are only very few people, if any, whom we can trust when it comes to our most intimate and private issues. But to these persons, whom we *trust*, we *entrust* those most confidential issues. These issues usually concern pieces of information that we consider to constitute our identity and personality. Should this information, however, be disclosed by government officials and thereby be disclosed to the public, this would precipitate a crisis to the relevant identity; the individual would become vulnerable.

However, not all types of information about an individual should be considered equally vulnerable. Therefore, not all information ought to be denoted as confidential and placed under the protection of privacy. What kind of different effects on the autonomous individual, then, does the recording of genetic information have, as it was intended for the Icelandic Health Sector Database?

In order to find an answer to this question, we must go back to another consideration. First we have to make clear that the planned Database in Iceland is intended to store at least three different kinds of data, with varying needs for confidentiality. Only one of these three databases contains genetic data (see McInnis 1999, 236), but the other two databases, containing genealogical data and medical records, are obviously also highly in need of protection. The latter data, especially, obviously have to be treated in a very sensitive way. Up to this point, the keeping of medical data has usually been entrusted to medical staff, but in the Icelandic project, the data were to be transferred to officials of the Ministry of Health and to a private enterprise called deCODE genetics. In addition, deCODE genetics was to gain monopoly rights to the databases. Even when these data are one-way coded, the data are not anonymous in the sense of the Directive 95/46/EC, issued by the European Parliament and the Council on the protection of individuals with regard to the processing of personal data, and the free movement of such data (European Union 1995). Anonymous data are neither personally identifiable, nor is there any decoding key to connect them with previously collected data. Thus confidentiality is not guaranteed anymore in these instances. With regard to the genealogical database called Íslendingabók, however, the most common argument employed to defend the establishment against worries of data protection, is that in a community as small as the Icelandic one, with merely 285,000 individuals, everyone is able to investigate the interconnection between someone's relatives, if they cared to do so. The same argument, by the way, is used against the objections to the storage of medical records, as if the details of someone's health were as public as their genealogical data, even though they should be treated in the confidential physician-patient relationship. Databases with genealogical and medical data raise a tremendous number of questions concerning the effects on autonomy; this paper, however, will focus on the impacts of genetic data.

What kind of information do DNA data yield?

First we have to make clear that there is a difference between data and information. The pure quantitative collection of data does not contain any information. Data have to be interpreted by someone in a certain way in order to become information. This information then has to meet the needs of a recipient who must be able to read and understand the information.

DNA is something all individuals have. But even when you isolate it by means of the most modern forms of high technology, almost no one is able to get any information out of the data. You need an interpreter, someone who has been initiated into this knowledge. In this respect there is hardly any difference between DNA records and medical records: it is virtually impossible for a non-initiated person to evaluate the quality of a newly gained piece of information. This fact is the basis for an argument frequently used for defending elitist positions, e.g. that science is too complex for average people to understand its problematic issues. But even if this were true, it does not in any way make people's feelings of uncertainty and anxiety disappear, nor does it lessen the necessity for communication with or explanation to these people. Quite the opposite is true; the existing feelings of uncertainty and anxiety imply the democratic necessity to communicate the complexity to the people.

To whom does the DNA belong?

Each individual's DNA is different, of course. But does this fact imply that the DNA is the individual's property? We can also consider it a newly assembled set-up of two former DNA chains, namely of those of our parents. Taking this idea further, one single DNA is merely a part of a whole DNA-network and the individual part is only fully understandable within the general framework of the whole DNA network. Although every DNA chain is unique and belongs to one individual, it becomes nevertheless an essential part in understanding the genetic compositions of our closest relatives, and so on. Thus we could propose at least three different owners of the DNA (i) the individual, (ii) the closest relatives, and (iii) the whole ethnicity of which that individual is part. This circumstance explains the absolute necessity to evaluate all those claims of ownership with regard to their effects on the individual's autonomy.

When is the collection and the interpretation of the DNA information a breach of autonomy?

The answer to this question seems to be quite simple. Whenever the collection happens against the individual's free will, i.e. without an informed consent, this constitutes a breach of autonomy. But as a first step we should consider the following question: to what extent is the collection of genetic data already a violation of the physical integrity? This can obviously be the case if we assume that (i) the DNA is an essential component of what we consider a human body to consist of and (ii) that the collection is experienced in a harmful way by the donor.

On the one hand, we do not consider the DNA to be part of the body, as we do for example the heart, a leg etc., since the DNA is invisible and somewhat abstract to us. Although scientists have proclaimed DNA to be a mirror of our unique identity, we give our hair, for example, much more attention. We are only aware of the bearer of genetic information, not of the DNA itself.

On the other hand, the collection of genetic data cannot be treated as a breach of physical integrity, as it causes no harm to the donor and the integrity of the body is never violated; it is sufficient just to take a used teaspoon or a stubbed-out cigarette end in order to gain the DNA. Therefore, there are pressing new problems of defining physical integrity, considering that the body leaves its marks widely in time and space. We should seek a new concept of physical integrity, which takes into account the fact that the body constantly leaves its marks everywhere.

Collecting genetic data while respecting the principle of autonomy

If we regard autonomy as the human capacity for self-determination and as the ultimate source of the individual's values, the collection of genetic information should only be allowed, if (i) the donor's privacy is protected and if (ii) the donor expresses an informed consent. In order to protect his or her privacy, we must respect the patient's right to have control over his or her own data, and to accept or refuse a suggested treatment. When we are collecting data as sensitive as genetic data, the individual has to be informed in the best possible way in order to respect his or her autonomy. The concept of informed consent and informed refusal is based on the prior condition of autonomy, and the principle of confidentiality presupposes an informed consent. Even the Icelandic act on the rights of patients follows this argumentation by stating that "no treatment may be given without the prior consent of the patient … The consent shall be in writing whenever possible and indicate the information the patient has been provided with and that he has understood the information" (Icelandic Parliament 1997, article 7). Logically and consistently, the Act also demands an informed consent prior to scientific research: "A patient shall give his formal consent prior to participation in scientific research … It shall be explained to the patient that he can refuse to participate in scientific research and that he can cease participation at any time after it has commenced" (article 10).

However, while writing a bill on a Health Sector Database in 1998, the Icelandic legislative assembly perverted these rights by replacing the informed consent with a presumed consent, leaving an opting-out possibility for those who fill in the respective forms (see Sigurdsson 1999). Furthermore, the legislature claims to have respected the informed consent according to the patient's right. It is mentioned in the notes to the bill that the staff of the Health Sector Database should "ensure that citizens are aware of what the centralised database involves, and their rights to refuse to participate, so that they can make an informed decision." (Icelandic Parliament 1998, sector III, 2, 11). Instead of informed consent, ensuring that the people in question support the offered treatment, people are now to make an informed decision, i.e. they have not to consent, but to opt out actively by sending a withdrawal letter to the Directorate of Health. This results in the principle of autonomy being severely violated, as there is no guarantee anymore that the donors of genetic data have really received all the necessary information, that the consequences have been explained to them, and that they have given their free informed consent to the acquirement. Only with those persons listed as opted out, can we be certain that they have been informed. This is a sarcastic perversion of the patient's rights and is a paternalism legitimized by the Icelandic government, which is completely unacceptable in respect to the principle of autonomy.

Conclusion

The current discussion about the effects of conflicting interests on the principle of autonomy shows problematic issues with regard to our concepts of the individual's rights. Democratic societies are obliged to protect those individual's rights without reservations, and, if necessary, must forgo prospective economic or political benefits. Political paternalism in health care must never lead to the disregard of civil rights, as happened in the Icelandic case. The only acceptable political paternalism should be measures to avoid possible violations and abuses.

However, there are still many unanswered questions requiring further research. In the whole debate about the health-sector database and deCODE genetics, no one has investigated the legal situation of deCODE's results being kept by the Swiss enterprise Hoffmann-La Roche. Although both Iceland and Switzerland have ratified the same European directive on data protection (95/46/EC), the collection of data in Iceland has never fulfilled the Swiss requirements on data protection. It is highly disturbing that only the transfer of personal data, but not the commercial goods (e.g. medication) resulting from the processing of that personal data, is so well-protected by international laws.

References

European Union. 1995. Directive 95/46/EC of the European Parliament and of the Council of 24 October 1995 on the protection of individuals with regard to the processing of personal data and on the free movement of such data. *Official Journal of the European Communities* 23/11/1995, L. 28:31–50.

Icelandic Parliament. 1997. Act on the rights of patients no. 74/1997.

Icelandic Parliament. 1998. Act on a Health Sector Database no. 139/1998. Notes to the bill.

McInnis, M. G. 1999. The assent of a nation — Genetics and Iceland. *Clinical Genetics* 55:234–239.

Sigurdsson, S. 1999. Icelanders opt out of genetic database. *Nature* 400:707–708.

28

Do biobanks promote paternalism?
On the loss of autonomy in the quest for individual independence

by Bjørn Hofmann

Introduction

IN ORDER TO handle the moral challenges with respect to modern biobanks, and to avoid paternalism, we tend to turn to the moral principle of respect for autonomy, e.g. to procedures of informed consent. Traditional understandings of autonomy refer to persons' intentionality, understanding, and voluntariness. More precisely, autonomy is defined in terms of making choices. A person acts autonomously only if that person acts intentionally, with understanding, and without controlling influence (Beauchamp and Childress 2001; Faden and Beauchamp 1986, 238).

However, the appeal to autonomy with respect to handling the ethical challenges with modern biobanks runs into a series of problems. Some are related to the special characteristics of biobanks, whereas others are related to the profound challenges to the concept of autonomy. This paper will address the first while investigating the latter. The point is that in our quest for autonomy we may violate its prerequisites.

Intentionality, understanding, and voluntariness?

Many patients requiring health care are in a state of pain, despair, or suffering. Some patients have reduced consciousness. The point is that their deliberative faculties are reduced, and hence, they cannot act autonomously. Correspondingly, patients' ability to understand their situation and all relevant aspects of their health care options can be strongly reduced.

Moreover, many patients are in crisis or misery, which strongly reduces their voluntariness. Patients have little time to make decisions; they are under the influence of relatives and physicians. Many feel that they "have no choice" other than to do what is expected of them. Hence, many patients by definition have reduced or no autonomy. In other words, appealing to a principle of autonomy will not save them from paternalistic tendencies.

Here we could argue that this critique is irrelevant with respect to biobanks. The question of whether one will contribute to a biobank can be posed when a person is not ill, and is as safe, sound, and autonomous as possible. As in everyday life our actions can be autonomous with respect to deciding whether our DNA, blood, tissues, cell serum, or plasma can be entered into a biobank.

Although we may be in a situation in which our actions are completely deliberative and voluntary, it is far from obvious that we are able to understand the situation.[1] This is of course because we cannot foretell the future research results or applications of our contributions to the biobank. Hence, the understanding of what biobanks can do is at best reduced. And hence, according to the traditional conception of autonomy, our ability to make autonomous choices with respect to biobanks is limited.

Autonomy as capacity

We could of course argue that other conceptions of autonomy would be more appropriate than the traditional liberty-based framework. One compelling alternative is to define autonomy in terms of different aspects of human ability, capacity or, personal character, e.g. to define it as the ability to be authentic, independent, in command, and consistent (Benn 1988), or by self-determination, rationality, or the ability to make voluntary choices.

Gerald Dworkin argues that autonomy instead is based on a person's reflective capacity. Autonomous persons have the capacity to "reflect critically upon their first-order preferences, desires, wishes, and so forth and the capacity to accept or attempt to change these in light of higher-order preferences and values" (Dworkin 1988, 108). According to Dworkin it is important to distinguish autonomy from liberty and privacy. A person can be autonomous even if he has reduced liberty. When Odysseus was tied to the mast of his ship so that he could hear the siren song, it certainly restricted his liberty, but not his autonomy. Correspondingly, deception can reduce autonomy, without restricting liberty or privacy (Dworkin 1988, 104–106).

However, this does not seem to solve the problem raised by biobanks. Although we might be able to act according to first-order preferences corresponding to second-order preferences formed under critical reflection in general, it is hard to know our preferences with respect to biobank decisions. We cannot foresee future results, and we certainly do not know what our preferences will be then. It is well known that our preferences are strongly context-dependent; prospective medical possibilities that we prefer, may very well be rejected when they become real, and *vice versa*. Hence, although we may be autonomous in general according to such a capacity conception of autonomy, autonomy does not seem relevant, as we have no access to future preferences. Another possibility would be to define autonomy in terms of personal control. However, it does not make the case easier.

Autonomy as control

An autonomous person has control over his life including control over his own actions, how he is treated by others, and over the obtaining and dissemination of personal information about him. Robert Young has captured this conception of autonomy in his state-

---

1 For the issue of competence, see Welie and Welie 2001.

ment that "the fundamental idea in autonomy is that of authoring one's own world without being subject to the will of others" (Young 1986, 19).

How one is to control future obtaining and dissemination of personal information and how one is treated by others in the context of biobanks is hard to see. Again, we do not have the slightest control over what information about us biobank research is going to bring to light.[2] Hence, being autonomous with respect to biobank decisions according to a conception of autonomy as control is hard.

Consequentialistic autonomy

Another option is to argue that autonomy is a value in itself, and that we should find the institutions and formal rights that best maximize autonomy (Gundersen 1990, 262). According to such a view, the best way to maximize autonomy is by establishing a web of rights. However, challenges appear when we try to specify what autonomy is, so that it can be maximized. Then we tend to find ourselves back at square one.

Furthermore, it is hard to define the rights that will maximize autonomy with respect to biobanking, because we do not know what these rights should entitle us to or protect us against.

Some general challenges to autonomy

There are other more general arguments against autonomy which are relevant to whether it can guide moral decisions with respect to biobanking. The principle of autonomy has been heavily criticized for being too individualistic. It has been reproached for not taking into account the social nature of individuals and the impact of individual choices and actions on others, for neglecting emotions, and for highlighting legal rights and ignoring social practices (Jennings 1998).

Autonomy in modern bioethics is tied to the enlightenment project within North-American liberalism based on universal principles of justice and reason. "Worth comes from the particularity and culturally situated nature of the self's relationships and commitments, and not from the abstract universality of having and exercising freedom or choice as such" (Jennings 1998, 265). Accordingly, it is argued that to meet the moral challenges in modern health care, a viable bioethics has to be based on difference liberalism. Although this may resolve the challenges with respect to "intentional ignorance," it still does not solve the problems related to the person who insists on knowing about, or being in command and control with respect to, biobank decisions.

Others have argued that principlism in general, and the principle of autonomy in particular, mirror certain aspects of American common morality and are therefore untransferable to other contexts and other societies (Holm 2000, 333). Additionally, (practical) interpretations of autonomy (informed consent) in different countries are quite different, and there are few cultural patterns to explain these differences (Lie 2003). It is pointed out that professional forces are stronger than cultural forces with respect to how

---

2 It may also be argued that one could not control whether one would like to have this information, or whether a third party could gain access to such information.

the autonomy-model is implemented. Thus, if autonomy is highly contextual, and only relevant within North-American common morality, it is not viable as a moral foundation for regulating the global formation of biobanks.

No autonomy within biobanking?

The point here is that many of the traditional conceptions of autonomy are challenged by biobanking, due to non-specific research aims (if any) and uncertainty with respect to research results.

One could of course argue that all the conceptions of autonomy discussed above are viable, if one is obliged to ask the contributor for an informed consent for every new test that is run, every new research question, or every new knowledge gained. Then the person would be able to make liberal choices, would be able to reflect on his or her (first-order) preferences, and would be in control. However, this does not seem to be feasible. Biobanks are established to be able to address tomorrow's research questions with access to today's biomaterial. Having to poll the population for every new use of biobanks will be cumbersome and in some cases even impossible. The research results obtained from biobanks established with our biomaterial and consent may become available after we are dead, and be of vital relevance for our children or grandchildren, who for obvious reasons have not consented.

Identified, identifiable, anonymized, and anonymous

At this point it becomes urgent to address the question of whether a differentiation with respect to information would be able to rescue autonomy. If information is anonymized, or anonymous, would that not save autonomy? Could we not then autonomously contribute to the achievement of knowledge important to others, without any fuss? In principle this appears to be so.

However, there are some practical problems. To be able to make the best of biomaterials with respect to research outcomes, one needs as much information as possible. The coupling of biobanks and health registers of all kinds, as well as medical records, is necessary. Within this context it becomes questionable whether information really can be anonymized or anonymous.

Autonomy-based paternalism

The focus on autonomy, in particular in North-American bioethics, is itself an expression of paternalism (Gundersen 1990; Lie 2003). Proponents of the prominence of autonomy are more concerned with what people should want than with what they actually want (Schneider 1998). Autonomy can be interpreted as a mandatory duty to choose, which opposes ordinary people's ideas.

This is highly relevant with respect to biobanks, as the knowledge of future research results and their significance to an individual and his or her family is unknown. The urge to enter a biobank under glorious prospects of future treatments can in itself become paternalistic.

This also goes for the claim that autonomy is based on what reasonable persons would agree upon. Such (contractarian) conceptions of autonomy tend to be a *petitio principii*, as the effort to find what a reasonable person would agree upon in is itself paternalistic. In other words, justification of anti-paternalism tends to be based on paternalistic assumptions (Lie 2003).

Autonomy as a formal straitjacket

This also becomes evident in the insistence on informed consent with respect to biobank research. As the information has to be limited or uncertain, the formalism of informed consent may lead to two situations. The first is where there cannot be a "real" informed consent, as information is insufficient, and the second is where the information is so general that the consent becomes a waiver. Both these alternatives appear to be morally controversial. If moral issues related to biobanks are believed to be handled by autonomy and informed consent, they may well become paternalistic.

Conclusion

Biobanks pose ontological, epistemological, and ethical challenges, and autonomy does not seem to be the appropriate measure to meet the ethical ones. The concept of autonomy faces a series of theoretical and practical problems that are highlighted in the case of establishing biobanks and biobank research. In particular, the limited understanding of the future use of biobanks and associated research restricts autonomy's relevance. Even if we regard ourselves as autonomous with respect to biobank research, future generations may find the research results just as strongly restrictive to their life projects and flourishing.

Although autonomy appears to have its weaknesses in general, and with respect to regulating biobanking in particular, this does not imply that it is useless. Respect for patients' or subjects' autonomy appears to be an important moral perspective in all health care. However, this does not mean that it can be applied as a regulative principle.

The autonomy-paternalism dichotomy is a useful analytical tool, which can guide our analysis of morally challenging situations, even if it does not prescribe their solutions. Biobanks challenge us at so many levels, and until we have found a theoretical perspective that is capable of integrating all of them, we will have to apply existing analytical means: autonomy being *one* of them.

References

Beauchamp, T. L., and J. F. Childress. 2001. *Principles of biomedical ethics*. New York: Oxford University Press.

Benn, S. 1988. *A theory of freedom*. Cambridge: Cambridge University Press.

Callahan, D. 1984. Autonomy: A moral good, not a moral obsession. *Hastings Center Report* 14:40–42.

Dworkin, G. 1988. *The theory and practice of autonomy*. New York: Cambridge University Press.

Faden, R., and T. L. Beauchamp. 1986. *A history and theory of informed consent*. New York: Oxford University Press.

Gundersen, M. 1990. Justifying a principle of informed consent: A case study in autonomy-based ethics. *Public Affairs Quarterly* 4:249–265.

Holm, S. 2000. *Developing a European biomedical ethics: Can American biomedical ethics be transferred without modification?* PhD Diss. Univ. of Copenhagen.

Jennings, B. 1998. Autonomy and difference: The travails of liberalism in bioethics. In *Bioethics and society: Constructing the ethical enterprise*, ed. R. DeVries and J. Subedi, 258–269. New Jersey: Prentice Hall.

Lie, R.K. 2003. The ethics of the physician-patient relationship: The Anglo-American approach in the European context. In *Healthy thoughts: European perspectives on health care ethics*, ed. R. K. Lie, P. T. Schotsmans, B. Hansen, and T. Meuenbergs. Leuven: Peters.

Schneider, C. E. 1998. *The practice of autonomy: Patients, doctors, and medical decisions*. New York: Oxford University Press.

Welie, J. V. M., and S. P. K. Welie. 2001. Patient decision making competence: Outlines of a conceptual analysis. *Medicine, Health Care and Philosophy* 4:127–138.

Young, R. 1986. *Personal autonomy: Beyond negative and positive liberty*. New York: St. Martin's Press.

Monozygotic autonomy and genetic privacy

by Anne Maria Skrikerud

MONOZYGOTIC TWINS ARE popular in genetic research, especially in research attached to biobanks where their special genetic setup is important to detect new genes and environmental factors that cause hereditary illnesses. The standard analysis of the right to genetic information relies on the concepts of "autonomy" and "right to information." The case of monozygotic twins does, however, raise challenges to the standard analysis, and may lead us to a deeper understanding of the potential problems in "autonomy" and the "right to information." Monozygotic twins are characterized by having the same DNA but different consciousness. "Autonomy" and the "right to information" are features of the consciousness, but in genetic research they apply to the DNA or that which monozygotic twins share.

Let us consider a constructed case with two monozygotic brothers. One of their parents recently died of colon cancer. No hereditary gene was found during the treatment, but since colon cancer is a common disease in the brothers' family there is reason to believe that the cancer was hereditary. The brothers are offered the opportunity to participate in a research project to try to detect new genes that might cause colon cancer (Möslein et al. 2003; Meincke et al. 2003). In return the company in charge of the project offers to give the brothers information about any risks detected. One of the brothers, brother A, wants to participate, while brother B does not.

First, we must take for granted that the brothers believe that participating in the research project is the only chance to receive knowledge about their future. The medical information that might come out of their participation is not available in standard clinical practice. Second, we must also take for granted that even if we might have a situation of conflict, the brothers will still stay in touch, so that giving information to one brother is the same as giving it to the other. How are the two concepts "autonomy" and "right to information" related to this case?

The right to information versus the right to remain ignorant

Brother A uses his *right to information* and says that not only does he want to participate; he wants to be informed about the research findings. Brother B on the other hand wants to use his *right to remain ignorant*, and does not want to participate either. Let us first consider the following opposite concepts, "right to information" versus "right to ignorance." Brother A and brother B both have the same genetic setup but they are not the same person. They have made opposite decisions. A exercises his right to know. Häyry and Takala have interpreted this in three ways:

(a) A has no duty to remain ignorant.
(b) Others have a duty not to interfere with A's quest for information.
(c) Somebody has a positive duty to assist A in [his] quest for information (Häyry and Takala 2001).

Häyry and Takala understand the three interpretations of the right as three subsequent rights. (a) is a right that is more easily understood as a *license*; (b) is a *negative claim-right*; and (c) is a *positive claim-right* (Häyry and Takala 2001).

But Brother B does not want to know whether he is carrying a gene for colon cancer. Häyry and Takala characterize the "right to ignorance" accordingly:

(d) B has no duty to know.
(e) Others have a duty not to inform B against his will.
(f) Somebody has a positive duty to assist B in remaining in ignorance (Häyry and Takala 2001).

Now let us go back and look at the case of brother A and brother B. What if these two brothers are not only brothers but each others' best friends and they find themselves disagreeing for the first time? To expect brother A to actively not say a word about a possibly lethal gene to B, but maybe undergo a colectomy might be conceptually possible, but considering their strong relationship, brother B is surely certain to understand what is going on. In that situation, *everything is not done* so that B *remains in ignorance*. If A has a *positive duty* to assist B in staying uninformed, A would have to refrain from a medically advised colectomy. But for a *true innocence* B would have to know in advance that in both cases A's actions would be the same.

On the other hand, if only the "no duty to know" is valid, characterized by Häyry and Takala as only a license, it means practically nothing: as soon as A learns about a positive outcome of the research project he will go to the doctor. If (e) is valid A has a duty not to inform B against his will. This does not mean that A has to hinder B from seeing any written result in his home.

From brother A's viewpoint, he does not have a duty to remain ignorant. To ask of him that he should refuse information in respect for his brother would be to interfere with both (a) and (b), (b) saying in this case that brother B has a duty not to interfere with A's quest for information. That means that if brother B wants to remain in ignorance, this wish cannot stop brother A's quest for information.

How is (c) to be interpreted? If (c) is fulfilled then someone must *be obliged to do a positive act* so that A will get the information he wants. This "someone" cannot be counted as the research company since it, after all, takes the initiative and one cannot see any moral duty from the company's viewpoint, any *moral raison d'être*. One must assume the company has reasons of its own to be committed to research. In this example there is no one who has a *positive* duty to assist brother A in his "right to information." But the company has an obligation through the arrangement with each participant, to inform those participants once the medical results are at hand. Accordingly, it would be wrong to promise information about any medical results, and then hold that information back at the end of the research project. The company has a *quasi-contractual* obligation toward brother A. The company is more committed than the duty in (b) not to interfere with A's quest for information, yet its situation is not covered by Häyry and Takala's obligation to do a positive act.

In the European Convention on Human Rights and Biomedicine, article 10(2) says: "Everyone is entitled to know any information collected about his or her health. However, the wishes of individuals not to be so informed shall be observed" (Council of Europe 1997; Harris and Keywood 2001). If this article in the European Convention is to provide any guidance concerning what the brothers ought to do, it might seem as if the right of brother A is equally important as the right of brother B. But the rights are only negative rights. In the European Convention no one is obliged to assist someone else in a quest for information or in remaining in ignorance. How may this article be interpreted? It is clear that although A has no *positive claim-right to information* he may surely ask for a *negative claim-right to information*. So B has no right actively to hinder A. But B's negative claim-right to remain ignorant should mean that A has a duty not to inform B against his will (Häyry and Takala 2001). The European Convention is not applicable to this case.

Autonomy

"Autonomy" is important because this concept describes the feature that enables persons to choose between the *right to information* and the *right to ignorance*. And "autonomy" is often characterized as being related to choice, although Harris and Keywood (2001) claim that "autonomy is not simply the exercise of choice but of control—literally self government." Harris and Keywood do not consider the choice that includes *the right of ignorance* to be an *autonomous* action. It may be a right of liberty to act in a way that includes the deprivation of knowledge from further action. But they say that to take an autonomous choice one has to be informed. A possible interpretation of this statement may be that if someone has some kind of illness and does not want to be informed, he will be treated without consent. And that would of course be against the law. "Furthermore, failure to provide information relating to the nature of any proposed intervention will invalidate patient consent and give rise to an action for battery before the civil or criminal courts" (Harris and Keywood 2001). But that is of course not necessarily the case, especially if the *information* has come from a research project. We can hardly imagine a reasonable, healthy adult being told by researchers that "we are just going out for a coffee," and ending up having a colectomy. When B refuses information he of course says no to any prophylactic treatment, and we must expect him to be aware of this fact.

Harris and Keywood claim that to accept ignorance is the same as to refuse *control* of one's life, and control or self government is what characterizes autonomy. The right to ignorance is thereby characterized as a liberty-claim. Harris and Keywood (2001) see that their argument is limiting autonomy: "It might be based on a simple claim of a right to liberty where liberty includes the right to make free, but non-autonomous decisions or autonomous but autonomy-limiting decisions." They argue that when in doubt one should always prefer to inform the patient. And they refer to a House of Lords decision when saying that "'reasonable patients' would wish to know, rather than not know, information about their health status" (Harris and Keywood 2001; *Sidaway v. Board of Governors of the Bethlem Royal Hospital and the Maudsley Hospital and Others* [1985]). But it does not have to be like that. For instance, a research project in Germany has proven that predictive information to people in families with Huntington's chorea has a negative effect. Of those who got a positive result many turned suicidal or depressed (Meincke et al. 2003). And Chapman (2002) has been able to show that of all those who would be relevant to test for Huntington's chorea, only between 5% and 25% wish to have a predictive test. Harris and Keywood discuss autonomy and the right to information from a clinical perspective, and consequently their argumentation is not automatically transferable to research cases as discussed in this paper.

Häyry and Takala (2001) have another way of understanding autonomy. Inspired by John Stuart Mill they say that "individuals are entitled to make their own decisions on whatever ground they wish, as long as they do not inflict harm on others by these decisions." When it comes to the right to information versus the right to ignorance, they interpret Mill accordingly:

(j) Individuals do not have a duty to know about their genetic makeup, if they do not harm others by remaining in ignorance.
(k) Others have a duty not to force-feeding them with the information, if third parties are not harmed by the ignorance.
(l) Somebody, presumably the government, has a duty to prevent others from force-feeding them with the information, if third parties are not harmed by the ignorance (Häyry and Takala 2001).

How are these three norms related to the case of the two monozygotic brothers? The first norm, the license, would mean that B would not be forced by law to be informed about his genetic makeup, unless this ignorance may harm someone else—in this case, his brother A. If A is kept uninformed because of B's preferences, is that such a harm, that B has a duty to be informed? (k), on the other hand, would protect him from others force-feeding him with information, as long as no one is harmed by his ignorance. Here A may claim that he is harmed by B's ignorance. Is A's claim important enough to warrant force-feeding B with information he does not desire? The duty in (l) would in this case practically mean the same as in (k).

Conclusion

How should this case be solved? To begin with, it is important to hold on to one thing, namely that no one has the right to participate in research projects. In this case that means that if the research company prefers to refuse A when B has said "no," they may do so. The company is not a clinical practice, and it is not in general to be concerned with other people's health. Is there a reason why it still should accept A if it wanted to? In this case it is assumed that colon cancer may be hereditary in the family, and the only way to learn if one carries a gene is to participate in a research project. Information about a possible gene may be important considering the prophylactic treatment that may be done to prevent the illness; consequently I consider it as justified if the company accepts A as a research subject. B would have to learn about his genetic makeup, to prevent harm coming to his brother by his ignorance. It would be different if the illness were not colon cancer but a disease like Huntington's chorea, where there is no known prophylactic treatment. B's wish to remain ignorant should be emphasized more, since no harm can be caused directly by his ignorance (Häyry and Takala 2001). A possible conflict between two monozygotic twins concerning the right to information and the right to ignorance may vary from one case to another. The consequence of the information must be considered first.

References

Chapman, E. 2002. Ethical dilemmas in testing for late onset conditions: Reactions to testing and perceived impact on other family members. *Journal of Genetic Counseling* 11:351–367.

Council of Europe. 1997. *Convention on human rights and biomedicine*. ETS No. 164, 4 April. Available online at:
http://conventions.coe.int/treaty/en/treaties/html/164.htm

Harris, J., and K. Keywood. 2001. Ignorance, information and autonomy. *Theoretical Medicine* 22:415–436.

Häyry, M., and T. Takala. 2001. Genetic information, rights, and autonomy. *Theoretical Medicine* 22:403–414.

Meincke, U., Ch. Kosinski, K. Zerres, and G. Maio. 2003. Psychiatrische und ethische Aspekte genetischer Diagnostik am Beispiel der Chorea Huntington. *Nervenarzt* 74:413–419.

Möslein, G., S. Pistorius, H. D. Saeger, and H. K. Schackert. 2003. Preventive surgery for colon cancer in familial adenomatous polyposis and hereditary nonpolyposis colorectal cancer syndrom. *Langenbecks Archive of Surgery* 388:9–16.

Sidaway v. Board of Governors of the Bethlem Royal Hospital and the Maudsley Hospital and Others [1985]AC 871.

Privacy in public

by Salvör Nordal

DURING THE SEVERAL years I have been working on the concept of privacy, I have grown more and more skeptical about the ways in which it has been used and referred to in the scholarly literature. One of the main reasons for this skepticism is the problem we face in understanding and defining the term—both its scope and its content. Judith Jarvis Thomson (1975) stated in a famous paper that "perhaps the most striking thing about the right to privacy is that nobody seems to have any clear idea what it is," and 30 years later this statement is still uncomfortably true. This is unfortunate to say the least. For the past few decades the interest in the right to privacy has not only been acknowledged in US courts but found its way for instance into the European Declaration of Human Rights, and from there into the constitutions of many European countries. How can we take a right seriously if it can mean almost anything, or nothing for that matter?

The title of this paper is *privacy in public*. For some this may sound like a contradiction. Traditionally, privacy has been defined within the private sphere; what we regard as private we do in private and keep away from others. The fact is, however, that there is a growing concern for privacy in the public sphere, especially with the development of information technology and genetic databases. Therefore we need to ask if privacy in public does make sense. In order to answer this question, I will examine the interrelations between privacy and the private, and argue in the end that the traditional way of defining privacy has been misdirected.

What is privacy?

There is no unified definition of privacy, and in fact many scholars have argued lately that privacy is a cluster concept referring to distinct concerns. These concerns have for instance been labeled as: 1) *informational privacy*, the interest of protection of personal information; 2) *decisional privacy*, the interest of deciding on personal issues; 3) *physical privacy*, the interest of having access to person and personal spaces; and 4) *proprietary privacy*, the interest of appropriation and ownership of human personality (see Allen 1988; Allen 2003; DeCew 1997).

What these different concerns have in common is that there are some personal affairs that an individual should be able to keep to him or herself. This indicates some kind of haven or sanctuary where others are not allowed to interfere with personal lives and decisions—a sanctuary where we should be let alone and able to restrict access to ourselves. In our private sphere and with regards to our private affairs we should be in control. On the other hand, if we decide to go into the open, into the public realm, we give up our privacy, i.e. privacy does not apply anymore. In the public sphere we allow people to observe us, and similarly we can observe others.

This understanding of the private and privacy is quite common in the literature, and seems to coincide with our intuition. The word "privacy" has after all the same roots as "private." This view has, however, been criticized for being too limited (Nissenbaum 1998). Here I want to draw attention to three problems:

First, this understanding seems to rest on the assumption that we can make a sharp distinction between the private and the public sphere. It indicates, furthermore, that we have a common understanding of these two spheres and that we are well aware of when we are acting in one or the other. This seems however highly problematic.

Second, and closely related to the first point, is the assumption that privacy concerns are necessarily linked with private things of the private sphere. What happens in public is public, and we are not able to demand a privacy protection for public affairs. This might sound convincing, but again it rests on a sharp distinction between the private and the public, and more importantly on the notion that particular issues or affairs belong to one or the other.

Third, I want to draw attention to the emphasis on individual control in the privacy literature. The focal point in the privacy literature seems more often than not be that individuals have and should have control of their privacy, i.e. that individuals control when they enjoy privacy and when they do not. This seems to me to be a rather naïve understanding of privacy and, more importantly, if we stick to this demand many of the privacy worries of today fall outside the scope of privacy.

Is the public/private distinction a myth?

It is an important fact of human existence that we can withdraw from the eyes of others into our private space. Some things we want to do in private and some things we are required to do in private. In most societies there are conventions that draw such a boundary between the private and the public and indicate what behavior is or is not acceptable in public. In this context, "in public" refers to places open to or accessible to everyone, such as the marketplace or public streets. The private sphere on the other hand is a sphere where access is restricted. In public streets we want to be let alone with our business without interference or nuisance, we do not want to be exposed to certain kinds of behavior (Guess 2001).

A different way of making a distinction between the public and the private is that of contemporary liberalism. There the distinction is used to set a limit to governmental intrusion into the private sphere. The private sphere is seen as highly valuable for human flourishing.

Yet another way of drawing a distinction is found in Aristotle's work, where a boundary is drawn between the domestic life of the household, which serves bodily functions, in contrast with the political or the public sphere which is, in his theory, and contrary to liberalism, the locus of human flourishing.

There are many more ways of describing or defining the public and the private, such as in terms of public or private interests, or public or private goods. It is not even necessary, as Raymond Geuss points out, that a definition of the public sphere has a residual private sphere, and *vice versa* (Geuss 2001, 109).

Consequently, if privacy rests on the distinction between the public and the private, we need to be clear about what we mean by the words "public" and "private," since that may change fundamentally the meaning of privacy. In liberal terms privacy is the realm of human flourishing, while in the Aristotelian sense the public sphere is the realm of human flourishing and privacy has a rather different function.

Privacy and the private

From what I have said so far it is safe to conclude that the public/private distinction does not give a ground for a unified understanding of privacy. Rather, it seems to obscure our understanding of this concept. Now I want to take a closer look at the relation between privacy and the private, given that we can define such a sphere, and ask if it is essential to define a private sphere to get a clear understanding of privacy. Of course, this depends on how we understand the private, or how we draw the distinction.

It is well known that feminists have targeted the private and the private sphere. For them, the subordination of women is reflected in the distinction between the public and the private, where women's role were defined in terms of domestic work, which denied them access to the public. Furthermore, the idealized version of the private realm, in terms of liberalism, has also been unconvincing to many feminists, since women have found it hard to enjoy privacy within the private sphere, due to domestic work and responsibilities (Allen 1988).

Instead of regarding the private as a sphere where we should be let alone and without public scrutiny, feminists have argued that personal matters should be made political and discussed in the open. One main theme in this criticism, a theme that matters for my purpose, is the rejection of the individualism that portrays persons as unrelated to others. Many feminists argue, on the contrary, that the private sphere is a locus of important personal relations where our actions and behavior affects many others and to whom we are accountable. A recent book by Anita Allen, a well known privacy scholar, discusses privacy within the private sphere and the many cases in which we are accountable to others for our personal matters, such as in the case of drug use, sexual behavior, and regarding our health condition (Allen 2003). Her point is not that we are always accountable for these matters—far from it—but that these issues are in reality very complex and cannot easily be categorized as either private or public, but can be both depending on the context and circumstances. We may for instance be required to provide information concerning our health, which is normally regarded as private, to our relatives or even in some cases to authorities.

Here I want to pause for a moment. It is often argued that we should be able to protect personal information from others and appeal to privacy. Information on health and genetic makeup is a case in point. But why appeal to privacy? Is it because genetic information is private in any sense, and if so what does it mean to say so?

Genetic information may certainly contain sensitive information such as the fact that the individual is HIV positive, but it also contains non-sensitive information anyone can observe from seeing us, such as sex, hair color, etc.

Is genetic information sensitive because it is difficult to obtain. Again the answer is both "yes" and "no." We might need to use force to get a blood sample from someone, but on the other hand he or she may well leave a sample of DNA around, in a hair sample, for example.

Is genetic information private because it does not affect anyone—i.e. is no one's business? Again, that depends. Genetic information may contain information relevant to family members and, moreover, we are not in full control of our genetic information because we share almost all of it with our family. This is especially evident in the case of informational privacy, to which I now turn.

Privacy in public — the cyber sphere

We have seen the difficulties involved in establishing the scope of the private, and its interrelation with privacy. The difficulties with categorizing issues such as personal information as strictly public or private should also be evident. I want to turn now to the issue of privacy in public.

As I mentioned earlier, it seems to be a conceptual error to talk about privacy in public: privacy seems to be located in the private, if anywhere. And I have already discussed some of the difficulties with that view. Another argument against privacy in public concerns our own choices and control (see Nissenbaum 1998). The public sphere is seen as open and accessible to everyone. Moreover, when we are acting in the public sphere, we give away personal information freely. When we walk on public streets and buy in public shops, those around us can see what we do and how we behave, what we are wearing, what we are buying, and so on. We have thereby given away our control of personal information. More accurately, we are revealing it to others. A person who complains of privacy loss in the public sphere has misunderstood something essential about that sphere.

Or so the story goes. With recent technological developments, this simple description of the public sphere is changing dramatically. Our movements and actions in open public space are not only open for others to see and observe; it may possibly be monitored, stored, and kept in one of the new modern databanks. It used to be the case that we had anonymity in public. It is well known that in crowded streets we are seen by many but observed by none. This is changing. And this change is so fundamental that we may well describe it as a transformation. Cities have monitors on their down town streets, shopping behavior is well listed when we use our credit cards, the local videostore has a list of our movie preferences, and so on.

Is this a privacy loss? After all we know we are observed in public places and, moreover, information of our walking in the streets is non-sensitive anyway. Who cares what cheese I buy and whether I buy meat once or three times a week? Still, something is wrong. It is true that we give this information freely, we pay with credit cards, we walk in the streets, and so on. And it is true that our daily habits are not categorized as highly sensitive. The problem we face is that this information is stored and can be used in new and different contexts. Along with information on our health, finances, and so on, it is possible for instance to build up a profile. Instead of being seen by many and observed by none in the public sphere, as has been the case, we are now scrutinized by some, and our activities are recorded and stored for purposes we do not know, and at present cannot even imagine (Nissenbaum 1998). In this sense it is safe to say that information technology is transforming the public sphere. And this transformation has given us good grounds for privacy concerns in the public realm.

Privacy interests

I have shown that it is misdirected to define privacy in relation to the public/private distinction, and to put too much emphasis on individual control and private affairs. Rather, we should think again of why privacy is important and what privacy is meant to protect. It is important to remember that the privacy issues were brought up to begin with as a reaction to new technology: a technology that could publish photographs of individuals, tap phones, and so on. In a world of information technology, I think the fear of misuse of information, stigmatization of groups, and unjustified interference are the real fears behind privacy claims. In the public sphere we give information freely—by simply walking the streets we are consenting to being monitored. But still we fear how this information will be used. In this respect we cannot appeal to individual control, because the technology does not give us control.

How then can we protect privacy? Instead of focusing on the nature of personal information, we should look at the context and circumstances and ask how we can build safeguards and protect individuals from misuse and deception. This means that it must be clear why information is stored, how it is used, and so on. Instead of emphasizing individual rights as we have in the past, we need to look at the interests those rights are meant to protect. In this way, we will achieve a fuller understanding of privacy.

Acknowledgements

This paper is a product of the ELSAGEN project (Ethical, Legal and Social Aspects of Human Genetic Databases: A European Comparison), financed 2002–2004 by the European Commission's 5th Framework Programme, Quality of Life (contract number QLG6-CT-2001-00062). Research for this paper was also aided by participation in workshops funded by the Nordic Academy for Advanced Study (NorFA), through the NorFA Network "The Ethics of Genetic and Medical Information." I gratefully acknowledge the support of the European Community and NorFA.

References

Allen, A. 1988. *Uneasy access: Privacy for women in a free society*. Totowa, NJ: Rowman and Littlefield.

Allen, A. 2003. *Why privacy isn't everything: Feminist reflections on personal accountability*. Lanham, MD: Rowman & Littlefield.

DeCew, J. W. 1997. *In Pursuit of Privacy: Law, ethics and the rise of technology*. Ithaca and London: Cornell University Press.

Geuss, R. 2001. *Public goods and private goods*. Princeton, NJ: Princeton University Press.

Nissenbaum, H. 1998. Protecting privacy in an information age: The problem of privacy in public. *Law and Philosophy* 17:559–596.

Thomson, J. J. 1975. The right to privacy. *Philosophy and Public Affairs* 4:295–314.

Genetic screening, eugenics, and commodification

Genetic screening, prospective parenthood, and the internal perspective

by Peter Herissone-Kelly

WHY ARE GENETIC samples and data stored in biobanks? One of the primary reasons, of course, is to facilitate research into hereditary conditions. For each such condition, either that condition can be cured, or it cannot. But for as long as a cure for a condition remains unavailable, and where that condition can be detected by pre-implantation genetic diagnosis (or "PIGD"), it might seem acceptable to use this technique to select only "healthy" embryos—embryos unaffected by the condition—for implantation.

That it *is* acceptable has been hotly disputed, especially by representatives of the disability rights movement, who have condemned the practice as discriminatory. However, a number of bioethicists (including John Harris and Raanan Gillon), have vigorously rebutted these claims (Harris 2001; Gillon 2001). And some (amongst them Julian Savulescu and, again, John Harris) have gone further, arguing not only that the practice is ethically acceptable, but that, where available, it is morally obligatory (Savulescu 2001; Harris 2001). That is, it has been held that, when faced, say, with the choice of implanting one of two embryos, it would be positively wrong for a prospective parent not to choose that which would be most likely to develop into the person with the better life (for ease of reference in what follows, I shall call such an embryo a "better-life embryo," and its companion a "worse-life embryo"). And, so long as we are convinced that this claim is not discriminatory, it may appear to be convincing.

I will not, in this paper, address the important question of whether the practice of PIGD and embryo selection is discriminatory. Nor will I ask—though again I think it is a crucial question—whether it is really the case that all those without impairments are even *likely* to lead better lives than all those with impairments (though I will state, purely in passing, that this is not at all obvious to me). Instead, I will focus solely on the claim that it would always be wrong for a prospective parent to select a worse-life embryo in preference to a better-life embryo. I want to suggest that this claim is incorrect. I will argue (with one qualification) that when a person stands in the relation of prospective parenthood to a better-life embryo and a worse-life embryo, there is something about his or her standing in that relation that makes it equally legitimate for him or her to select either embryo.

For the sake of clarity, let me restate the claim against which I am arguing:

(BL) Whenever a prospective parent is faced with a choice of implanting one of two embryos, it is obligatory for him or her to select the better-life embryo, and so it would be morally wrong for him or her to select the worse-life embryo.

As I have already conceded, (BL) is at least *prima facie* compelling. What, then, could be wrong with it? Well, there is one possible criticism which, I think, ultimately fails, but which can nevertheless push us in the direction of a more decisive criticism. It will do this by introducing us to a distinction between what I call "the internal perspective" and "the external perspective."

The criticism attacks the notion of better and worse lives on which (BL) depends, by asking *for whom* a better life is better. Suppose that I have before me two embryos, A and B, and that I have to decide which ought to be implanted. Suppose too that it is abundantly clear that A is the better-life embryo—that is, that A is likely to develop into the person with the better life. But *for whom* will this life be better? Our natural reaction might be to say that it will be better *for* the person (call him Jim) who will develop from A. But this appears to suppose that Jim could have lived either of the two possible lives available, and fortunately got to live the better one, and this is not the case. Had B been implanted, the worse life would have been lived, but it would not have been lived by Jim.

Similarly, the worse life would not have been worse for the person who would have developed from B—call her Jane. To say that the worse life would have been worse *for* Jane again implies that Jane could have lived either of the two lives available, and unfortunately got to live the worse. And again, this is not the case. If A had been implanted, the better life would have been lived, but it would not have been lived by Jane.

This criticism of (BL) appears to rest on two assumptions. First, it appeals to an intuition that it is somehow illegitimate to appeal to the notion of a better life that is not better *for* anyone, but is simply better, *period*. Someone making the criticism may think either that this sort of free-floating betterness is incoherent, or, if coherent, cannot ground the sort of moral obligation that (BL) asserts. Second, it assumes that "better for" judgments can only be made from what I will call the "internal perspective." Precisely what this means I will explain shortly.

At first sight, it might seem that (BL)'s notion of better and worse lives is indeed a notion of lives that are not better and worse *for* anybody. And it does seem possible to question the coherence of this sort of "betterness, period." Oddly, however, at the same time it seems possible meaningfully to talk of better and worse lives in this way. That this is so is demonstrated by an example given by Derek Parfit (1984, chapter 16) in his modern classic *Reasons and Persons*.

In broad outline, the example runs as follows. Suppose that one of two possible social policies could be adopted. The first would be a policy of depletion of natural resources, which would significantly improve the quality of our lives for the next couple of centuries, but then make for a significantly lower (though not unbearable) quality of life for the generations that followed. The second would be a policy of conservation of resources, which would cause our quality of life to remain where it is, and keep it constant for all future generations. Now, importantly, Parfit thinks it safe to assume that adoption of the depletion policy will alter future events to such an extent that existing people will meet other partners, conceive children at different times, and so on, than they would have done under the conservation policy. Within a couple of centuries, a wholly different set of people will exist than would have existed had the conservation policy been chosen.

Will these people have worse lives because the policy of depletion was chosen? The answer has to be "no," since had the policy not been adopted, they would not have existed, and so would not have had better lives. Now suppose that the conservation policy is chosen. Will the people in existence several hundred years later have better lives as a result of its adoption? Again, the answer is "no," since those people would not have existed to have worse lives had the depletion policy been adopted. Thus, the better lives produced by the conservation policy would not be better *for* anybody, and the worse lives produced by the depletion policy would not be worse *for* anybody. And yet, crucially, it does seem eminently possible to judge that better lives will result from the conservation policy, and worse lives from the depletion policy. There is nothing incoherent about this judgment. But if the better or worse lives are not better or worse *for* anybody, then the better lives must be better, period, and the worse lives worse, period. Therefore, the notion of "better, period" cannot be incoherent.

I have said that our critic of (BL) assumes either that the notion of "better, period" is incoherent, or that it cannot create moral obligations (the implication being that only "better *for*" judgments could give rise to obligations). Parfit's example has made it look as if the notion of "better, period" is not incoherent. But crucially, it also appears to show that it can ground obligations. That is, if we were in a position to choose between the conservation policy and the depletion policy, we would be obliged to adopt the former: the policy that would produce lives that were better, without being better *for* anyone. As Parfit says (1984, 363), "the great lowering of the quality of life must provide *some* moral reason not to choose depletion."

This can seem puzzling. On the one hand, we can readily agree that the policy makers in Parfit's example ought to adopt the conservation policy. On the other hand, we can remain mystified as to why they should be obliged to do something to produce lives that are not better *for* anyone. I think the solution to this puzzle is that the policy-makers are in fact *not* obliged to do this, even though they *are* obliged to adopt the conservation policy. This is because I think that any coherent "better, period" judgment must be analyzable into a distinctive sort of "better for" judgment. In short, it must be analyzable into the sort of "better for" judgment appropriate to what I call "the external perspective." It is time I explained what I mean by "the internal perspective" and "the external perspective."

In thinking about the life of a possible future person A, we adopt the internal perspective when we (1) "imaginatively inhabit" that life, attempting to gain a sense of what it would be like to be A, and (2) make the sort of "better" and "worse" judgments that we would make about A's life *if we were A*. All the "better for" judgments made from the internal perspective will be judgments about what would be better *for* A. It should be clear, then, that our critic of (BL) is taking up the internal perspective, when she argues that the better and worse lives mentioned in (BL) are not better or worse *for* anyone.

259

We are unmoved by (BL)'s talk of better and worse lives when we view it from the internal perspective, since such talk belongs only to the external perspective. In thinking about the life of a possible future person A from the external perspective, we (1) imaginatively inhabit that life, attempting to gain a sense of what it would be like to be A; (2) imaginatively inhabit the life of another possible future person B, attempting to gain a sense of what it would be like to be B; and (3) draw back from the perspective of both A and B, in order to make a judgment about which life is better. Since this will not be a judgment about which of the two lives is better *for* A or *for* B, we can if we like say that it is a judgment about which life is better, period. But we should understand that what we have done in taking the external perspective, is in fact to have made a different sort of "better for" judgment. That is, *we have judged that, say, the life that A would lead would be better for A than the life that B would lead would be for B*. That the "better, period" judgments we make from the external perspective are after all analyzable into a type of "better for" judgment is, I think, what rescues them from incoherence.

The upshot is this: our critic of (BL) is right to assume that "better, period" judgments are incoherent, only when those judgments are not analyzable into "better for" judgments. And she is wrong to assume that "better for" judgments can be made only from the internal perspective, and so only about a single person. From the external perspective, what we might call *transpersonal* "better for" judgments are possible. It is, I submit, a transpersonal "better for" judgment that is made in Parfit's example.

The external perspective is, it seems to me, a wholly fitting one to take when making decisions that will affect which of a range of possible future people will be born. Judging that a not-yet-existent person A's life would be better for A than another not-yet-existent person B's life would be for B, is a compelling reason to do what will bring A into existence. But if I believe this, why do I not believe (BL)?

My position is that a prospective parent, *qua* prospective parent, is entitled (though probably not obliged) to take up the internal perspective when faced with a decision about which of two embryos to implant. Taking up the internal perspective will give him or her no compelling reason to select either embryo in preference to the other. The one exception is where an embryo is likely to develop into a person for whom it would be better not to have been born; thinking about this possible future person's life from the internal perspective, we are led to the conclusion that the embryo that will develop into her ought not to be implanted.

Before I expand on these points, I want to anticipate an objection. It might be protested that it is never legitimate to take up the internal perspective when thinking about future possible lives, since doing so would involve imaginatively inhabiting the life of someone who does not exist, and so has no interests. My response to this is that taking up the *external* perspective in order to judge which of two lives would be better—which is something we must do if we are to be guided by (BL)—*also* involves imaginatively inhabiting the lives of people who do not exist or have interests. And yet we take their interests into account when we judge that A's life would be better for A than B's life would for B, since this, it seems to me, is to judge that the interests that A would have would be better met by A's life, than the interests that B would have would be met by B's life.

Suppose, then, that a prospective parent is trying to decide which of two embryos should be implanted. To do this, she must imaginatively inhabit both future possible lives, in order to compare their quality. (I say "she" and "her" in what follows only to avoid cumbersome disjunctions—I am aware that prospective parents can be of either sex!) Now, when she, as prospective parent, does this, she is imaginatively inhabiting the life of her future offspring, and it is therefore appropriate that she should relate to that future offspring's life in a way that is proper for a parent. This, I would argue, will involve an extremely close identification with the possible subject of the life she is imagining, so close that it will include, and take seriously, the sorts of assessments that the offspring would be likely to make of his own life. In short, it will include just those features from which the external perspective prescinds; it will be the internal perspective. The only "better for" judgments available from this perspective will be about what would be better for the person who would develop from that embryo.

Now, it might be urged that I have said only what a prospective parent will *as a matter of fact* be likely to do—that she will probably take up the internal perspective. And this, it might be objected, does not entail that it is permissible for her to take up that perspective, nor that taking up that perspective will lead her to act in a permissible way. But my claim is actually stronger—I have not simply said something about what it would be *natural* or *understandable* for a prospective parent to do. Rather, I have said it is *appropriate* that she should imaginatively inhabit each future possible person's life in a manner *proper* for a parent. She may in fact not do this at all, and there may be no fault in her not doing it. But if she does, she will be acting appropriately.

The role of "parent" is not merely biological—it is in very large part a moral role. There is a correct way for a parent to relate to her child: a good parent is one who is able to see things from her child's perspective, who makes the child's interests her own, who takes seriously the child's view of his own life. Of course, when the prospective parent makes a decision about which embryo to implant, there is no child, and *a fortiori* no child's interests, and the prospective parent is not yet a parent. Even so, to make a decision at all the prospective parent must, as we have seen, imaginatively inhabit the lives of each future possible person. And in this process of imaginative inhabitation and decision-making, she is not a detached, neutral person. She is someone who, in undertaking the process in the first place, is preparing for the role of parent, and whichever embryo she selects will, if all goes well, become her child. It is not only understandable, but perfectly proper, that she should take the internal perspective in thinking about the possible lives of her potential children.

The prospective parent will doubtless be able to form an opinion about which embryo will lead the better life. But from the internal perspective this sort of external "better" judgment will have no force. She will have no compelling reason to select the better-life embryo in preference to the worse-life embryo, or, for that matter, to select the worse-life embryo in preference to the better-life embryo. She will do nothing morally wrong whichever she chooses. The external perspective may well be the default position from which to make decisions about which future people should exist, but it is a default that can be overridden when those people will be our children.

References

Gillon, R. 2001. Is there a "new ethics of abortion"? *Journal of Medical Ethics* 27:ii5–ii9.

Harris, J. 2001 One principle and three fallacies of disability studies. *Journal of Medical Ethics* 27:383–387.

Parfit, D. 1984. *Reasons and persons*. Oxford: Clarendon Press.

Savulescu, J. 2001. Procreative beneficence: Why we should select the best children. *Bioethics* 15:413–426.

The ugly curve — genetic screening into the 21ˢᵗ century

by Asterios Tsioumanis, Konstadinos Mattas and Elsa Tsioumani

Introduction

DURING THE *ancien regime*, term that Lewontin (1982) uses for the time period prior to the democratic bourgeois revolutions in Europe and North America, social hierarchy and inequality were attributed to God's will. However, after the stated objectives of the new era, *social practice was, for the first time, in direct contradiction to the stated political values of the society*.

In contrast to the proclamations of liberty, equality, and fraternity, stark inequalities between individuals and hierarchical societal structures have not ceased to exist. "Biological determinism has been built over the last two hundred years as a solution to this socio-political paradox" (Lewontin 1982), claiming that the existence of dominant and subordinate groups is due to natural and intrinsic differences between individuals and groups.

History reconsidered

The first systematic attempt to link biological characteristics to behavior was made during the 19th century. Cesare Lombroso, a well-known criminologist in his time, claimed that a person with a predisposition to criminal activities could be identified by the shaping of his skull (Lombroso 1876). As opposed to cephalometry, defined as the scientific measurement of the head, craniometry applies to measurement of the dead skull, and is defined as the scientific measurement of the dimensions of the bones of the skull and face. This kind of measurement of cranial features has been used in order to classify people according to race, criminal temperament, or intelligence. The underlying theory linked the size of the skull to the size of the brain, thus addressing brain capacity, which apparently corresponded to a person's mental ability. Female inferiority was taken for granted, and the relative skull sizes of men and women were taken to confirm this theory. Paul Broca (1824–1880), a French neurologist, concluded that mature adults are more intelligent than the elderly, and that men are more intelligent than women. Other measurements were used to "prove" Western European supremacy.

During the 20th century the French psychiatrist Alfred Binet developed the first widely accepted intelligence test. Binet himself designed the method as a test for learning disabilities in order to draw appropriate remedial programs. However, Binet scales have been adapted and extensively used in the USA for quite different purposes. Dr Arthur Sweeney (quoted in Rawat 1995), addressing the House Committee on Immigration and Naturalization in 1923, cannot be outclassed in the description:

> We have been overrun with a horde of the unfit ... We shall degenerate to the level of the Slav and Latin races ... pauperism, crime, sex offenses ... we must apply ourselves to the task with the new weapons of science ... it is now as easy to calculate one's mental equipment as it is to measure his height and weight ... this new method ... will enable us to select those who are worthy and reject those who are worthless.

Unfortunately, similar views were adopted by the US authorities with tragic social consequences. "Over several decades, 60,000 American citizens and native Americans were deemed mentally retarded (i.e. genetically inferior) and forcefully sterilized for race hygiene purposes" (Rawat 1995). Sterilization laws were applied under the pretext that heredity plays the most important part in the transmission of crime, idiocy, and imbecility. The state of Iowa legislated for *the prevention of the procreation of criminals, rapists, idiots, feeble-minded, imbeciles, lunatics, drunkards, drug fiends, epileptics, syphilitics, moral and sexual perverts, and diseased and degenerate persons*. Until 1931, 24 states adopted similar laws and 34 forbade mixed marriages.

From 1910, the tests were administered to immigrants under severe conditions of scientific bias, resulting in the Immigration Restriction Act of 1924, introducing quotas for the number of accepted immigrants based on their country of origin. The Act favored the "Nordics" of northern and western Europe over the "undesirable races" of eastern and southern Europe (Higham [1955] 1988).

The Bell Curve…

Arthur Jensen stirred the IQ debate in 1969, arguing in his article *How much can we boost IQ and scholastic achievement?* that intelligence is a highly heritable trait, and that differences in intelligence across races are quite possibly genetic. As a result, it was held that compensatory education programs could not prove effective in removing differences in social status between races. When Charles Murray and the late Richard Herrnstein published their book *The Bell Curve: Intelligence and Class Structure in American Life* in 1994, few expected the range of reactions it would occasion. While Brimelow (1994) glorifies the book, comparing it, in terms of an intellectual event, to Darwin's *Origin of Species*, other critics present it as nothing more than a pseudo-scientific racist screed.

The Bell Curve, a 550-page textbook (accompanied by 300 pages of additional material), addresses a variety of considerations and can be regarded as rather complex. Although much of its commercial success may be due to the critics, it is a fact that the book stayed at the *New York Times* bestseller list for 15 weeks. However, unless a dramatic cultural cleavage has occurred between Europe and the U.S., which passed unnoticed by cross-continent research during the last decade, it is not easy to conceptualize how the average reader can associate a value judgment on the text's integrity and scientific value without possessing the necessary knowledge, which includes economics, psychology, biology, political sciences, statistics, survey research methodologies, and psychometry in an indicative list.

Part 1, entitled "The Emergence of a Cognitive Elite," argues that cognitive stratification is increasing. This is accurately stated as "our thesis is that the twentieth century has continued the transformation, so that the twenty-first will open on a world in which cognitive ability is the decisive dividing force" (Herrnstein and Murray 1994, 25). Part II presents a series of empirical analyses observing statistical associations between various social behaviors in 1990 and measured IQ in 1979. The analysis focuses on the partial effect of "cognitive ability" after controlling for measured parental socioeconomic status (SES). The study includes items such as the probability of being in poverty, the determination of schooling or labor supply, illegitimacy, becoming a chronic welfare recipient, and other topics. The story heats up at Part 3 in which, under the title "The National Context," ethnic differences in cognitive ability are addressed. Finally, Part 4 can be regarded as Herrnstein and Murray's vision of America's future as "something resembling a caste society" (509) because "cognitive partitioning will continue. It cannot be stopped, because the forces driving it cannot be stopped" (551).

...And the Bell Curve Wars

A thorough critique of *The Bell Curve* cannot be performed in this article, as the task would on the one hand be extensive, and on the other hand call upon the cumulative work of scientists from many disciplines in order to fulfill its purpose. A few notes however, may prove useful to show that the scientific evidence that Murray and Herrnstein employ can hardly qualify as indisputable.

After stating that "IQ is substantially heritable ... the genetic component of IQ is unlikely to be smaller than 40 percent or higher than 80 percent ... For purposes of this discussion, we will adopt a middling estimate of 60 percent heritability, which, by extension, means that IQ is about 40 percent a matter of environment" (Herrnstein and Murray 1994, 105), the analysis focuses on the partial effects of cognitive ability and parental socio-economic status on a number of cases. A number of conclusions such as "cognitive ability is more important than parental SES in determining poverty" (135), "low cognitive ability is a much stronger predisposing factor for illegitimacy than low socioeconomic background," (167) or "a white mother's IQ is more important than her socioeconomic background in predicting the worst home environments" (222), are drawn, and presented together with a series of graphs and charts.

However, in order for empirical evidence to be utilized, cognitive ability and socioeconomic background have to be measured. Cognitive ability, a term that is used almost interchangeably with IQ throughout the book, is approached as a normalized transformation of the respondent's score on the Armed Forces Qualifying Test (AFQT), which was administered in 1979 to virtually all NLSY (National Longitudinal Survey of Youth) respondents. The socioeconomic environment is measured by the SES Index, which accounts for education, income, and occupation.

While Herrnstein and Murray use the whole of Appendix 3 to prove that the AFQT qualifies as a widely accepted IQ test, they dismiss concerns that the test results were influenced by educational environment because of the respondents' age (15–23 years). The actual differences that can be encountered when using tests from age 7 or 8 are attributed to random error rather than to an experiential effect. On the contrary the authors make no value judgments about the adequacy of the SES Index as a means of portraying socioeconomic environment. After some criticism on the issue Murray states: "measures of SES are a staple in the social sciences ... to avoid controversy, we deliberately constructed an SES index that uses the same elements everybody else uses: income, occupation, and education ... in the jargon, our measure of SES is robust, and as valid as everyone else's has been" (Murray 1995).

Even if a full SES Index had been constructed—which was not the case in *The Bell Curve* as out of the 12,000 NLSY respondents (including non-whites and supplementary samples) only 7,500 reported data on all four components of the index (Herrnstein and Murray 1994, 574–575)—it is still a great conceptual leap to accept that this single variable can carry "the burden of expressing all aspects of the child's upbringing, from family structure to sibling relationships to neighbourhood characteristics" (Goldberger and Manski 1995). Nisbett (1995) points out a 16-item indicative list of variables that the literature on the effects of social background on children's academic achievement has identified as having significant influence on children's performance in addition to income, education, and parental occupation. Thus Goldberger and Manski (1995) can hardly be disregarded when they state that the casualness with which Herrnstein and Murray treat socioeconomic environment is astonishing.

The problems do not end there. Assuming that AFQT determines cognitive ability and the SES index depicts social environment, the two items still have to be compared. Lewontin (1982) makes the following remark: "the height of a person is a natural attribute of a real object. The average of the heights of a group of people is not an attribute of any real object ... It is a mental construct." Lewontin extends this idea to factors.

It is important to point out that the distinction between mental constructs and natural attributes is more than a philosophical equivoque, even when those constructs are based on physical measurements. As the normalized AFQT score and raw SES index are measured in qualitatively different units, Herrnstein and Murray transform each variable x into the standardized form $(x - m)/s$, comparing apparently similar units: standard deviation. While standardization as it is applied (essentially using "beta weights") is a common practice in social sciences, it is very rarely encountered in economics. "The reason is that standardization accomplishes nothing except to give quantities in non-comparable units the superficial appearance of being in comparable units" (Goldberger and Manski 1995). As scaling depends on the range of values for the variables in the data, the predictor with the largest standardized beta weight cannot be regarded as the most important.

Furthermore, Herrnstein and Murray assume in their logistic regression model that there is no IQ-SES interaction. They seem to confuse that assumption when writing that "socioeconomic status is also a result of cognitive ability, as people of high and low cognitive ability move to correspondingly high and low places in the socioeconomic continuum. The reason that parents have high or low socioeconomic status is in part a function of their intelligence, and their intelligence also affects the IQ of the children via both genes and environment" (Herrnstein and Murray 1994, 286–287). As they are probably right in their quotation and the two variables actually are correlated, the problem of collinearity in linear regression arises. Weisberg (1985, 196) notes that "when the predictors are related to each other, regression modelling can be very confusing. Estimated effects can change magnitude or even sign depending on the other predictors in the model."

Ignoring these shortcomings along with many others that can be encountered across their book, Herrnstein and Murray draw a series of ultimate conclusions. However, their beliefs cannot influence the timeliness of Darwin's quotation, as given on the title page of Gould's (1981) *The Mismeasure of Man*: "if the misery of our poor be caused not by the laws of nature, but by our institutions, great is our sin."

Challenges ahead

Murray thinks that his book "is filled with 'on the one hand … on the other hand' discussions of the evidence, presentations of competing explanations, cautions that certain issues are still under debate, and encouragement of other scholars to explore unanswered questions" (Murray 1995), and what he has actually said regarding the relationship of genes to race differences in intelligence is that "it is scientifically prudent at this point to assume that both environment and genes are involved, in unknown proportions; and, most importantly, that people are getting far too excited about the whole issue" (Murray 1995).

However, this is a rather lax way of describing Murray's standpoint. The claim that natural and intrinsic inequalities between individual human beings at birth are determinative of eventual differences in their status, wealth, and power is the defining property of biological determinism. According to this theory, "differences in ability between individuals and groups will always be translated into hierarchical social structures with dominant and subordinate groups because the tendency to form such hierarchies is coded in the human genome" (Lewontin 1982). In 1975, the evolutionist E. O. Wilson wrote in the *New York Times Magazine* that all societies of the future, no matter how egalitarian, would always give a disproportionate share of power to men because they are genetically advantaged, thus portraying the sexism present in modern human societies as both natural and unavoidable,

Peter Brimelow (1994), one of Murray's supporters, states that "Psychometrics, the measurement of mental traits including intelligence, was a rapidly developing science earlier this century. But then came the savagery of Nazism. The pendulum swung. Any talk of inherent differences became taboo."

As history teaches us, the Nazi atrocities were not the only example of humans being segregated according to existing or non-existing, perceived, imagined, or tangible biological differences. Recent advances in genetic and genomic science may create new bases for illegitimate discrimination between individuals. While potential discrimination in employment and in the provision of insurance are important considerations, challenges do not cease. "The possibilities of genetic manipulation might eventually allow allegedly 'undesirable' genetic traits to be progressively eliminated. This gives rise to concern about which genetic traits are so 'undesirable' that their elimination should be encouraged or allowed" (United Nations 2002). There is an urgent need for all responsible members of society to engage in a fruitful dialogue, safeguarded by the use of bioethics, in order to determine future applications of emerging human knowledge. Otherwise, once more, technological leaps may prove to be tortuous for humanity, and influenced by factors not inherent to them.

References

Brimelow, P. 1994. For whom the bell tolls. *Forbes* October 24:153–163.
Goldberger, A. S., and C. F. Manski. 1995. Review article: *The Bell Curve* by Herrnstein and Murray. *Journal of Economic Literature* 33:762–76. Available online at: http://www.ssc.wisc.edu/irp/featured/bellcurv.htm
Gould, S. J. 1981. *The mismeasure of man*. New York: Norton.
Herrnstein, R., and C. Murray. 1994. *The Bell Curve: Intelligence and class structure in American life*. New York: Free Press.
Higham, J. [1955] 1988. *Strangers in the land. Patterns of American nativism, 1860–1925*, 2nd ed. New Brunswick, NJ: Rutgers UP.
Jensen, R. 1969. How much can we boost IQ and scholastic achievement? *Harvard Educational Review* 39:1–123.
Lewontin, R. C. 1982. *Biological determinism*, the Tanner lectures on human values, delivered at the University of Utah on March 31 and April 1.
Lombroso, C. 1876. *L'uomo delinquente. In rapporto all'Antropologia, alla Giurisprudenza alla giurisprudenza ed alle discipline carcerarie*. Turin: Fratelli Bocca Editori.
Murray, C. 1995. "The Bell Curve" and its critics. *Commentary* 99 (5): 23–30.
Nisbett, R. E. 1995. Letter to the Editor. *Commentary* 100 (2): 17–18.
Rawat, R. 1995. The return of determinism? The pseudoscience of the Bell Curve. *Cornell Science and Technology Magazine* 2 (1): 10–13. Available online at: http://www.rso.cornell.edu/scitech/archive/95sum/bell.html
United Nations. 2002. *Conclusions of the Group of Experts on human rights and biotechnology*. Geneva: Office of the United Nations High Commissioner for Human Rights. Available online at: http://www.unhchr.ch/biotech/conclusions.htm
Weisberg, S. 1985. *Applied linear regression*. 2nd ed. New York: Wiley.
Wilson, E. O. 1975. Human decency is animal, *New York Times Magazine* October 12.

Categorizing genes: Commodifying people?

by Donald Bruce

Introduction

GENETICS IS A reductionist means of inquiry into the human person and other organisms. Within the confines of science, this is fair enough, as far as it goes. But genetic knowledge comes with the implicit assumptions of a particularly focused and reduced way of seeing living organisms. These assumptions do not necessarily map across to other disciplines or domains of understanding. Indeed they may create conflicts if this is attempted. If one is unaware of this, there is a danger that these assumptions unconsciously exert an undue or inappropriate influence in other spheres. The carry over of genetic assumptions beyond scientific inquiry into medical, social, and spiritual spheres can raise significant problems. For example, if it is accompanied by the philosophical belief that this genetic way of seeing humans is a more fundamental explanation of human being, compared with more traditional, common sense, or religious understandings of humanity, conflicts will arise.

In this context, the rhetoric which has typically accompanied the public presentation of the Human Genome Project and its results—expressions such as "the book of Life," "blueprint of life," or "the DNA Code"—is rather suggestive. In the first two, here is a new way of reading what a human being is, and indeed what all species are. With the third, genetic science will unravel something that is currently a mystery. All such framings make value assumptions. The problem occurs if these assumptions then override wider representations of human beings found in existing ethical frameworks. This short paper explores some of the implications of the assumptions of genetics in functionality, prediction, and human enhancement, and how these challenge and are challenged by some assumptions from Christian ethical reflection.

Functionality

Genetic ways of describing human beings express the functions which particular genes or combinations of genes have in terms of local physical effects in the body or larger characteristics. Human beings are regarded in terms of the functions of the body. This is most clearly seen in the notion of those dysfunctions which are now explained as having genetic origins, for which the term "genetic defect" is commonly used. Thus some forms of emphysema may be described in terms of a defect in one gene which fails to produce enough of the protein AAT, resulting in damage to the lung wall. The genetic code presents humans as individuals who are a collection of functions, most of which work more or less correctly (otherwise we would presumably not have survived to birth), but some of which are or go wrong. One implication is that we should correct all the aspects that are seen as wrong. Another is that what humans are is seen primarily in functional terms. This would represent a tendency to commodify the human person, in terms of what is useful, instead in terms of who and what a human being is.

These contrast with some assumptions of Christian ethical systems, which see all human beings as created by God and as "made in the image of God." (Gen. 1:26–27). Because of this each one is seen as special and precious in God's sight, whatever bodily and functional imperfections they may have. The value of a person is not connected with their correct functioning, or in how far they approximate to some notion of physical perfection or a theoretical perfect genetic blueprint. A Christian understanding of human being is also relational not just individualistic. Humans are individuals who are also persons, whose meaning and value is inextricably tied up with those with whom they have relationships—first of all to God, then to each other, and also to the non-human world and its creatures. We would see therefore humanity as persons in relationship, whose fulfillment lies not in individual perfection of function but in moral and spiritual development and in success in relationality (Bruce 2002).

This raises some concern about a deterministic and functional streak in some discourses about genetics, especially in the popular media and, through its influence, in society at large. Thus the title of the book "The God Gene: How Faith Is Hardwired Into Our Genes" (Hamer 2004) was chosen for its sales appeal, but in doing so plays on a misplaced notion of determinism which misleads both scientifically and theologically.

The medical practices of pre-implantation genetic selection and prenatal diagnosis for serious diseases carry a risk of unintended moral change by gradual clinical steps. The very act of repeating these procedures and regarding them as clinically normal may gradually promote a philosophical view which values humans in terms of correct genes. Similar questions are prompted by a recent case in England in which a late termination of pregnancy on the grounds of a cleft palate has been challenged as legally unjustified. A primarily genetic understanding runs the risk of stigmatizing any people whom society comes to regard as departing from what it considers normal (Church of Scotland Board of Social Responsibility 1995). In the field of genetic databases, the search for correlations at a population level, to aid in the interpretation of genetic function, needs to be moderated by a continuous reminder to take explicit account of the fact that the genetic "book of life" is not the full story of what human beings are.

Prediction

Genetic knowledge is also an inherently predictive and usually probabilistic way of interpreting the human condition. An otherwise healthy person is told, "because you have a defected gene which does this, you carry an especial risk that at some point in the future your body will do that, or that you will get such and such a disease." Even if it is only probabilistic, there is the sense of genetic knowledge foretelling the future in a way which differs from normal human experience hitherto. Even in families where a particular condition was prevalent, until some symptoms appeared a family member did not know if he or she would "get it too." Molecular genetic knowledge pre-empts this situation, and usually does so in a fatalistic way, because the identification of the gene is not normally accompanied by the immediate availability of an effective clinical treatment. For the person who has been tested, the future is no longer open and contingent, but in some sense determined. The human predicament of knowing one's future has been much explored in literature and the arts, but it is generally portrayed as something which humans do not handle well.

This contrasts with the role of the prophet in biblical times, which was generally as a warning to people to change behavior because of the serious spiritual, moral, and political consequences which would follow. It was usually conditional. Thus in the story of Jonah, when the city of Nineveh repented, God did not bring the destruction that Jonah had been told by God to proclaim. There is a sense in which the situation can still be redeemed. With genetic information, in some cases changes of diet or lifestyle may make a difference to the severity or onset of the disease, but sadly so far it is rarely the case that the outcome which is predicted by the presence of the gene can be prevented. Genetic information has an apparently redemptive motivation in so far as it seeks to understand disease and thereby to help alleviate human suffering, but in practice thus far it usually lacks the redemptive power to deliver it. Eventually, it may prove possible to offer cures for some of the diseases which we can now only test for, but there will usually be a substantial gap, perhaps of several decades, between identification and cure. In some cases, cure may never be possible. Thus for significant numbers of people, the effect of a genetic test may be to feel one is reading one's death certificate in advance. Except for those who believe in an after life, the knowledge of one's own genetic predispositions may be counter-redemptive (Conference of European Churches 2003).

The probabilistic nature of genetic knowledge also focuses human thinking on an uncertainty and an accompanying fear which again goes beyond the normal uncertainties of human life. Thus a woman in the UK will be aware in a general sense that she runs a risk that she will contract and eventually die from breast cancer, simply because this has become one of the more common causes of death. Perhaps it might happen to her. If she discovers, however, that she carries one of the identified breast cancer genes, she now knows that she has a much higher likelihood of developing breast cancer, but it remains uncertain. She carries with her the shadow over her life of a fear of something which may yet never happen. All that genetic knowledge has told her is that she is much more likely to have this awful disease than the average woman of her age.

The predicament of knowing but not knowing may be harder than facing up to the diagnosis of the disease itself. Christian teaching recognizes a proper time for things in one's life—in Greek thought *kairos*, or in the Old Testament the observation that "to everything there is a season, and a time for every purpose under heaven" (Eccles. 3:1–8). Genetic information can infringe that right moment in a life because it is driven only by the logic of what Jacques Ellul (1964) called *la technique*. Here is perhaps a case for restraint in the use of genetic testing, certainly until the development of wider knowledge about the disease and what can be done about it. Although it certainly opens up new fields of future enquiry for the medical researcher, in knowing the gene one still knows so little.

Enhancement and the cult of the perfect function

The emphasis on seeing the human being in terms of genetic functions also moves subtly away from accepting the givenness of the human condition, at a genetic or a familial level. Potentially it is now something which we can manipulate to our own design, if we think we can do better. This is one consequence of isolating individual characteristics from an account of the whole human being. Instead of life being a gift to be handled well, as in traditional morality, a genetic framing of knowledge can transform it into something to be designed, or as something accepted only conditionally, in so far as it meets with our specification. In Christian thinking and spirituality, the reality of a person's character lies in what they do with their "lot" in life, and much less in the lot itself. Even though Christian principles have been the motivation for very large efforts toward the relief of suffering, the ability to cope with unrelieved suffering has always remained a key concept, framed by the perspective of eternity and of God's own suffering for humanity.

In contrast, the notion of performing a technical fix on human characteristics carries the risk of mistaking the worth of human beings with limited notions of functional perfection. This is not helped by the trend in contemporary western culture to take the proper celebration of individual creativity and achievement out of proportion into a cult of celebrity, focused on a few perfect achievers. In sport this would be the Olympic champion, in the performing arts the Oscar winner or rock idol, in popular magazines the glamorous celebrity, and so on. A success orientation applied to genetic functions in any sphere of human excellence would exalt those with the most perfect genes as the ideals toward which all should strive. This presents a very fragile view of human beings—as individuals acceptable only while their success lasts. It has no place for the sports star whose body wears out prematurely because of an extreme focus on a few bodily functions. It exalts youth and beauty over age and normality. It also glorifies the winners and has no satisfactory account of the underclass who do not reach the mark.

There are also the troubling lessons of history in which genetic knowledge was turned either into a social philosophy that society should not accept anything less than a perceived ideal, or into elevating evolution from science to ideology. The extension of the biological notion of the survival of the fittest to produce correlative social ideas about which members of the human race are seen by society as fit to survive and who should be enhanced genetically remain deeply problematical. Even if the likelihood is very small that characteristics could ever be manipulated genetically in anything but an extremely crude way, that aspiration apparently exists for some.

From the point of view of Christian ethics, the notion of genetic enhancement seems misplaced for a more fundamental reason. In theological terms, the issue of what is wrong with the human condition is not in terms of lack of intelligence, beauty, athleticism, art, science, or even education, but in the moral and spiritual shortcomings of humanity individually and collectively. These are beyond any amount of genetic knowledge or enhancement to redeem, but require instead a spiritual solution. Christians see this as found in having a relationship to Jesus Christ. In whatever sense these deeper issues are cast, genetic functional enhancement is, by comparison, a distraction.

Conclusion

This paper has briefly considered some problematical underlying ethical trends in the framing of humanity in primarily genetic terms. The impact of such an understanding on a Christian ethical framework has been critically assessed. Concerns are raised about seeing human beings too much in terms of genetic functions, and about the potential manipulation of human genetic characteristics in search of a misplaced idea of individual functional perfection. This represents a commodification of the human being. People are valued in terms of functional usefulness instead of wider moral, ontological, and theological categories. Some difficulties brought by the predictive aspect of genetic knowledge are also highlighted, especially in the uncertainty of "knowing but not knowing" and its insensitivity to the times and seasons of a person's life.

Some of the insights offered would also be shared by other faith communities and by some secular ethical systems, but these lie beyond the scope of this short paper. If in Europe the thinking about the nature of the human person has hitherto drawn significantly on a Christian understanding, it is important to identify places where the pursuit of an exclusively genetic knowledge would challenge that understanding.

There is much to be celebrated about genetic knowledge, but it must also be seen in relation to other factors of humanity. Many geneticists and clinicians are aware of the limitations of a genetic framing, and may take care in their professional lives not to lose sight of wider views and understandings of human life. But those devising genetic databases and their accompanying ethical, legal, and social programs would do well to be aware of the unintended effects which genetic knowledge may have on the perceptions and beliefs of the wider society about this emerging science.

References

Bruce, D. M. 2002. Stem cells, embryos and cloning — unravelling the ethics of a knotty debate. *Journal of Molecular Biology* 319:917–925.

Church of Scotland Board of Social Responsibility. 1995. *Human genetics*. Edinburgh: St. Andrew Press.

Conference of European Churches. 2003. *Genetic testing and predictive medicine*. Report of the Working Group on Bioethics. Strasbourg: Conference of European Churches.

Ellul, J. 1964. *The technological society*. Trans. John Wilkinson. New York: Knopf.

Hamer, D. 2004. *The God gene: How faith is hardwired into our genes*. New York: Doubleday.

Genetic medicine and genetic research

34

Interpretation of genetic data for medical and public health uses

by Paul A. Schulte

THE GROWTH OF genetic data presents various resources that may be of use to clinicians and public health practitioners. However, before there can be effective and safe application of genetic information various interpretive issues need to be addressed. These include: (1) the need for population thinking about genetic information; (2) the extension of the biomedical paradigm to include it; and (3) attention to issues in the application of genetic information to qualitative and quantitative risk assessments. These issues have scientific, ethical, and social aspects.

Genetic databases are derived from various types of clinical, laboratory, and field studies involving single or multiple genes, expression products, and covariate information. These are derived from a range of efforts including microarray experiments and examination of expression profiles, and assessments of case series or etiologic epidemiologic studies. Epidemiologists, worldwide, often establish genetic databases when conducting epidemiologic studies (Steinberg et al. 2002). Sometimes these databases are not the primary objective, but since the investment to conduct population-based field studies and obtain data and specimens from subjects is so large, it is cost effective to collect and bank DNA for the current and future research.

Research on genetic factors in people can be conceived on a continuum that has been described as leading from basic research through population research to medical and public health practice (Schulte 2004). Regardless of whether a genetic factor has been analytically and clinically validated, the mere collection of DNA and the subsequent assay and results are de facto genetic tests and ethical, legal, and social issues of genetic testing need to be considered. There is a rich literature on these topics (e.g., see Grody 2003; Laberge and Knoppers 1992; Davison, Macintyre, and Smith 1994).

For the most part, the discussions of the ethical and social issues in genetic testing have involved one or a few genes. The ability to assess hundreds or thousands of genes or gene products in one microarray amplifies issues that have already been identified for individual genes, and creates some situations with new implications (Grody 2003; Schulte 2004). These implications involve how scientists and nonscientists conceive of variability. To clearly define these terms, there is a need to apply the disciplines of population science—the understanding that individual research subjects or patients are part of population groups that can be defined by genetics and other factors (Knoppers 2000). Finally, the applicability of genetic information to the socially desirable practice of risk assessment has various implications that should be considered.

Need for population thinking

Genetic analysis at the macro and micro levels has provided an opportunity to quantify variability in populations. "Variation" has been the topic that sparked creative thinking in biology and epidemiology. In the view of Mayr (1982), Western thinking for 2000 years after Plato was dominated by "essentialism," or the belief that there were a limited number of immutable essences. Much of epidemiologic (and for that matter, biomedical scientific) thinking to date has been essentialist. For example, the early statistics used in public health by Graunt (1620–1674) and Quetelet (1796–1874) attempted to calculate true values to overcome the confusing effects of variation. Quetelet, who attempted to use a mathematical function of height and weight as a biomarker, hoped to calculate characteristics of the average person. To him and like thinkers, variations were nothing but "errors" around mean values. In the nineteenth century, new ways to view nature began to spread; a concept called "population thinking" was developed that stressed the uniqueness of everything in the organic world. To population thinkers there is no "typical" individual, and mean values are considered abstractions (Mayr 1982).

Lloyd (1998) has elegantly set the stage for further discussions of variation by recognizing that there is a diversity of theories and models from the different scientific disciplines involved in the Human Genome Project that provide a unique challenge to producers and consumers of DNA-sequencing information. While it is generally thought that science provides guidance as to variation, there are actually social views and constructs intertwined with the level of scientific analysis that can lead to divergent interpretations of genetic data. Specifically, science deals with observation of events under various circumstances. However, depending on the scientific model, the interpretation may change. For example, for a mechanistic model such as the "biochemical causal-pathway model," the emphasis is usually on function and the model presents, in detail, a picture of normal or proper functioning. As Lloyd (1998) notes, there is no room for simple variation in the explanatory scheme of such a model. Variation is seen as nonfunctioning. For example, with sickle cell anemia, the goal of the causal model is to explain at least one causal chain from DNA in its initial state arranged on a chromosome to the ultimate state involving iron molecules arranged in hemoglobin carrying around oxygen in the body. There is no variation in function that is addressed. In contrast, very different types of descriptions are used when populations of people are considered. Population thinking would suggest tri-level analysis: individual, propagating family, and community. At each level, in population sciences, a great deal of information is needed to delineate the variation. These include: (1) the distribution and frequencies of the types of genes, phenotypes, and functions associated with these genes, and (2) the range of environmental variables and the related parameters of reaction (Lloyd 1998). This kind of information is not provided in biochemical causal models that are prominent in genetics. Therefore, depending on the level of scientific analysis, it is possible to get interpretations of the role of genetic factors that are either deterministic or probabilistic. These interpretations have different implications for users and consumers of genetic information. Too often genetic reductionism leads to flawed thinking about the role of environmental factors in genetic diseases. Of the more than 7,000 genetic disorders known, only a few have been studied for environmental factors in their phenotypic expression (Samuels 2003).

The definition and range of variation that can occur in microarray studies or that will be stored in databases poses a rich, but problematic, challenge to medical and public health practitioners. Beyond the issue of standardization of approaches and scales are the issues of the extent to which the people in the database represent those who are not in the database. That is, are they representative in terms of genetic and ethnic factors as well as in terms of the various other host or environmental factors? Of concern is whether various ethnic groups have similar opportunities to be in a database. If not, characterization in a database can make one ethnic group appear more or less "susceptible" than another ethnic group without the same opportunity for characterization.

Utilization of genetic bases for public health requires that variation in the population be accurately categorized and that the concept of abnormal be thought of more in terms of susceptibility, then determined appropriately, and interpreted probabilistically. For public health purposes, there is a need to define concepts such as susceptibility on a population level. As Lloyd (1998) concluded "public and scientific misconceptions of susceptibility are probably one of the most prominent problems facing those interested in the development of genetic medicine."

Extension of the biomedical paradigm

Predictive genetic testing is a departure from the traditional paradigm of laboratory medicine (which involves laboratory testing or confirmation in a patient with signs or symptoms) in that disease or high risk can now be diagnosed years or decades before the clinical appearance of signs or symptoms (Grody 2003). Such predictive knowledge forces us to re-examine the definition of disease or pre-existing condition and the way society deals with people in those states of health. Increasingly, there will be a need for pre- and post-test counseling. Press and Burke (2000) have suggested that this type of predictive genetic testing and information be assessed in terms of how well it fits into the progression of biomedicine that has increasingly moved from bringing people into the medical system at the onset of symptoms to bringing them in as part of a prevention and early detection paradigm in which risk factors, including genetic information, trigger monitoring or intervention. The concept of "high risk management," however, needs to be based on epidemiologic research that links the genetic marker or markers to risk and, ultimately, to diseases (Samuels 2003). The mere presence of an allele in a test is not a sufficient indicator of risk or disease. Before a predictive genetic test can be useful for clinical or public health purposes, the analytical and clinical validity and clinical utility should be determined (Burke et al. 2002). At present, much of the data in databases derived from microarray studies have not been validated or linked to diseases or risks in populations. Additionally, such testing extends beyond the immediate person being tested and may impact all with that germline. Microarray technology will make such testing too easy and may further bring a departure from traditional laboratory medicine (Grody 2003).

Finally, one of the vexing problems that limits use of genetic information is the difficulty in translating genetic information on groups to individuals. Epidemiologic research pertains to group risk assessment, not to individuals in the group. Information about the presence or absence of variants of genetic factors in groups may not be useful for individual classification without research specifically designed for such translational purposes (McCanlies et al. 2002).

Risk assessment

One area for application of genetic data is in risk assessment. The term "risk assessment" has different meanings in different countries. In some countries, it is an effort to provide society with estimates of the likelihood of illnesses and injury as a consequence of exposure to various hazards. Risk assessments are needed when social policy decisions are in dispute, when alternative policies in question are not subject to direct measurements, and when scientific analysis of a hazard is not complete (IPCS 2001). Risk assessment can be qualitative or quantitative at the individual or population level. The term quantitative risk assessment describes the response associated with a specific level of exposure. The promise is for a more refined assessment of risk through the identification of gene-gene and gene-environment interactions, and also for the focusing of prevention and control programs for high-risk individuals. The perils of using genetic information in risk assessment include ethical and social issues such as stigmatization, discrimination, and the misconception that removing a susceptible person from the exposure scenario without reducing exposure opportunities will reduce risk effectively, when it may not, on a comparative basis (Vineis and Schulte 1995).

Most of the research on genetic modifiers of exposure-disease associations has involved single genes which are polymorphic for a particular enzyme. These have been referred to as "metabolic polymorphisms." They have also been referred to as susceptibility markers. One example of the use of susceptibility markers examined how changing glutathione-S-tranferase theta (GSTT1) genotype frequencies would impact cancer risk estimates from dichloromethane by the application of Monte-Carlo simulation methods in combination with physiologically based pharmacokinetic (PBPK) models (El-Masri, Bell, and Portier 1999). The investigators reported that average and median risk estimates were 23% to 30% higher when GSTT1 polymorphism was not included in the models.

Although the scientific press has been replete with promises of the use of genetic information in risk assessment, the use of even a single metabolic polymorphism in quantitative risk assessment is rare. More recently, similar pronouncements have been made for toxicogenomics; however, there is a large amount of work necessary to devise effective risk assessments based on these technologies. The ability to interpret data sets from toxicogenomics and other microarray studies will be quite difficult due to the large number of potential combinations of gene expression states (Morgan et al. 2002). Before more use is made of this type of information and extended to include multiple genes from microarray generated databases, there is need for new thinking on how to incorporate the information in risk assessments. Using the results of human studies and animal data, risk assessors define environmental exposures that may be linked to disease in a portion of the population (Waters et al. 2003). Often these determinations are based on the use of default assumptions to reflect limitations in knowledge. When used in risk assessment, genetic information can provide increased mechanistic insight and replace default assumption in species and other extrapolations. There is also need for consideration of how assessments that identify differential risks in populations will be incorporated into laws, regulations, or practices in ways that do not have untoward effects. If protective measures based on genetic risk assessments are to be provided to particular groups or individuals, criteria should be established for how those groups are identified and how their rights will be preserved (Hornig 1988).

References

Burke, W., D. Atkins, M. Gwinn, A. Guttmacher, J. Haddow, J. Lau, G. Palomaki, et al. 2002. Genetic test evaluation: Information needs of clinicians, policy makers, and the public. *American Journal of Epidemiology* 156:311–18.

Davison, D., S. Macintyre, and G. D. Smith. 1994. The potential social impact of predictive genetic testing for susceptibility to common chronic diseases: A review and proposed research agenda. *Sociology of Health and Illness* 16:340–371.

El-Masri, H. A., D. A. Bell, and C. J. Portier. 1999. Effects of glutathione transferase theta polymorphism on the risk estimates of dichloromethane to humans. *Toxicology and Applied Pharmacology* 158:221–30.

Grody, W. W. 2003. Ethical issues raised by genetic testing with oligonucleotide microarrays. *Molecular Biotechnology* 23:127–38.

Hornig, D. F. 2002. Conclusion. In *Variations in susceptibility to inhaled pollutants*, ed. J. D. Brain, B. D. Beck, A. J. Warren, and R. A. Shaikh. Baltimore, MD: The John Hopkins University Press.

International Programme on Chemical Safety (IPCS). 2001. Environmental Health Criteria 222: Biomarkers in risk assessment; Validity and validation. Geneva: World Health Organization.

Knoppers, B. M. 2000. From medical ethics to "genethics." *Lancet* 356, Suppl. 1:s38.

Laberge, C. M., and B. M. Knoppers. 1992. Rationale for an integrated approach to genetic epidemiology. *Bioethics* 6:317–30.

Lloyd, E. A. 1998. Normality and variation: The human genome project and the ideal human type. In *The philosophy of biology*, ed. D. L. Hull and M. Ruse. Oxford: Oxford University Press.

Mayr, E. 1982. The growth of biological thought: Diversity, evolution, and inheritance. Cambridge, MA: Harvard University Press.

McCanlies, E., D. P. Landsittel, B. Yucesoy, V. Vallyathan, M. L. Luster, and D. S.

Sharp. 2002. Significance of genetic information in risk assessment and individual classification using silicosis as a case model. *Annals of Occupational Hygiene* 46:375–81.

Morgan, K. T., H. R. Brown, G. Benavides, L. Crosby, D. Sprenger, L. Yoon, H. Ni, et al. 2002. Toxicogenomics and human risk assessment. *Human and Ecological Risk Assessment* 8:1339–1353.

Press, N., and Burke W. 2000. "Genetic exceptionalism" and the paradigm of risk in US biomedicine. Paper presented at "A Decade of ELSI Research" Conference, January 16, Washington, DC.

Samuels, S. W. 2003. Occupational medicine and its moral discontents. *Journal of Occupational and Environmental Medicine* 45:1226–33.

Schulte, P. A. 2004. Some implications of genetic biomarkers in occupational epidemiology and practice. *Scandinavian Journal of Work Environment and Health* 30:71–79.

Steinberg, K., J. Beck, D. Nickerson, M. Garcia-Closas, M. Gallagher, M. Caggana, Y. Reid, et al. 2002. DNA banking for epidemiologic studies: a review of current practices. *Epidemiology* 13:246–254.

Vineis, P., and P. A. Schulte. 1995. Scientific and ethical aspects of genetic screening of workers: The case of the N-acetyltransferase phenotype. *Journal of Clinical Epidemiology* 48:189–197.

Waters, M. D., J. K. Selkirk and K. Olden. 2003. The impact of new technologies on human population studies. *Mutation Research* 544:349–60.

35

Perceptions of risk and human genetic databases: Consent and confidentiality policies

by Timothy Caulfield

I Introduction

ENSURING THAT DNA databanks have legally and ethically appropriate consent processes remains an enduring challenge. It has been the focal point of policy discussions and social commentary (for example, see Laurie et al. 2003). Various policy options have been proposed, most of which involve some altering of the traditional approach to obtaining consent in the research setting.

In this brief discussion paper, I consider the relevance of the concept of "minimal risk" in the context of DNA databanks and consent policy. We will see that it should not be presumed that the public will always view DNA databank research as involving minimal risk, even if the data is de-linked. This conclusion will impact on the application of existing ethics policy. Though much of the analysis relies on Canadian law and research ethics policies, I believe the conclusions will have relevance to policy development throughout the world.

II Minimal risk and DNA databanks

The concept of "minimal risk" has become an important part of the consent policy debate. First, it is often argued that research involving health and genetic information poses minimal risk and, therefore, a lower standard of consent may be justified—including the use of blanket consent or a complete waiver of consent (Tu et al. 2004). For example, it has been suggested that DNA databanks usually involve research on low penetrance genes, and the information generated by the research will likely have little relevance to individual participants (Beskow et al. 2001). As a result, the information created by the research is unlikely to be harmful, even if it is inadvertently disclosed. In addition, because DNA databanks usually have a very large number of participants, the significance of any one sample is likely to be relatively minor.

Second, and more importantly, the concept of minimal risk is central to formal research policies and emerging laws that allow the "waiver of consent" or the altering of the consent process. There are both research policies and legislation that permit research involving health information without consent (or with the provision of minimal information) if the risk associated with the research can be deemed minimal in nature. In general, it is left to a research ethics board (REB) to make the determination as to whether a given research project meets this standard.

For example, the concept of minimal risk pervades the governing Canadian research ethics document, the Tri-Council Policy Statement (TCPS). Article 2.1 of the TCPS states that an REB may approve a consent procedure which does not include, or which alters, some or all of the traditional elements of informed consent or may waive the requirement to obtain informed consent if, *inter alia*, the "research involves no more than minimal risk to the subjects" (Medical Research Council of Canada et al. 2003). Likewise, emerging health information legislation, such as Alberta's *Health Information Act*, allows REBs to waive or alter the consent process in relation to research involving health information, including genetic data.[1] Minimal risk is an implicit criterion in this context.

III Assessing the risk of DNA databanks

While there are legitimate arguments that DNA databanks are not as risky as research involving more direct clinical interventions, such as new pharmaceuticals, there is a growing body of evidence that the public has strong views about genetic privacy, consent and the use of personal health information in research (Caulfield and Outerbridge 2002).[2] In fact, one study, conducted by Jon Merz, found that individuals view the risks associated with "genetic research with identifiable human tissue as substantially higher than minimal" (Merz 1997).[3] In addition, there are numerous studies that have found that the public, rightly or not, has specific concerns about the privacy of genetic information.[4] One recent Canadian study, for example, found that it was "clear that for many, genetic information is more personal and more fundamental to identity" than general health information. In addition, "most people indicated they would be more disturbed if their personal genetic information was inadvertently made public than if their health

---

[1] *Health Information Act*, R.S.A. 2000, c. H5, s. 50(1), which requires an ethics committee to consider whether "the proposed research is of sufficient importance that the public interest in the proposed research outweighs to a substantial degree the public interest in protecting the privacy of the individuals who are the subjects of the health information to be used in the research." The magnitude of risk to the individual participant may be an implicit part of this assessment.

[2] It is also important to note that although genetic research may not pose the same physical risk as clinical interventions, they have been associated with psychological and social concerns, such as the discrimination and the stigmatization of communities.

[3] Though there is no Canadian data on point, it is worth noting that available evidence indicates that the public wants to re-consent. See generally Caulfield and Outerbridge 2002; and Robling et al. 2004.

[4] Pollara Research and Earnscliffe Research and Communications 2003. The study involved focus groups and a telephone survey of 1200 Canadians. See also, Dennis Bueckert, "Hands off my genes, Canadians say" in the *Globe and Mail* on March 14, 2004.

records had been" (Pollara and Earnscliffe 2003; see also HGC 2001). This finding is consistent with a 2000 survey that found that 90% either strongly agree (61%) or agree (29%) that genetic information is different and rules governing access should be stricter than for other forms of personal information (Pollara and Earnscliffe 2000, 51).

Whether genetic data is, in fact, different from other forms of personal information remains a highly contested issue. Indeed, it has been noted that this view is based on a scientifically inaccurate notion of genetic essentialism—that is, the belief that genetics plays a more simple and direct role in the development of human disease and personal characteristics than is actually the case. Regardless of its accuracy, this view has greatly influenced recent national and international policy development. For example, article 4 of UNESCO's (2003) International Declaration on Human Genetic Data declares that human genetic information *is* special because it can be used to identify genetic predispositions, has relevance to biological relatives and may have cultural significance for persons or groups. As a result, it is recommended that "[d]ue consideration should be given, and where appropriate special protection should be afforded to human genetic data and to the biological samples." Similarly, in late 2004, the World Health Organization released a report, entitled *Genetic Databases: Assessing the Benefits and the Impact on Human and Patient Rights*, that develops strict criteria for when and how genetic databanks should be developed (Laurie et al. 2003).

In the context of risk, many commentators and policy documents also make a distinction between linked and de-linked data. However, even here, it is unclear that the public makes the same distinction (see Willison et al. 2003; also Gibson 2003). Even though one can pose reasonable rationales for why DNA databank research should, on an objective basis, be considered minimal risk (though not all would agree), the public may not view the research in this light. Likewise, there is a growing body of national and international policy statements that have called for special protections in the context of DNA databases. These documents further emphasize, at least for the public and policy makers, the potentially sensitive nature of the research.

IV Who should quantify the risk?

The fact that the public may view research involving DNA databanks as posing more than minimal risk has a number of significant policy implications. While more in-depth research on public perception is certainly needed, the available evidence challenges the presumption that human genetic databases pose minimal risk to research participants. This is not to say that the public is not supportive of genetic databanks. On the contrary, most in the public seem keen to participate—as evidenced by the high recruitment rates for both large and small databanks.[5] However, we should be careful not to conflate a willingness to participate with perceptions of risk. Indeed, existing research suggests that waiver of consent or, even, a blanket consent approach cannot be justified on the basis

---

5 See McQuillan et al. 2003 where it is noted that 85% of eligible participants in the U.S. National Health and Nutrition Examination Survey consented to having a blood sample stored in a national repository for future genetic studies.

of minimal risk alone. As noted by Robling et al. (2004) in their study of attitudes about research on health care records, "Public acceptability regarding the use of medical records in research cannot simply be assumed."

In some jurisdictions, public perception of risk may also have legal implications. In many countries with a common law legal tradition, the standard of disclosure for consent is viewed through the lens of the "reasonable patient" (Dickens 2002; Picard and Robertson 1996). In Canada, for example, the standard of disclosure is anything that a reasonable person in the research participant's position would want to know.[6] As such, whether a research activity poses minimal risk must be determined from the perspective of the research participant. This is not a wholly objective standard to be determined by the available evidence regarding risk. In jurisdictions with a patient-centered standard of disclosure, the question is whether a participant may consider the risk as more than minimal—the view of the researchers or, even, the research ethics board is not the determinative consideration. If a reasonable person in the participant's position would want to be provided with a particular bit of information, that information must be disclosed. Applying this participant-focused standard would mean that it may be difficult to meet the minimal risk justification for altering the traditional consent process.

Interestingly, Canadian research ethics policy has a similar, participant-focused, view of minimal risk. For example, the TCPS definition of minimal risk is as follows: "[I]f potential subjects can reasonably be expected to regard the probability and magnitude of possible harms implied by participation in the research to be no greater than those encountered by the subject in those aspects of his or her everyday life that relate to the research then the research can be regarded as within the range of minimal risk" (Medical Research Council of Canada et al. 2003, 1.5).

It should also be noted that there is evidence that particular activities or funding sources may influence how the public views the risks associated with a DNA databank. For example, the involvement of a commercial entity is a factor a reasonable person in the patient's position may want to know about. A recent study found that the Canadian public was far less willing to contribute samples to a private company. "Only 22% would deny medical researchers access to Canadians' genetic information while 49% would do so for health care companies" (Pollara and Earnscliffe 2003).

V Genohype and science agenda paradox

Over the past few decades, a vocal segment of the scientific community has actively promoted genetic research. This has included advocating for large research initiatives, such as the Human Genome Project and the UK Biobank, to government funding entities (Cook-Deegan 1994).[7] Inevitably, this requires an emphasis on the speculative, near-

---

6 *Reibl v. Hughes* (1980) 114 D.L.R. (3rd) 1 (S.C.C.).

7 See Pallab Ghosh's article "Will Biobank Pay Off?" on *BBCNews* on September 24, 2003, (http://news.bbc.co.uk/2/hi/health/3134622.stm): "It's hoped that the information collected will be the research equivalent of gold dust—containing the secrets of how genes and environmental factors conspire to make us ill."

future and practical implications of the research. These messages, which are amplified by the increasingly commercial nature of the research environment, are picked up by the media and transmitted to the public (Bubela and Caulfield 2004). Some commentators have speculated that these messages amount to an inappropriate "hyping" of science and have facilitated a belief in genetic essentialism (Don't Feed the Hype! 2003; see also Condit 1999).

For researchers, the selling of the genetic research agenda has created a policy paradox. On the one hand, the public has been consistently told that genetic information holds the "key to life" and is as precious as "gold dust."[8] It has even been called the "language of the Gods."[9] In such an environment, it is not surprising that the public now views genetic information as unique and worthy of special protection. On the other hand, the research community often seeks to characterize DNA research as low risk. Can researchers have it both ways? To some degree, the research community may be a victim of its own marketing success.

VI Conclusion

Given the importance of maintaining public trust, understanding and respecting public perceptions of risk seems essential. DNA databanks provide an exciting opportunity to do valuable research. But in most jurisdictions, navigating the consent issues remains a critical and unresolved issue. Categorizing research involving DNA databanks as minimal risk may make some of these consent issues easier to address—particularly in countries that have laws and research ethics policies that allow the consent process to be altered for minimal-risk research. Available data suggests, however, that the public may view databank research as involving substantial risks. In part, this perception may be a result of the belief that genetic information is unique.

Acknowledgements

I would like to thank Nola Ries for her insight and Genome Prairie and the Alberta Heritage Foundation for their continued support.

References

Beskow, L. M., W. Burke, J. F. Merz, P. A. Barr, S. Terry, V. B. Penchaszadeh, L. O. Gostin, M. Gwinn, and M. J. Khoury. 2001. Informed consent for population-based research involving genetics. *JAMA* 286:2315–2321.

---

8   See Ghosh, note 7 above.
9   "Language of the Gods" in the *Globe and Mail* A1 on June 27, 2000; see also *Time Magazine: Special Issue: The Future of Medicine* on January 11, 1999.

Bubela, T. M., and T. A. Caulfield. 2004. Do the print media "hype" genetic research? A comparison of newspaper stories and peer-reviewed research papers. *Canadian Medical Association Journal* 170:1399–1407.

Caulfield, T., and T. Outerbridge. 2002. DNA databanks, public opinion, and the law. *Clinical and Investigative Medicine* 25:252–256.

Condit, C. M. 1999. How the public understands genetics: Non-deterministic and non-discriminating interpretations of the "blueprint" metaphor. *Public Understanding of Science* 8:169–180.

Cook-Deegan, R. 1994. *The gene wars: Science, politics and the human genome*. New York: Norton and Company.

Dickens, B. 2002. Informed consent. In *Canadian Health Law and Policy*, 2nd ed., ed. J. Downie, T. Caulfield and C. Flood. Toronto: Butterworths.

Don't Feed the Hype! 2003. Editorial. *Nature Genetics* 35:1.

Gibson, E. 2003. Is there a privacy interest in anonymized personal health information. *Health Law Journal* Special Edition: 97.

Human Genetics Commission (HGC). 2001. *Public attitudes to human genetic information*. A study conducted by MORI Social Research for the HGC. London: HGC. Available online at: http://www.hgc.gov.uk/business_publications.htm

Laurie, G., F. Dekkers, A. Kent, and C. Shalev. 2003. *Genetic databases: Assessing the benefits and the impact on human and patient rights*. A report prepared for the European Partnership on Patients' Rights and Citizens' Empowerment, a network of the World Health Organization Regional Office for Europe. Geneva: World Health Organization.

McQuillan, G. M., K. S. Porter, M. Agelli, and R. Kington,. 2003. Consent for genetic research in a general population: The NHANES experience. *Genetics in Medicine* 5:35–42.

Medical Research Council of Canada, Natural Sciences and Engineering Research Council of Canada, and Social Sciences and Humanities Research Council of Canada. 2003. *Tri-council policy statement: Ethical conduct for research involving humans, 1998 (with 2000, 2002 updates)*. Ottawa: Public Works and Government Services Canada. Available online at: http://www.pre.ethics.gc.ca/english/policystatement/policystatement.cfm

Merz, J. 1997. Psychosocial risks of storing and using human tissue in research. *Risk: Health, Safety and Environment* 8:235–248.

Picard, E., and G. Robertson. 1996. *Legal liability of doctors and hospitals in Canada*. Toronto: Carswell.

Pollara Research and Earnscliffe Research and Communications. 2000. *Public opinion research into biotechnology issues, third wave*. Report prepared for the Biotechnology Assistant Deputy Minister Coordinating Committee (BACC); Government of Canada, December.

Pollara Research and Earnscliffe Research and Communications. 2003. *Public opinion research into genetic privacy issues*. Report prepared for the Biotechnology Assistant Deputy Minister Coordinating Committee (BACC); Government of Canada, March.

Robling, M. R., K. Hood, H. Houston, R. Pill, J. Fay, and H. M. Evans. 2004. Public attitudes toward the use of primary patient data in medical research without consent: A qualitative study. *Journal of Medical Ethics* 39:104–109.

Tu, J. V., D. J. Willison, F. L. Silver, J. Fang, J. A. Richards, A. Laupacis, and M. K. Kapral. 2004. Impracticability of informed consent in the registry of Canadian Stroke Network. *NEJM* 350:1414–1421.

UNESCO. 2003. *International declaration on human genetic data*. Adopted on 16 October.

Willison, D. J., K. Keshavjee, K. Nair, C. Goldsmith, and A. M. Holbrook. 2003. Patients' consent preferences for research uses of information in electronic medical records: interview and survey data. *BMJ* 326:373–377.

36

Individual boundaries and the impact of genetic databases upon collectivist cultures: Molecular, cognitive and philosophical views

by Janet K. Brewer

Introduction

THE HUMAN GENOME has become a battleground for cultural ways and belief systems. Western scientists have historically conceptualized the body as a collection of distinct and separable parts, as detachable, as things-in-themselves, and in some cases as ownable, whether they are blood, organs, or human gametes (see Lock 1999, 100). In some ways, the pursuit of scientific knowledge has become a capitalistic venture. Opposition to the commodification of DNA is vigorous and widespread among the 5,000 groups of peoples in the world recognized as indigenous. Clearly, the concerns raised regarding molecular research into human genetic diversity is most pronounced in those parts of the world formerly subject to colonization. As human blood, cells, and genetic material are understood simply as things-in-themselves to which monetary value can be attached, their worth as culturally significant entities, as the basis and affirmation of human life in a specific time and place, may be overshadowed (Lock 1999, 100).

*The concept of "gene" in western and indigenous culture:
Belief and the boundary of self*

The characterization of genetic material by Westerners as a tangible, ownable object has several common law underpinnings. The writings of Lord Edward Coke offer the earliest recorded treatise on the subject of property rights in the human body: "the burial of the [c]adaver (that is caro data vermibus) is nullius in bonis, and belongs to [e]cclesiastical cognizance" (Coke 1644, 203). This statement became the foundation for the Anglo-American law that a property right cannot be vested in human body parts (Boulier 1995, 705). No one questioned Coke's assertion for many years.

    The 1980 landmark decision, *Diamond v. Chakrabarty*, is the case charged with giving birth to the now flourishing biotechnology industry. The Supreme Court justified vesting property rights for a bacterium in a researcher. In *Diamond*, through a gene-splicing process, Dr. Ananda Chakrabarty developed bacteria with the genetically enhanced ability to break down crude oil. The US Supreme Court held that live, human-created bacterial microorganisms are patentable pursuant to 35 U.S.C. § 101. The Court deemed the genetically altered living organisms "compositions of matter" because they were not natural occurrences. In the five to four decision, Chief Justice Warren Burger held that "anything under the sun that is made by man" is patentable. The majority determined that human intervention is the crucial factor in patentability (see US Supreme Court 1980).

A decade later, *Moore v. Regents of the University of California* brought the commodity candidacy controversy into the human domain. In *Moore*, Dr. Golde of UCLA Medical Center removed the spleen of a patient John Moore and took blood, bone marrow, and sperm from this patient being treated for hairy cell leukemia. Dr. Golde then developed a cell line from Moore's T-lymphocytes and obtained a patent. The patent had an estimated three billion dollar earning potential due to its capacity to generate a variety of rare pharmaceutical products. At no time during Moore's treatment did Dr. Golde disclose to Moore the extent of his research and economic interests in his cells. Subsequently, Moore sued for conversion, and claimed a proprietary interest in each of the products created from his cells or from the patented cell line. The California Supreme Court determined that Moore did not own the genetic material already removed from his body. The court reasoned that the patient's rights were adequately protected via remedies for breach of fiduciary duty and lack of informed consent (see California Supreme Court 1990). Ownership rights in human body parts and genetic material continues to evolve.

In stark contrast to the traditional Western view, the very notion of owning genes clashes fundamentally with indigenous culture and belief systems. For most indigenous people, the human gene is their genealogy (see Whitt 1998). In many indigenous cultures, a human is not reduced to a clump of atoms and charged particles; rather, it is imbued with a "life spirit" that has been handed down by ancestors and is passed on to future generations. It is an element of the heritage of families, communities, tribes, and entire indigenous nations. It is not and cannot be the property of individuals. As a representative from the Paiute tribe stated: "We don't view our genes as protein actions ready to be interpreted; for us our genes are sacred" (Whitt 1998).

The traditional Western view toward vesting property rights in genes is inconsistent with many indigenous conceptions of property because it fails to acknowledge belief in group ownership and collective symbolism. Essentially, indigenous and Western scientific philosophies differ on one fundamental point: the difference in understanding of the origin of humanity, the responsibility of individuals, and the safety of future generations which sits so firmly at the core of indigenous opposition to gene banking. Gene banking therefore interferes in a highly sacred domain of indigenous history, survival, and commitment to future generations (see Whitt 1998).

In addition to contradicting the indigenous views regarding ownership of DNA, the very concept of gene banking flies in the face of the culture and belief systems of many indigenous peoples. There are genomics research programs that use gene banking to determine the precise origins of mankind and of culturally distinct peoples, building on the premise that mankind originated in Africa. The idea that mankind migrated out of Africa many thousands of years ago is an abhorrence to many indigenous peoples. This account is in conflict with their own account of historical events. In fact, there are as many stories of creation as there are groups of indigenous peoples. For example, the Pueblo Indians of the southwestern United States worship in round earthen structures called kivas, which represent the mythic realm of the underground from which humans emerged.

But Westerners themselves have religious beliefs about world creation that contradict scientific fact. Can't indigenous peoples separate out their culture and belief systems from the scientific underpinning of an endeavor like gene banking? The answer is that the word "religion" in and of itself inadequately defines the culture and belief systems of indigenous peoples. In truth, the Indian equivalent of the word does not appear in any of the hundreds of languages and thousands of dialects spoken in North America. The word implies that the various aspects of life can be segmented into the sacred and the secular. According to traditional Indian thinking, there is nothing that can be seen or touched, living or inanimate, that does not have a spirit. Spirituality and the ordinary life are as interconnected as the strands of a tightly woven rug. A pipe or a pair of moccasins that may appear quaint or mundane to a non-Indian may have enormous spiritual significance to an Indian. A carving or a painting that the non-Indian observer considers ornamental art may be part of a sacred process. The cycles that are dominant in the natural world—the path of the sun moving across the sky, the change of the seasons, the germination of seeds, and the birth, growth, and death of all creatures—evolve in an orderly fashion under the control of unseen forces. In essence, while indigenous beliefs are expansive, the traditional Western characterization of genetic material is inherently reductionist in nature. Researchers harboring this view assert that the human body can be divided into many parts and studied as tangible entities.

The origin of modern humans: The new molecular view

Nowhere is the clash of Western and indigenous belief more evident than in the context of the Human Genome Diversity Project. This markedly aspiring project is aimed at revolutionizing our understanding of human development and the expression of both our normal traits and our abnormal traits, such as disease. The purpose of the gene banking was to understand the origins and evolution of mankind, human migration, reproductive patterns, adaptation to various ecological niches, and the global distribution and spread of disease (Roberts 1991; Cavalli-Sforza et al. 1991). The originators boldly proclaimed during the first organizational meeting in 1992 that the project would ultimately discern "who we are as a species and how we came to be." Highest priority was to be given to groups defined as unique, historically vital populations that are in danger of dying out or being assimilated. To that end, the DNA would be kept in a database for use by all molecular scientists—molecular epidemiologists, molecular anthropologists, population geneticists, etc. The original plan was to obtain blood samples from twenty-five individuals who were members of specific populations. These samples would then be preserved as "immortalized" cell lines for future analysis, thus ensuring that there would be no further need to return for more blood (Roberts 1991; Cavalli-Sforza et al. 1991). By 1992, a total of 722 groups of people had been singled out from among the 7,000 groups under consideration (Christie 1996, 35). These groups quickly learned they had been deemed "genetic isolates." As they understood it, their blood cells would be stored as "cell lines" in American laboratories, and for a small fee, scientists could gain access to the stored blood for experimental purposes. Opponents of the gene banking project immediately labeled the endeavor "The Vampire Project" (Christie 1996).

In an effort to make gene banking more inclusive and sensitive to indigenous people's rights, the North American Regional Committee of the Human Genome Diversity Project (1997) published a proposed Model Ethical Protocol ("Protocol"). One of the major points addressed in the document is that, if any financial reward accrues from a specific analysis, a mechanism should be in place whereby individuals or populations who donated the blood can receive fair monetary compensation. The Protocol also states that permission must be obtained from both communities and individuals before samples can be taken. Community and individual permission would also be required before applications could be made for patenting or the marketing of products. In addition, the Protocol recommends that a respected international body such as the United Nations Educational, Scientific, and Cultural Organization should be made a trustee in connection with negotiations. Finally, the Protocol suggests that HGDP participants should have the right to ask for their samples to be withdrawn and destroyed at a later date if they decide not to participate further in the project.

The impact made by the Protocol is questionable, as the collection of blood continues with variable levels of consideration given to those from whom the blood is drawn. In October 1995, the United States Patent and Trademark Office granted the National Institutes of Health a patent on a human T-cell line developed from the genetic material of an indigenous person of the Hagahai people from Papua, New Guinea's remote highlands (see Aharonian 1995). The Hagahai, immune to the deadly cancer leukemia, are a 260-member hunter-horticulturist tribe who consented to the genetic research. Essentially, the antigens and other components of the Hagahai native's blood enabled the NIH to create a cell line that would further the formulation of "vaccines, bioassays and kits for the detection and diagnosis of infection with and diseases caused by HTLV-I and related viruses" (Aharonian 1995). The patent was later withdrawn.

Globalization, ethics, and law

PRIVACY AND INFORMED CONSENT: THE CONUNDRUM OF NATURAL LAW

Within the context of genetic research, individualistic ideals, not collectivist, are typically embodied in the notions of informed consent and the right to privacy. Informed consent in medical research was spawned by the Nuremberg Code which itself is derived from Natural Law theory. Within the Nuremberg Code, holding primacy above all others, both in number and importance, was the necessity of obtaining authentic, uncoerced *individual* informed consent from human subjects participating in scientific research (*see* Proctor 1992, 25). In 1953, the Clinical Center of the National Institutes of Health (NIH) produced the first US federal policy regarding research on human subjects, adopting for its research program the Nuremberg Code's emphasis on the use of healthy, competent volunteers in clinical research (see NIH 1995, Appendix 1). Today, all codes and regulations concerning the rights of human research subjects, both ethical and regulatory, import this notion of informed consent. But whether or not inalienable, God-given, or "natural" human rights are bestowed *individually* or *collectively* poses a challenge in population-based genetic research. Like the Nuremberg Code, the notion of informed consent appears first among the guidelines set forth in the Protocol published by the North American Regional Committee of the Human Genome Diversity Project (1997). The Protocol states that "[th]e discussion of informed consent that follows is the longest part of the Model Protocol—not just because of its undoubted complexity, but because of its importance." The Protocol shows how personal autonomy is of paramount concern in the context of scientific research—that no human should be subjected to participation in research through coercion and all subjects should knowingly give consent. Unlike the Nuremberg Code, the Protocol recognizes that many indigenous people do not conceptualize themselves as being distinct from the group, and advocates that informed consent be obtained from the group population, in addition to the individual.

Extending Western Philosophical Thought to Collectivist Cultures

Though the dilemma between Native American and traditional Western philosophies concerning collective rights is apparent, Congress has recently resolved a conflict in favor of Indian nations with respect to the Indian Child Welfare Act (ICWA; 25 U.S.C. SS 1901-1963 (1983)), recognizing that in Native American culture, children belong to the group as a whole (see Goldsmith 1990, 3). Thus, in an attempt to reflect tribal values, the ICWA establishes federal standards for Indian child welfare proceedings that give preference to the placement of the child with a member of the child's extended family, another Indian family, or a family chosen by the child's tribe (Goldsmith 1990). Since the passage of ICWA, Congress has continued enacting legislation premised on group-rights philosophies for Native peoples. The year 1990 marked a landmark legislative victory for Native Americans when Congress adopted the Native American Graves Protection and Repatriation Act (NAGPRA; 25 U.S.C. SS 3001-3013 (1991)). The Act, which sets out to protect human remains, sacred objects, and cultural patrimony of indigenous peoples, reflects a growing consensus in the United States that "[t]he sacred culture of Native American and Native Hawaiians is a living heritage. This culture is a vital part of the ongoing lifeways of the United States, and as such, must be respected, protected, and treated as a living spiritual entity—not as a remnant museum specimen."

Conclusion

In the context of the gene banking, it remains clear that a number of questions have yet to be satisfactorily answered: Who "owns" genetic material? Individuals? Communities or tribal groups? Corporate organizations? Or humankind? The notion of group versus individual rights is at the heart of the gene banking controversy, because what is truly at stake with regard to whether or not to further or abandon such projects is the basic question of what the purpose is of humanity itself. The notions of group rights versus individual rights involve contradictory views of the human situation, human life, and the purposes for which humans have been created. Respect for the interests of cultural groups as a whole involves a fundamentally different alignment of humans toward life itself from that which seeks to protect individual rights and freedoms. Protecting the interests of cultural groups requires viewing individuals as members of cultural groups, thereby rejecting the individual as distinctly separate from a cultural group. The individual is, therefore, conceptualized as a product of a single belief system. It is the cultural beliefs of the group, rather than the individual, that hold spiritual significance. Thus, deference is due a common set of ideas, rather than choices of individuals.

References

Aharonian, G. 1995. Aboriginal person is claimed in a U.S. patent. *Internet Patent News Service* Nov. 1.

Boulier, W. 1995. Sperm, spleens, and other valuables: The need to recognize property rights in human body parts. *Hofstra Law Review* 23:693–762.

California Supreme Court. 1990. *Moore v. Regents of the University of California* 793 P.2d 479 (Cal. 1990).

Cavalli-Sforza, L. L., A. C. Wilson, C. R. Cantor, R. M. Cook-Deegan, and M. C. King. 1991. Call for a worldwide survey of human genetic diversity: A vanishing opportunity for the human genome project. *Genomics* 11:490–491.

Christie, J. 1996. Whose property, whose rights? *Cultural Survival Quarterly* 20:34.

Coke, E. 1644. *Institutes of the laws of England*. Volume 3. London.

Goldsmith, D. J. 1990. Individual vs. collective rights: The Indian Child Welfare Act. *Harvard Women's Law Journal* 13:1–12.

Lock, M. 1999. Genetic diversity and the politics of difference. *Chicago-Kent Law Review* 75:83–111.

National Institutes of Health (NIH). 1995. *Guidelines for the conduct of research involving human subjects at the National Institutes of Health*. Bethesda, Md: National Institutes of Health. Available online at: http://ohsr.od.nih.gov/guidelines/graybook.html

North American Regional Committee of the Human Genome Diversity Project. 1997. Proposed model ethical protocol for collecting DNA samples. *Houston Law Review* 33:1431–1473.

Proctor, R. N. 1992. Nazi doctors, racial medicine and human experimentation. In *The Nazi doctors and the Nuremberg Code: Human rights in human experimentation*, ed. G. Annas and M. Grodin. New York: Oxford University Press.

Roberts, L. 1991. A genetic survey of vanishing peoples. *Science* 252:1614–1617.

US Supreme Court. 1980. *Diamond v. Chakrabarty*. 447 U.S. 303 (1980).

Whitt, L. A. 1998. Indigenous peoples, intellectual property, & the new imperial science. *Oklahoma City University Law Review* 23:211.

Research biobanks

37

Mapping the language of research-biobanks and health registries: From traditional biobanking to research biobanking

A project presentation

by Jan Helge Solbakk, Søren Holm, Paula Lobato de Faria, Jennifer Harris, Anne Cambon-Thomsen, Marit Halvorsen, Camilla Stoltenberg, Roger Strand, Bjørn Hofmann, Anne Maria Skrikerud, and Jan Reinert Karlsen

Background and introduction

Biobanking, i.e. storage of biological samples or data emerging from such samples for diagnostic, therapeutic, or research purposes, has been going on for decades in many European countries and elsewhere. However, it is only since the mid 1990s that these activities have become the subject of considerable public attention, concern, and debate. This shift in climate is probably due to several factors:

a) increased knowledge among the general public about the existence of biobanks and health registries and their potentials for research;
b) public concern about health insurance discrimination due to "risky" genetic profiles;
c) the proliferation of initiatives from public as well as private research-institutions with the purpose of establishing *new* biobanks for conducting population-based research that involves genetics, or with the purpose of converting biological samples stored in diagnostic or therapeutic banks into "biocurrency" subject to such kinds of research;
d) the emergence of new regulative initiatives in many countries concerning research biobanking; and finally
e) rising interest among private and public sponsors in the possibilities of commercial utilization of research biobanks and research results acquired from samples and data stored in research biobanks and health registries.

The project is the result of an interdisciplinary collaboration between epidemiology and genetics with ethics, law, and philosophy of science, and involves researchers from seven different institutions in Norway, France, Portugal, and the UK. The aim of our project is to investigate some of the ethical, legal, and social challenges raised by research biobanking in its different modern forms and formats. Besides analyzing current legislation, other forms of regulation, and public debates concerning biobanks in European and other relevant jurisdictions, the project also includes an "in-vivo" study of ethical, legal, and social issues likely to emerge from two new genomics/functional genomics projects, BIOHEALTH-NORWAY administered by the Division of Epidemiology at the Norwegian Institute of Public Health, and GENOMEUTWIN, an EU project, the Ethics Core of which is the responsibility of the Division of Epidemiology at the Norwegian Institute of Public Health (see later).

We describe here briefly these two genomics/functional genomics projects with which our project on research biobanking is linked:

(A) BIOHEALTH-NORWAY:

The overall purpose of this project is to provide a nationwide network and resource for studying the molecular nature of disease and to explore gene-environment interactions. The ultimate goal is to improve prevention and treatment of disease through this new molecular knowledge. This research platform includes data from two large cohorts:

1. CONOR — a cohort of 200,000 subjects, including blood samples and standardized health and exposure data.
2. The Norwegian Mother and Child Cohort Study (MOBA) — a cohort of 100,000 pregnant women, 100,000 children and 70,000 fathers, including biological samples and standardized health and exposure data.

(B) GENOMEUTWIN:

The second project that will serve as a template for analyzing the ethical, legal and social issues raised by research biobanking is GENOMEUTWIN. This project is a genetic epidemiological, multinational, multi-center study. It was developed in response to the EC Quality of Life Programme, Area 8.5: Integrated Projects in Functional Genomics Relating to Human Health. It was funded by the EU as one of three centers of excellence in Genomics, and started on October 1, 2002. It is an integrated project including components in research, networking and training, and mobility. Genome-wide analyses of DNA from European twin and population cohorts will be conducted to identify genes involved in common diseases. The twin cohorts include 680,000 pairs of twins from Denmark, Norway, Sweden, Finland, The Netherlands, and Italy.

GENOMEUTWIN is organized around five intellectual cores (database, epidemiology, DNA extraction and genotyping, statistical issues, and ethical and social issues). The ethical core operates across the entire project.

The wider objectives of GENOMEUTWIN are to:

1) develop a framework facilitating the development and use of novel strategies to utilize maximally the unique features of twin cohorts, including the availability

of longitudinal data and ample information about lifestyle and environmental factors, in European and global biomedical research;
2) guarantee the access of European researchers to the epidemiological, phenotypic, and genotypic information on European population cohorts; and
3) create a unique infrastructure for research into common diseases and for training of scientists in quantitative biology.

GENOMEUTWIN offers a unique platform for analyzing the ethical, legal, and social impacts of population-based research involving genetics. Besides, this multinational "data-bank" provides an excellent opportunity to test cultural differences, and explore how differences in national standards and legislation affect subjective understandings and public perceptions of population-based research involving genetics.

Research biobanking: Four main clusters of issues

The ethical, legal, and social issues raised by research biobanking in its modern form can be divided into four main clusters of issues:

1. Issues concerning how biological materials are entered into the bank;
2. Issues concerning research biobanks as institutions;
3. Issues concerning under what conditions researchers can access materials in the bank, problems concerning ownership of biological materials and of intellectual property arising from such materials;
4. Issues related to the information collected and stored, e.g. access-rights, disclosure, confidentiality, data security, and data protection.

The first cluster of issues has been much discussed. Relevant problems are, for instance, what kind of consent should be given by persons who give material to a research biobank, under what conditions can material in diagnostic or therapeutic biobanks be "converted" into research materials, and under what conditions can materials obtained without consent or against the will of the "donor" be "converted" into research materials. Other problems in this cluster of issues concern exactly what rights the donor gives to the bank and what rights the donor retains, questions about incentives for giving, grounds for withdrawal from the bank, as well as renewed consent from children with stored tissues when they reach the age of legal maturity.

The second cluster of issues is concerned with the biobank as an institution. What kind of institution is it? Under what conditions can it be sold, merged with other biobanks, exported, divided, or destroyed? These issues are much less discussed in the literature, but may be of importance for the two other clusters of issues (is the distinction between public and private biobanks for instance important when regulating consent procedures?).

The third cluster of issues raises questions concerning research ethics governance of the use of stored biological materials, as well as questions concerning how a biobank should set priorities among a number of competing research projects. This cluster is also concerned with ownership and intellectual property issues, including various modes and levels of profit sharing, if any.

The fourth cluster of issues concerns the long term relations between researchers and users of the biobanks on one side, and the sampled population on the other. It includes access to results on an individual or a global level, ways of dissemination of information about biobank use, and data protection and confidentiality issues.

There is a considerable interplay between the ethical and legal issues in each of the described clusters. If, for instance, relatively liberal rules are implemented concerning the entry of materials into biobanks, stricter rules concerning the use of these materials are likely to be needed, and vice versa.

Research biobanking: The traditional approach

The traditional approach to the ethical and legal issues raised by research biobanks has been to extend the informed consent and other research ethics procedures that are already in place, and to supplement them by measures directly transferred from the area of data protection. This is the case with current Norwegian legislation (LOV 2003-02-21 nr 12: Lov om biobanker). There are, however, reasons to believe that such a direct application of the traditional approach creates more problems than it resolves. Informed consent was originally developed in the context of the doctor-patient relationship, and later extended to the researcher-research participant relationship, but still mainly in the clinical setting (Beskow et al. 2001; NBAC 2001, 13). In this setting it is possible to inform the potential research participant about the exact nature and purpose of the research project in considerable detail. However, in the context of research biobanking this level of specification is hardly achievable, because the research performed on banked materials is, by in nature, open-ended. We cannot know how the materials stored today will be used in 20 years' time, because we have no idea what will be possible in 20 years' time. Furthermore, it is practically impossible to obtain actual informed consent for each new use of the stored materials.

This problem of specification might be an indication that current consent procedures are insufficient to provide the donors of biomaterials with adequate protection of their rights.

In the biobank setting consent is required not only for a specific research procedure, but also for a transfer of some or all of the rights of control over the actual material and its use. If this is conceptualized as a "transfer of ownership," informed consent suddenly looks like a very odd procedure for such a transfer, since ownership is usually transferred by means of contracts, and based on the advice of lawyers, not on information given by medical doctors. If, on the contrary, one considers this not as ownership, but as "right of control" over its use without proprietary rights, then the direct consent or the consent to transfer this right over use to a body or another person looks appropriate. In any case, the clarification of this issue seems to be crucial.

Even if the traditional approach is problematic, it is nevertheless important to continue analyses along these lines, since they are the currently acknowledged legitimate basis for biobanking. Such analyses are, however, unlikely to sufficiently resolve the major ethical and legal issues actualized by biobanking.

Primary project objective: Biobanking — do we need a new conceptual approach?

Thus, the primary objective of this project is to critically assess the traditional regulatory approach to research biobanking, and develop alternative ways of conceiving of and regulating research biobanks. Our main hypothesis is that instead of attempting to directly extend the analysis of the traditional approach described above, it will be fruitful to investigate the conceptual potential of analogies from a range of areas outside medical research, where people transfer something to a common institution. The analogies we propose are not intended as templates or models to be followed, but as tools for analysis. Examples of such analogies are ordinary commercial banking, associations, clubs (e.g. book clubs) or unions, libraries, military conscription, taxation, and management of art pieces.[1] By developing these analogies, analyzing their implications, and identifying their limitations, we believe it will be possible to:

a) achieve a deeper understanding of the structural arrangement of research biobanking;
b) critically assess the vocabulary prevailing in the field of biobanking, in particular the labels employed and the roles, rights, and duties ascribed to different parties affected by or involved in biobanking (donor, "biobanker," researcher, research ethics committees, the sampled population); as well as
c) make recommendations regarding different ways in which a biobank could be structured as a social institution.

Secondary or specific project objectives

Based on the considerations mentioned above, our project has the following specific project objectives:

* To compare legislation and other regulation concerning biobanks in European and other relevant jurisdictions;
* To study the public debate preceding and following regulation of biobanks in selected European jurisdictions (Denmark, Finland, France, Italy, Portugal, Sweden, The Netherlands, UK, and Norway);
* To analyze the extent to which traditional, individual informed consent conceptions and procedures can usefully be employed to "solve" the ethical and legal issues concerning entry of materials into research biobanks;
* To explore a range of analogous social institutions where people transfer something to a common institution and draw out possible implications for biobanking. This will (among many other aspects) include:
    o Investigating to what extent different forms of incentives employed by *associations, clubs, and unions* to recruit new members might be of help in resolv-

---

1 This last analogy plays a prominent role in the commercial world (some art pieces are considered "invaluable," are protected, have a cultural and symbolic dimension etc.).

ing the question about appropriate forms and uses of incentives in relation to donors of biological materials for research purposes;
- o Investigating the possibilities provided by the notion of *currency exchange* in addressing the problem of converting materials stored in already existing diagnostic banks and therapeutic banks (e.g. blood banks, sperm banks, and IVF-banks) and health registries into *"valuta"* (bio-currency) for research purposes;
- o Studying whether commercial banking notions such as *savings account, savings, saver or depositor, small saver, interest rates, loan*, etc. might be of help in clarifying the relation between the right of ownership and the right of disposal of materials for research purposes stored in bio-banks and health registries;
- o Addressing the question whether notions from commercial banking might help clarify the issue of commercial utilization of research bio-banks, as for example the issue of whether individual donors of bio-currency for research purposes should be attributed economical rights analogous to the rights of small savers with savings accounts in ordinary banks;
- o Analyzing whether the notions of *conscription* and *taxation* might be of help in resolving the question whether there could be a duty to donate biological materials for research under certain circumstances;
- o Investigating the possibilities provided by the notion of *loan* employed in the running of libraries, in addressing the question of researchers' rights and duties in relation to materials "borrowed" from research biobanks;
- ★ To further explore alternatives to individual informed consent procedures, identified through the use of analogies. In particular, to critically analyze the ethical and legal justifications of employing collective information and consent-procedures in relation to converting materials stored in diagnostic banks, therapeutic banks, and health registries into materials for research purposes;
- ★ To analyze the issue of valid consent procedures in relation to future research projects not yet conceived and/or not sufficiently specified:
- ★ To study the institutional context of biobanking and its implications for the legal and ethical regulation of biobanks;
- ★ To outline different possibilities for resolution of the problems in the four clusters of issues on the basis of the analysis performed.

Acknowledgements

The project is funded by The Research Council of Norway and The Norwegian Institute of Public Health for the period of 2004–2007.

References

Beskow, L. M., W. Burke, J. F. Merz, P. A. Barr, S. Terry, V. B. Penchaszadeh, L. O. Gostin, M. Gwinn, and M. J. Khoury. 2001. Informed consent for population-based research involving genetics. *JAMA* 286:2315-2321.

National Bioethics Advisory Commission (NBAC). 2001. *Ethical and policy issues in research involving human participants*. Vol 1. Bethesda, Md: NBAC.

Norwegian Parliament. 2003. LOV 2003-02-21 nr 12: *Lov om biobanker* (biobankloven).

38

UK Biobank: Social and political landscapes

by Helen Busby

Introduction

THE ORGANIZATION OF biobanks on a national basis—as opposed to a more local or indeed on a non-geographical basis—is a recent phenomenon. Previously collections of tissue and data organized by a range of commercial and non-commercial bodies attracted relatively little attention except from the scientists and professionals involved (Hirzlin et al. 2003). More recently though the social issues arising from developments in biobanks have been defined as ethical issues.

Much of this recent discussion has been characterized by a search for generalized principles and frameworks which may govern such initiatives across diverse contexts. For example, it has been suggested that "solidarity" might be a new framework within which developments in genetic biobanks could be considered (Chadwick and Berg 2001). But there has been less discussion of the ways in which discourses of solidarity have a distinctive character and meaning in different national and historical contexts. There is also an emerging body of literature exploring the particular and various ways in which historically informed notions of trust and responsibility, of civic virtue, shape views about biobanks (Pálsson and Hardardóttir 2002; Hoeyer 2004). The tension between these two approaches is to an extent reflective of the different stances taken by bioethicists and social scientists (Haimes 2002; Hoffmaster 2001). In this paper I shall explore some of the historically informed social and political landscapes which surround the development of a biobank in the UK.

UK government policies on genetics and biobanks

In the UK, the development of a national biobank has been under discussion for some time. We have recently seen the establishment of an organizational structure for the biobank, although some key aspects are still under review. Recruitment of participants is expected to begin later in 2004. At governmental level, the biobank is nested into a series of policy initiatives concerned with innovation and the UK's place in the global knowledge economy. These developments and their drive to synthesize data across a whole nation are firmly located within an agenda of modernization. At the same time, the Secretary of State for Health has confidently announced that the values on which the NHS is based historically—providing for all on the basis of need—"are particularly suited to capturing the benefits of genetics advances" (Department of Health 2003, 8). The Government White Paper, entitled Our inheritance, our future goes on to stress the concept of the social solidarity represented by the NHS; people can take genetic tests without fear because "everyone, regardless of risk, is 'insured' by the NHS." This kind of rhetorical emphasis on pooling and sharing risks is unusual in today's policy climate in the UK, which tends rather towards a stress on the limits of the responsibilities of the state toward its citizens. Indeed it evokes a post-war ethos, and perhaps intentionally echoes the rationale of the Beveridge report of 1942.

The relationship between the historical and contemporary experience of the British NHS and the stance taken toward the new biobank initiative is of particular interest here. Perhaps more than any other national institution, the NHS is widely considered to underpin the values of the post-war settlement and the common good of the nation. Nevertheless, and despite the mythical status of that settlement, class and place have continued to impact on the health of Britons in significant ways (Szreter 2003). Thus the appeal to solidarity in contemporary UK policies on genetics is made across those divisions of place, class, and health status, to the commonalities of nationality.

Blood donation and concepts of "gift" in the UK

In my quest to consider some of the social contexts which may shape the UK's biobank, I turn first to the practice of blood donation, which holds a special place in the history of the NHS. As is well-known, British blood donors are not financially rewarded for giving blood. Richard Titmuss' famous formulation of gift relationships has long been a point of reference for thinking about blood donation in the UK (Titmuss 1997). Written in the late 60s, Titmuss' study included a comparison of the systems in the US and the UK, and put forward a raft of moral, economic and social arguments against a market in blood. Titmuss famously saw voluntary blood donation as "a gift." For him, "the voluntary community donor is the closest approximation in social reality to the abstract concept of a 'free human gift.'" Without immediate reward or sanction, these donors would donate blood "for unnamed strangers without distinction of age, sex, medical condition, income, class or ethnic group" (Titmuss 1997, 140). This "British" donor type, then, symbolized Titmuss' ideals of mutual social provision for medical need, regardless of social standing (Rose 1981).

Despite the many changes in blood services in the UK and elsewhere, and the criticisms leveled at the original study, the image or metaphor of blood donation as a "gift to strangers" has remained influential in clinical and policy discussions about blood donation.

I shall refer here to my own research, which included interviews with contemporary NBS donors.[1] As I have discussed elsewhere, many of these donors see their donation of blood as a contribution to a "bank" which they themselves, their relatives, or strangers can draw upon (Busby 2004): Donors perceive themselves and their families

---

1 One hundred short interviews were conducted with NBS donors, approximately equal numbers of men and women, ranging in age from 18 to late 60s, but with most being in their thirties, forties and fifties.

as potential recipients in a direct sense. These concepts of reciprocity or mutuality contrast with notions of "pure altruism" which sometimes appear in the literature.

The gift as a rhetorical framework for donation to biobanks in the UK

Recently the image or metaphor of blood donation as a "gift to strangers" with a moral dimension has achieved a metaphorical resonance beyond its original context in public debates and policies on tissue donation for genetic research (Tutton 2002). In remarks on the launch of UK Biobank, its new Chief Executive, John Newton, referred to donated blood as a gift. Previously, the authors of ethical guidelines published by the Medical Research Council (MRC) on the collection, use and storage of human tissue recommend that blood donated for research refer to the "gift relationship" between participants and researchers, which is said to underline and protect "the altruistic motivation for participation in research" (MRC 2001, 8). The theme of altruism is echoed too in some other recent policy papers on biobanks which endeavor to harness the ideal of patient altruism to a rhetoric of progress and prosperity. Ideas about altruism and "gift," then, have had some currency in the various papers and statements about donation for biobanks which can together be taken to constitute emerging policy.

It is notable however that the references to the idea of "gift relationships" in discussions on blood and tissue donation are rarely supported with an understanding of the depth and range of Titmuss' work in this context. In brief, Titmuss' model has several implications: firstly, it ascribes certain motives to donors; secondly, it implies that the structure of blood donation, over time and across society, is reciprocal; thirdly, it asserts that for these and other reasons, policies should not allow for a market in blood: it puts forward a raft of moral, economic and social arguments against a market in blood. In contrast, the notion of "gift" as it has been deployed in the context of biobanks and genetic research elides the question of commerce and ignores the issues about the kinds of practical mutual provision which Titmuss was interested in.

Research, biobanks and the "imagined community" of the NHS

The influence of clinicians and the shaping of clinical norms are well-established dimensions of peoples' decisions to participate in medical research.[2] Here I suggest that cultural and historical aspects of loyalty to the National Health Service more broadly conceived are deserving of attention in this context. I shall refer here to my own interviews with volunteers who donated blood samples for an epidemiological genetic research project,[3] in which a sense of responsibility generated by a relationship with the NHS often featured. These donors generally expected that an increased knowledge from the research would feed back directly into treatments for NHS patients. At the same time, they raised concerns about their own and others' experience of the NHS.

---

2   Beginning with Fox ([1959] 1974), and see for example Corrigan (2003) for an elaboration of this theme in the context of more recent drugs trials.
3   A group of 27 interviews was undertaken with a group of donors from a research project concerned with possible genetic factors in psoriatic arthritis.

They cited instances in which their relatives or others whom they knew had been unable to get timely and adequate care for serious conditions. In these accounts, the cherished ideals of the imagined NHS were often not lived up to, a notable example being the "postcode lottery" in which patients in some parts of the country received treatments unavailable to those in another region.

We can think of this in terms of Anderson's concept of imagined communities: For Anderson, an essential dimension of the nation is that "regardless of the actual inequality and exploitation that may prevail in each, the nation is always conceived as a deep, horizontal comradeship" (Anderson 1991, 7). One important element of this analysis is the stress on people's imagining of a nation, and another is the role of institutions in shaping that imagination. I suggest that this analysis can inform our thinking about the ways that people conceive of their national health service—and by extension their participation in research such as that envisaged by the national biobanks.

Concluding remarks

At a time when public perceptions are seen as key to legitimizing developments in biotechnologies, the organization of biobanks on a national basis can be seen to be one route to command legitimacy of this kind. In the UK, it is likely that the national biobank will recruit subjects via the NHS, historically one of the common institutions through which national identity is mediated. I conclude by drawing out some of the implications of framing contemporary policies for genetic research and biobanks with reference to the touchstones of post-war solidarity and gifted blood. Without a tangible framework of regulation for the new forms of commerce and public-private partnerships entailed in these developments, these frameworks are likely to be more stylistic than political.

Acknowledgements

I would like to acknowledge the support of a doctoral training grant from the Wellcome Trust, and the assistance of my PhD supervisors Dr Paul Martin and Professor Robert Dingwall.

References

Anderson, B. 1991. *Imagined communities*. 2nd ed. London: Verso.
Busby, H. 2004. Blood donation for genetic research: How donor narratives challenge policy assumptions. In *Genetic databases: Socio-ethical issues in the collection and use of DNA*, ed. R. Tutton and O. Corrigan. London: Routledge.
Chadwick, R., and K. Berg. 2001. Solidarity and equity: New ethical frameworks for genetic databases. *Nature Reviews Genetics* 2:318–321.

Corrigan, O. 2003. Empty ethics: The problem with informed consent. *Sociology of Health and Illness* 25:768–792.

Department of Health. 2003. *Our inheritance, our future: Realising the potential of genetics in the NHS*. Cm 5791-II, HMSO.

Fox, R. [1959] 1974. *Experiment perilous*. Philadelphia: University of Pennsylvania Press.

Haimes, E. 2002. What can the social sciences contribute to the study of ethics? Theoretical, empirical and substantive contributions. *Bioethics* 16:89–113.

Hirzlin, I., C. Dubreuil, N. Preaubert, J. Duchier, B. Jansen, J. Simon, P. Lobato de Faria et al. 2003. An empirical survey on biobanking of human genetic material and data in six EU countries. *European Journal of Human Genetics* 11:475–488.

Hoeyer, K. 2004. Ambiguous gifts: Public anxiety, informed consent and biobanks. In *The genetic donation: Donating, collecting and exploiting human tissue in research*, ed. R. Tutton and O. Corrigan. London: Routledge.

Hoffmaster, B. 2001. *Bioethics in social context*. Philadelphia: Temple University Press.

Medical Research Council (MRC). 2001. *Human tissue and biological samples for use in research: Operational and ethical guidelines*. London: MRC. Available online at: http://www.mrc.ac.uk/index/publications.htm

Pálsson, G., and K. Hardardóttir. 2002. For whom the cell tolls. *Current Anthropology* 43:271–287.

Rose, H. 1981. Re-reading Titmuss: The sexual division of welfare. *Journal of Social Policy* 10:477–502.

Szreter, S. 2003. Health, class, place and politics: Social capital and collective provision in Britain. *Contemporary British History* 16 (3): 27–57.

Titmuss, R. 1997. The gift relationship. In *The gift relationship: From human blood to social policy*, ed. A. Oakely and J. Ashton. London: Routledge.

Tutton, R. 2002. Gift relationships in genetic research. *Science as Culture* 11:523–542.

Cell line research with UK Biobank

Why the new British biobank is not just another population genetic database

by Sebastian Sethe

Introduction: The UK Biobank

LET US BRIEFLY look at some of the main tenets of UK Biobank:[1]

(1) The project is funded by a private charity (the Wellcome Trust) and two public entities: the Medical Research Council (MRC) and the Department of Health. Revenue is expected from granting access to the resource.
(2) The study aims to recruit up to[2] or at least[3] 500,000 participants aged 45–69, who will provide information about their lifestyle and diet, their medical history, and a blood sample or other biological sample (UK Biobank, Frequently Asked Questions; UK Biobank, Website).
(3) There will be an ongoing link to the participant's medical record. Participants might be re-contacted periodically for renewed or further lifestyle information.
(4) Datasets will be encrypted (rendered reversibly anonymous) for the purpose of studies.
(5) Proposals to study the data will be subjected to peer review of their scientific quality, ethical review by an NHS Multi-Centre Research Ethics Committee, and review by UK Biobank.

It is important to note that all these aspects put UK Biobank in a unique position in comparison with other genomic and health databases (Austin, Harding, and McElroy 2003). However, on the face of it, there are at least some broad similarities between UK Biobank and other large-scale population genetic databases. However, there is one aspect that makes UK Biobank more than just the potentially largest resource of this kind in the world: the proposed proliferation of some samples into cell lines.

---

1 Further information is available on the UK Biobank Website; see Barbour 2003 and Wallace 2003 for a primer into some criticisms.
2 UK Biobank. Frequently Asked Questions.
3 Parliamentary Office of Science and Technology 2002.

Cell line science

The draft protocol states: "In order to provide an extra long-term source of DNA for controls, a random sample of 10,000 members of the cohort will have an extra 5ml of blood taken for cryopreservation of peripheral blood lymphocytes in a manner which will allow subsequent immortalisation of these cells" (UK Biobank 2003, Section 2.3.5.1).

This protocol proposes the creation of cell lines. A cell line is a continuously growing population of cells, which can be expanded in cell culture, and thus can be used for multiple appliances time and again. The draft protocol is silent on specifics, but "immortalisation" implies a method which ensures continuous proliferation. While certainly more routine today than even a few years ago, immortalization of lymphocytes is not a standard procedure (Stacey and MacDonald 2001). It mainly involves turning a normal "mortal" cell into a cancer cell. SV-40 is one example of a common viral oncogenic mediator, but there is a variety of potential methods to achieve this transformation (Rhim 2000). Some lines grow continuously, while others have a more limited lifespan. The most famous example might be the HeLa cell line that was derived from cells of a cancer patient by the name of Helen Lane in 1951. Half a century later, her cells continue to reproduce, and are now used in laboratories all over the world.[4] A more infamous example is that of James Moore,[5] whose spleen cells were proliferated into a commercial cell line without his knowledge or consent.

Given the cost and the difficulty of the procedure, clearly not each and every one of the "random samples" will be thus immortalized. The draft protocol confirms: "blood will be collected ... so as to allow the subsequent immortalisation of these cells, *if necessary*" (my emphasis). When, one might wonder, will such a procedure be deemed necessary? "For controls" suggests the draft protocol. How and when the creation of a cell line would be a valid measure to establish a "control" in a population genomic study is a scientific question. Another possible invocation of "necessity" could be to screen the sister samples for noteworthy genetic polymorphisms; and if one is found, to immortalize its frozen companion for further study and use.

In the light of this preciously tentative information, one might be tempted to suspect that the collection of samples for cell line purposes has been grafted onto the protocol in anticipation of future technological advances. It needs to be remembered that UK Biobank is a long-term project with the future built firmly into its design. As science progresses, there will likely be novel and more effective ways of immortalizing a sample and of utilizing a cell line. Only such developments, coupled with faster, easier, and more effective screening techniques, greater computing power, and enhanced genetic predictability, will enable the biobank to make effective use of the potential cell lines it accumulates in its freezers. In this sense, UK Biobank proves itself a true bio-bank, as opposed to simply a large genetic database.

---

4 A search for HeLa finds 42390 items in PubMed: http://www.ncbi.nlm.nih.gov/PubMed/
5 Moore v. Regents of the University of California, 793 P.2d 479, 271 Cal. Rptr. 146 (1990).

Cell lines are different

Interestingly, the sample immortalization feature has not featured prominently in many a discussion of the project. However, the feature was considered—briefly—by "The UK Biobank Ethics Consultation Workshop" two years ago (UK Biobank 2002, 9–10):

> The advantages of generating cell lines from a sub-set of the Biobank participants were discussed. It was agreed that misconceptions could arise since it was not clear whether this technique was well understood by the public. There was potential for confusion with other emotive research techniques such as cloning and stem cell research, which could lead to disquiet. It was therefore important to address people's concerns and possible misunderstandings regarding the generation of cell lines, and to explain as clearly as possible what was involved and why it was done. Further consideration should be given to the specific ethical issues that might be involved, such as perspectives on the human body in different cultures and their possible impact on willingness to participate. In any event, it was agreed that explicit consent had to be given by the Biobank participants for this procedure.

Let us turn to examine how the draft ethics and governance framework approached these suggestions, and how other particular issues not mentioned above arise.

INFORMATION

One might question whether explaining the procedure of cell line creation would necessarily have to be done with the dreaded reference to cloning and stem cell research that the workshop participants seem to suggest. However, the caveat they made about pitfalls in communicating the scientific essence of the procedure likely made an impression on the drafters of the ethics governance framework (Department of Health, MRC and The Wellcome Trust 2003): one searches in vain for an explicit reference to the procedure of sample immortalization, let alone for a proposal on how the procedure will be explained to participants.

When considering information provision to participants in this context, it is not sufficient to stop at the pre-donation stage. Will participants be told that their sample has been turned into a cell line? Will they be told if "their" cell line proves to be particularly noteworthy for some reason? It could be argued that for some donors it would be a fascinating and enriching experience to be able to see where copies of their "gift" are going and what scientific findings are derived from it.[6] This will likely not be possible for "normal" donors whose contribution is only ever considered as a tiny aspect in a large dataset, but the question must at least be asked why the selected few whose donations are immortalized should not be given such information if they wish. The dynamic link with the donor's medical record will ensure that even if the sample is only turned into a cell line after a considerable time, contacting the donor is unlikely to be a problem.

---

6 An analogy with sponsorship "adoption" schemes for trees or acres of rainforest comes to mind.

Consent

Consent has emerged as a concept central to medical ethics and medical research in particular. Usually, the term is used with the prefixes "informed," "full," or "voluntary." It has been pointed out that the consent obtained under the suggested biobank framework will to some extent be "blanket" consent of doubtful legal validity (Caulfield 2003). Since the potential creation of a cell line is a definite part of the proposed protocol, anything but explicit consent to cell line immortalization would clearly be insufficient.

However, the matter does not end there. Faced with an extensive debate about legal and ethical perspectives on how to handle *unanticipated* uses of a donation in research,[7] blanket consent in biobanks has been justified under the contention that it would be utterly impractical and prohibitively expensive to contact thousands of participants each time their samples are to be used for studies. These arguments apply to cell lines in a much lesser degree: there will likely be far fewer cell lines and instances of requested use. Once again, the individual nature of the cell line does not fit into the framework created for governing large data sets. While the latter make best use of the huge size of the proposed collection, the interest in a cell line will likely be tied to its very individuality. Thus, the (potential) consent of the individual gains more weight, while the notion that explicit consent can adequately be substituted by an appointed representative (ethics committee) that knows nothing of the individual's preferences becomes more dubious. Asking individuals whose samples will be turned into cell lines to narrow their consent to specific categories of use might be a compromise (Caulfield, Upshur, and Daar 2003). Nevertheless, in light of the anticipated duration of the project, the idea that we should be able to predict today what categories of research will be of particular interest in relation to a particular cell line ten years hence, seems problematic.

Privacy

For the purposes of research, biobank datasets will anonymomized in the sense that the link to the patient's name will be severed. But cell line samples bring about their own particular privacy issues. Stepping aside from debates surrounding the definition of anonymization, pseudo-anonymization, and exoneration (Beyleveld and Histed 1999), it can be argued that there is at least one notion of privacy that applies irrespective of how proficiently the link between the sample and the donor is severed. At least genetically, the sample and the cells derived from it remain something "of yours." It could rightly be argued that all biobank participants are reconciled to their genetic profile becoming the object of scientific study. But there remains a marked difference, and not only in degree, between having your contribution analyzed quantitatively as one among hundreds in a large dataset, and its becoming the unique object of in-depth study and manipulation, as would be the case with a cell line.

Commercialization

The debate about who—if anyone—can patent a product derived from the human body is far from over. To avoid some of these troubles, UK Biobank will demand that con-

---

7 I recommend Greely 1999.

tributors explicitly relinquish all potential rights they might hold in their sample. This sits uncomfortably with the impossibility of predicting who will exploit a donation in what way. In a world where specific commercial entities are increasingly recognized as having municipal power and responsibility, it cannot be automatically assumed that a donor would choose to renew her consent when all the proverbial cards are on the table.

This suggests the adoption of an authorization model as mentioned above. One could consider giving donors the ability to exclude their samples from any research that has a specific commercialization angle. For "normal" samples this might be regarded as impractical, since it would be hard to trace the specific contribution a specific donation makes to the research commercialization process. In cases of particularly remarkable cell lines, however, a closer involvement of the donor might be much more feasible. At a minimum, it would be possible to acknowledge certain broad specifications that the donor has given regarding venues of commercialization.

There is one other aspect: participants in the UK biobank are asked for an altruistic donation. Ownership of the data will remain with the three funding bodies, and knowledge gleaned from it will benefit commercial companies. In such a set-up, "it is reasonable to expect some of the benefits to be directed towards the wider public good." Recently, a US gene test patent was challenged on the ground that the donors of the tissue samples from which the patent was derived never consented to the commercial exploitation of their gift.[8] While this illustrates the importance of balancing commercial and community benefit, a closer examination of the case also reveals the importance of the individual's input. It can be speculated that this case and the partial victory (in settlement) for the claimants could only have come about because one donor family (Greenberg), could clearly be identified as the instigator of the whole database project and because the claimants benefited from the lobbying power of the particular disease-afflicted community. In a "normal" biobank project, such connections would likely be much weaker: for example, the discovery of the breast cancer gene BRCA1 relied on thousands of samples from affected and unaffected donors.

But a cell line can be traced back to an individual. In fact, its potential monetary value may rely mainly on the uniqueness of the donor's genetic makeup. The Mo cell line that was derived from Mr Moore was deemed commercially valuable precisely because Moore had a unique mutation. Cell lines that have rare features (such as not expressing a particular receptor) have a higher market value than "ordinary" ones.[9] How to manage such irregularities within a project that mainly attempts to chart the normal is a conundrum that biobank designers would be well advised to address, if it is not to come back to haunt them later.

---

8 Greenberg et al. v. Miami Children's Hospital Research Institute et al.; settlement announced on http://www.canavanfoundation.org/news2/09-03_miami.php

9 Compare the pricelist of InPro Biotechnology, Inc., as of March 2003 at: http://www.inpro-biotech.com/images/reagents/pdf_docs/Reagent_Price_List.pdf

Conclusion

The immortalization of selected samples—the factor that turns UK Genebank into UK Biobank—is not adequately addressed by the current regulatory framework and the discussions surrounding it.

I have charted a few points where the scientific, legal, and ethical considerations surrounding the immortalization of cell lines depart both in degree and in nature from the issues that are commonly considered in the context of population genetic databases. It has *not* been the purpose of this overview to argue against the inclusion of cell line proliferation in the protocol of UK Biobank. Nor should it be necessary to initiate a completely new debate surrounding this particular aspect. In the wake of the Moore case and the Alder Hey scandal in Britain, ongoing debates about ownership of human tissues and commercialization of research have fostered a substantial body of legal and bioethical literature that can—and ought to—be drawn upon for this purpose.

The challenge will be to join individual considerations arising with the immortalization of samples and the communal nature of genetic science[10] into a common governance framework that takes account of their conceptual difference.

References

Austin, M. A., S. Harding, and C. McElroy. 2003. Genebanks: A comparison of eight proposed international genetic databases. *Community Genetics* 6:37–45.

Barbour, V. 2003. UK Biobank: A project in search of a protocol? *Lancet* 361:1734–1738.

Beyleveld, D., and E. Histed. 1999. Anonymisation is not exoneration: R v. Department of Health, ex parte Source Informatics Ltd. *Medical Law International* 4:69–80.

Busby, H. 2003. Biobanks and imagined communities. Paper delivered at a meeting of the Postgraduate Forum for Genetics and Society (University of Sussex, 20–22 August).

Caulfield, T. 2003. Gene banks and blanket consent. *Nature Reviews Genetics* 3:577.

Caulfield, T., R. E. G. Upshur, and A. Daar. 2003. DNA Databanks and consent: A suggested policy option involving an authorization model. *BMC Medical Ethics* 4:1.

Department of Health, Medical Research Council (MRC) and The Wellcome Trust. 2003. *UK Biobank ethics and governance framework*. Version 1, 24 September. Available online at: http://www.ukbiobank.ac.uk/documents/egf-background.doc

Greely, H.T. 1999. Breaking the stalemate: a prospective regulatory framework for unforeseen research uses of human tissue samples and health information. *Wake Forest Law Review* 34:737–766.

Parliamentary Office of Science and Technology. 2002. The UK Biobank.

---

10 Critical: Busby 2003.

POSTnote July. Available online at: http://www.parliament.uk/post/pn180.pdf

Rhim, J. S. 2000. Development of human cell lines from multiple organs. *Annals of the New York Academy of Sciences* 919:16–25.

Stacey, G., and C. MacDonald. 2001. Immortalisation of primary cells. *Cell Biology and Toxicology* 17:231–246.

UK Biobank. 2002. The UK Biobank Ethics Consultation Workshop 25 April 2002. Available online at: http://www.ukbiobank.ac.uk/documents/ethics_work.pdf

UK Biobank. 2003. Protocol. 14 February. Available online at: http://www.ukbiobank.ac.uk/documents/draft_protocol.pdf

UK Biobank. Frequently asked questions. http://www.ukbiobank.ac.uk/FAQs.htm

UK Biobank. Website. http://www.ukbiobank.ac.uk

Wallace, H. 2003. A UK biobank: Good for public health? opendemocracy.net, 24 July. http://www.opendemocracy.net

Benefit-sharing, justice, and discrimination

40

Benefit-sharing and public trust in genetic research

by Graeme Laurie and Kathryn G. Hunter

Human genetic research, commercialization, and benefit-sharing

THE RAPID EXPANSION of commercialization in biotechnology science and research over the last two decades has given rise to widespread public concern, not only about the way in which genetic research is conducted, but also about what is done with the products of that research (Merz et al. 2002). The general effect has been a significant drop in public confidence in the research enterprise (Laurie 2002, 309; citing MRC 2000). At the same time, the historical assumption that subjects contribute to and participate in biomedical research solely for altruistic reasons is being challenged (Merz et al. 2002, 965), and there is a growing belief that lucrative commercial gains from research, whether through patents or other IPRs, should not flow only to researchers and investors. Patient and advocacy groups are becoming more active in clinical research and in the promotion of genetic studies (Lindee 2000), retaining, in some instances, ownership of IPRs in order to maintain influence over access to tests and medicines, costing, licensing, treatment research, and down-stream development (see Merz et al. 2002, 966).[1]

The problem, then, is that the traditional model of organizing medical research is breaking down. This model—which we might call the *altruistic gift model*—is but one option in a range of options available to us on how we should proceed in the face of growing public mistrust in researchers and their practices. Alternative models have been proposed. Gerard Porter, for example, offers a *contractual model* (see G. Porter, ch. 9 in this volume) whereby researchers and participants seek balance through negotiations of terms. There is, indeed, growing evidence of the popularity of this approach in the United States. This has arisen in large part precisely because of unacceptable behavior from researchers and notorious court rulings against participants' interests.[2] While there

---

1 Examples include: PXE International (http://www.pxe.org), the Alpha-1 Foundation (http://www.alphaone.org), and Patient Tumorbank of Hope — P.A.T.H. (http://www.stiftung-path.org/). For comment on the latter see Stafford 2003.
2 See *Greenberg et al. v Miami Children's Hospital Research Institute Inc. et al.*, settled action; for comment see Anderlik and Rothstein 2003.

is much to commend this approach, we offer yet another alternative in this paper, which may be termed the *benefit-sharing model*. Its advantage lies in addressing many of the underlying concerns while avoiding what we believe to be unnecessary and unhelpful legalistic models that can all too easily lead to polarized confrontation. Moreover, our experience of current initiatives in our own jurisdiction to establish genetic databases and tissue banks, *viz.* UK Biobank and Generation Scotland, afford the perfect opportunity to explore the practicalities of pursuing the benefit-sharing model in modern genetic research.

This having been said, successful benefit-sharing examples are few, particularly in relation to human genetic resources. Indeed, while benefit-sharing is an established principle of international law in relation to indigenous populations,[3] whether (and how) profits that are generated as a result of human genetic research should be shared with participants has been an on-going debate. Although the idea of benefit-sharing has generally been regarded as attractive, proposed practical models, which have evolved primarily in relation to the developing world to protect traditional knowledge and to prevent bio-piracy, have not been entirely successful, especially where there has been an inadequate consultation process (see, for example, Posey and Dutfield 1996; and Swiderska 2001). In addition, these models do not translate well to the developed world, where difficulties in formulating a rationale for benefit-sharing and in clearly defining "community" (Weijer 2000, 368) have presented significant challenges to the development of workable models. This endeavor has been further impeded by a resistance to benefit-sharing on the part of researchers, institutions, and investors, who adhere to the notion that research subjects do, and should, participate in genetic research for the sake of altruism alone.[4] This, however, fails entirely to recognize the potential interest that research subjects, advocacy groups, and foundations might have in the commercial benefits that may result from such research. It also fails to address the adverse impact that persistence with the one-sided altruistic gift model has on the motivation of these parties to become involved in research at all.

All of this distils to two issues: first, do we accept benefit-sharing *in principle*, and second, if we do, how might it work *in practice*?

Rationale for benefit-sharing

Many justifications have been advanced for benefit-sharing in the context of genetic research. These range from claims based in equity,[5] and on principles of justice and soli-

---

3   See HUGO (2000) *Statement on Benefit-Sharing* and HUGO (2002) *Statement on Human Genomic Databases*; see also the *Convention on Biological Diversity* (UNEP 1992), Articles 8j, 15(7) and 19(2).

4   See, for example, the Medical Research Council's (2001) *Human Tissue and Biological Samples for Use in Research: Operational and Ethical Guidelines*, which provide that samples will be treated as donations. The Human Genetics Commission (2002) similarly endorses the principle of altruism in its report, *Inside Information*.

5   Merz et al. 2002, 970; see also p. 969: "…fairness demands that profits be distributed among those who contributed to the research in an equitable manner."

darity (HUGO 2000), to those based on developing norms in international law (Sheremata 2003). It has also been suggested that a novel moral principle is required, namely, a principle of respect for communities (Weijer 2000, 368; see also Weijer 1999). While we find value in each of these, we would add more. A crucial feature that must be addressed is the lack of trust that the public in general, and research participants in particular, feel toward the research enterprise (Kaye and Martin 2000). Much of this stems from growing commercialization practices but also from feelings of disempowerment. The altruistic gift model treats the research subject as entirely passive. The truth is that those who participate as subjects and who provide vital genetic research material are the key components of the genetic research machine, and are crucial to its continued success. Whether they are represented by individuals or by communities, they are currently under-valued, under-respected, and undermined (Laurie 2002, 309). The way forward is to empower these parties to take a more equal role in the partnership that is formed when they participate in research (Greely 1997). This may come from a contractual model but, we believe, at the cost of confrontation and, we suspect, protracted legal dispute over time. The same may be true of a property model whereby research participants might be said to own their contributions to research (e.g. samples and/or information).

Benefit-sharing allows the normatively-appealing gift model to subsist while redressing the palpable imbalance of power in the researcher-subject relationship. For, within such a benefit-sharing model, the obligation of the researcher to the subject does not end with the transfer of the "gift." Rather, there is a continuing obligation to ensure that financial benefit (profit) to the researcher is shared *in some way* with the participant or her community. This overt commitment to contribute in the interests of those who, previously, have themselves contributed, may go far in restoring trust without the need for recourse to legal hubris.

One obvious retort is that there is sufficient benefit to the community from the fact that research is done at all. This discourse of promise (see W. Marsden, ch. 18 in this volume), however, grows unconvincing in the face of commercialization practices. The reality is that many, if not most, participants expect some more direct benefit from their participation.[6] Moreover, and unlike dedicated disease studies (where there is at least the possibility that the study will produce a direct health benefit, with a consequent incentive to participate), biobanks depend entirely on the goodwill of participants to act for the "greater good." The question, then, is whether abstract benefits—such as the resource itself or developments that are likely only to benefit future generations—are enough? It is suggested that they are not; at least, not when the realization of the promise is accompanied by commercialization which is directed away from, rather than toward, the community.

In the present climate, it is potentially self-defeating for biobank researchers and investors not to consider benefit-sharing arrangements with participants and the larger

---

6 On-going work of social science colleagues at Edinburgh University has uncovered some evidence to this effect. Details will be published in due course on the following sites: http://www.innogen.ac.uk and http://www.law.ed.ac.uk/ahrb/.

community. It is essential to the success of these projects that there are appropriate incentives in place in order to attract and retain participants, inspire public confidence in the venture, allay fears that the project will be exploited by private interests, and provide an acknowledgement that the on-going participation of individuals is both valued and of value on moral, scientific, and economic terms. We submit that a benefit-sharing model can contribute significantly to each of these ends.[7]

While it would be reassuring to find an "unambiguous justification" (Weijer 2000, 368) for benefit-sharing, it is unlikely that the search for a universally accepted principle will be fruitful. For present purposes we suggest that any, some, or all of the above reasons are sufficient to support the acceptance of benefit-sharing in principle. We turn now to the realities of the benefit-sharing model in practice.

Practicalities

The practical issues can be grouped under the following headings:

Who?

It has been suggested that a crucial stumbling block in this field is the need to define the community that will share in the benefits. We submit that in the particular context of genetic research, many of the relevant communities are self-defining. For example, if it is decided that the beneficiaries should be those who contributed to research, then, axiomatically, the research participants are the community. Other possibilities include communities that have mobilized themselves to encourage research on diseases affecting them.[8] Relatedly, the community may be those affected by the diseases that are the subject of the research, and sharing of benefits can be achieved through representative bodies or advocacy groups. Thus, we contend that *community* is not an impossible hurdle in this context.

When?

Benefit-sharing only becomes relevant when, and if, profits are generated. This raises the questions of how much profit, and what proportion, should be shared? There are no hard and fast answers: we support a casuistic approach. The additional task of tracing profits is complex, and requires, *ab initio*, clear intellectual property management agreements.

---

7 Winickoff and Winickoff (2003, 1181) point out the following: "The National Research Council has stated that 'in population studies, benefit to the population has become one of the critical issues in determining the ethical justification for the study itself, and sharing benefits with the population is critical in preventing exploitation' (National Research Council, Committee on Genome Diversity. Evaluating human genetic diversity. Washington, D.C.: National Academy Press, 1997). The obligation holds not only for groups defined on the basis of ethnicity or disease but also for those defined on the basis of geographic region or health care institution. In fact, one reason that subjects donate samples to a biobank is for the greater good."

8 See, for example, PXE International (above), Terry (2003, 166), and the Canavan Foundation: http://www.canavanfoundation.org/.

How?

What constitutes a "benefit" has been the subject of much debate. HUGO, for example, defines *benefit* as:

> a good that contributes to the well-being of an individual and/or a given community… Benefits transcend avoidance of harm… in so far as they promote the welfare of an individual and/or of a community. Thus, a benefit is not identical with profit in the monetary or economic sense (HUGO 2000).

What is striking here is that, while "benefit" may not be identical to "profit" *per se*, a benefit may, of course, be monetary, as contemplated by the *Bonn Guidelines*.[9] But the limits of the benefit-sharing model are set only by our own imaginations. We do not support a crude model whereby participants are merely paid for their trouble. Nor do we endorse undue inducement through compensation (HUGO 1996); and sharing profits from genetic research *directly* with individuals is both impractical and potentially harmful (Kerr 2003, 14). There are, however, a number of different ways in which profits can be shared with the public sector, which acknowledge the participation of individuals in genetic research. We discuss some of these options below.

In determining how, and with whom, profits are to be shared, it is important to bear in mind that different considerations will apply depending on context. Lessons may be learned from the developing world, but it will not be prudent to attempt to develop a single model for universal application. Nonetheless, we believe it is possible to discern key features of an equitable benefit-sharing model that can act as a template for later adoption and refinement. What follows are examples from the United Kingdom.

21st Century Genetic Health (21CGH)

21CGH is a multi-institutional, cross-disciplinary collaboration led by the University of Edinburgh, which addresses Scotland's three health priority areas: cancer, heart and stroke, and mental health. It provides the essential scientific infrastructure and social, ethical, and legal framework on which multi-disciplinary research can build to identify, evaluate, and utilize heritable genetic risk factors in early diagnosis, disease monitoring, treatment optimization, avoidance of adverse drug reactions, health care planning, and drug discovery (Genetic Health in the 21st Century).

21CGH builds on two major research initiatives: Generation Scotland (GS), which is aimed at combating disease through genetic studies in Scottish individuals and their families, and UK Biobank, a long-term project aimed at building a comprehensive database, which will include environmental and lifestyle information as well as medical data and biological samples from 500,000 participants aged 45–69, in order to support a diverse range of research into the causes, courses, and treatments of common severe diseases (UK Biobank 2003). Both projects will require samples of blood from participants and on-going access to medical records.

---

9 Appendix II of the *Bonn Guidelines on Access to Genetic Resources and Fair and Equitable Sharing Arising out of their Utilization* (Secretariat of the Convention on Biological Diversity 2002) enumerates 27 potential monetary and non-monetary benefits.

A unique feature of these projects is the approach to their establishment. Each has instituted robust measures to address the ethical, legal, and social issues ahead of the science. We are fortunate to be involved in these on-going processes, and we draw on our experiences in formulating the arguments in this paper.[10]

UK Biobank

UK Biobank does not contemplate benefit-sharing with participants or with the wider community beyond its stated purpose of generating and disseminating new knowledge (UK Biobank 2003). Any income that UK Biobank secures from access fees or intellectual property will be reinvested in the resource. We submit that the very *public* nature of this project—the funders are the Wellcome Trust, the Medical Research Council and the Department of Health—supports this approach. More radical benefit-sharing measures flow from profits that are diverted away from the public interest and toward private pockets. UK Biobank exists for, and will operate in, the public interest. It should be noted, however, that the luxury of such a public not-for-profit endeavor is rare; most biobanks and databases exist with a strong element of private investment and an expectation of profit. Other benefit-sharing measures are, therefore, required.

Generation Scotland

Unlike UK Biobank, the proposed framework for Generation Scotland (GS) acknowledges that the project's ambitious and important research concerns would not be able to proceed without funding from the private sector (UK Biobank 2003, 33). It is also recognized that there is great potential for GS's research activities to generate intellectual property and commercial advantages. We have recommended a need for that benefit to be shared in order to consolidate the partnership with those data subjects contributing to the research, and with the wider community as a whole (Laurie and Gibson 2003, 17). It is accepted by all involved in the project that its success depends entirely on the voluntary and on-going participation of individuals and their families; it is essential that the Scottish public has trust in the project.

Although individuals who participate in GS cannot expect any financial return for their involvement, nor will there be any direct health benefit accruing from their participation in the project, several benefit-sharing proposals have been identified.

As the project is still at the scientific development stage, these proposals have yet to be explored fully. But the early consideration of social, ethical, and legal issues means that there is every possibility that benefit-sharing might play a key role in the project. The initial focus could be on the various disease groups to be studied, thereby defining the community to benefit, but this may later be expanded to include a wider section of the population, if, for example, it is determined that a certain community or segment of the population in a specific area is "at risk" (of developing cancer, for example).

Our recommended approach is that benefit-sharing should occur with charitable institutions or similar groups that assist families or other disease communities involved

---

10 Laurie is a member of the UK Biobank Interim Advisory Group on Ethics and Governance and is responsible for the law and ethics research stream for Generation Scotland. Hunter provides research support for this stream.

in the project.[11] By these means, participants could themselves derive benefit in terms of support or improved services. We contend that this is the optimal means to acknowledge a participant's involvement. Benefits might be generated by various means, such as royalties, joint rights to intellectual property, one-off payments, on-going contributions, benefits in kind, or a combination of these (Laurie and Gibson 2003, 42). It is anticipated that GS will receive support from a number of different groups, such as the British Heart Foundation, the British Lung Foundation, Cancer Research UK, Genetic Interest Groups, The National Kidney Research Fund, and the Juvenile Diabetes Research Foundation (among others). Representatives from such groups, or participants themselves, could also have a role in deciding how benefits would be allocated, further empowering these communities. We favor further exploration of the trust model as a means of holding resources in stewardship for the community and distributing benefits accordingly (Gottlieb 1998). We have recommended that GS explore benefit-sharing arrangements with such bodies, and with patient groups, as a measure of goodwill, as a means of contributing directly to the research communities that are most valuable to GS, to participants, and as a way to inspire confidence in the project (Laurie and Gibson 2003, 43).

We acknowledge that much work is required on the procedural and legal aspects of any benefit-sharing model, but first steps have at least been taken. We remain at the stage of principle; but the case will be strengthened if it can be demonstrated that the practicalities are not insurmountable.

References

Anderlik, M. R., and M. A. Rothstein. 2003. Canavan decision favors researchers over families. *Journal of Law, Medicine and Ethics* 313:450–453.

Genetic Health in the 21st Century (21CGH). Fact sheet. http://www.erihost.com/gs/21CGH.htm

Gottlieb, K. 1998. Human biological samples and the laws of property: The trust as a model for biological repositories. In *Stored tissue samples: Ethical, legal and public policy implications*, ed. R. F. Weir. Iowa City, IA: University of Iowa Press.

Greely, H. T. 1997. The control of genetic research: Involving the "groups between." *Houston Law Review* 33:1397–1430.

Human Genome Organization (HUGO). 1996. *Statement on the principled conduct of genetics research*. HUGO Ethical, Legal, and Social Issues Committee Report to HUGO Council. Available online at: http://www.hugo-international.org/hugo/conduct.htm

---

11 A similar measure was suggested by Framingham Genomic Medicine Inc. in its proposal to establish a private genomics project using pre-existing local community health data. The company offered to pay 5% of its profits to a charity community development fund; and while the project stalled for other reasons, the precedent nonetheless remains, see Rosenberg 2001.

Human Genome Organization (HUGO). 2000. *Statement on benefit-sharing.* Prepared by the Ethics Committee of the Human Genome Organization. Available online at: http://www.hugo-international.org/hugo/benefit.html

Human Genome Organization (HUGO). 2002. *Statement on human genomic databases.* Available at: http://www.hugo-international.org/hugo/ethics.html

Human Genetics Commission (HGC). 2002. *Inside information.* London: HGC. Available online at http://www.hgc.gov.uk/insideinformation/

Kaye, J., and P. Martin. 2000. Safeguards for research using large scale DNA collection. *British Medical Journal* 321:1146–1148.

Kerr, A. 2003. The social and ethical aspects of medical genetic databanks. Report Prepared for the Generation Scotland Working Group.

Laurie, G. T. 2002. *Genetic privacy: A challenge to medico-legal norms.* Cambridge: Cambridge University Press.

Laurie, G. T., and J. Gibson. 2003. *Generation Scotland: Legal and ethical aspects.* Available online at http://www.law.ed.ac.uk/ahrb/publications/online/GSlawandethics.doc

Lindee, M. S. 2000. Genetic disease since 1945. *Nature Reviews Genetics* 1:236–241.

Medical Research Council (MRC). 2000. *Public perceptions of the collection of human biological samples.* London: MRC.

Medical Research Council (MRC). 2001. *Human tissue and biological samples for use in research: Operational and ethical guidelines.* London: MRC.

Merz, J. F., D. Magnus, M. K. Cho, and A. L. Caplan. 2002. Protecting subjects' interests in genetic research. *American Journal of Human Genetics* 70:965–971.

MRC see Medical Research Council.

Posey, D. A., and G. Dutfield. 1996. *Beyond intellectual property.* Ottawa: International Development Research Centre.

Rosenberg, R. 2001. Questions still linger on heart study access. *The Boston Globe*, 21 February, D04.

Secretariat of the Convention on Biological Diversity. 2002. Bonn guidelines on access to genetic resources and fair and equitable sharing of the benefits arising out of their utilization. Montreal: Secretariat of the Convention on Biological Diversity. Available online at http://www.biodiv.org/doc/publications/cbd-bonn-gdls-en.pdf

Sheremata, L. 2003. Population genetic studies: Is there an emerging legal obligation to share benefits? Paper presented at a HUGO Human Genome Meeting in Cancún, Mexico, on April 27–30. Abstract available at: http://hgm2003.hgu.mrc.ac.uk/Abstracts/Publish/WorkshopPosters/WorkshopPoster11/

Stafford, N. 2003. Patient-owned tumorbank opens. *The Scientist*, 20 November. Available at: http://www.biomedcentral.com/news/20031120/04.

Swiderska, K. 2001. Participation in policy-making for access and benefit-sharing: case studies and recommendations. *IIED Bio-Brief* 2, October.

Terry, S. F. 2003. Learning genetics. *Health Affairs* 22:166–171.

UK Biobank. 2003. *Setting standards: The UK Biobank ethics and governance framework.* September 24. Available at http://www.ukbiobank.ac.uk/ethics.htm

United Nations Environment Programme (UNEP). 1992. *Convention on biological diversity.* Available online at http://www.biodiv.org

Weijer, C. 1999. Protecting communities in research: Philosophical and pragmatic challenges. *Cambridge Quarterly Healthcare Ethics* 8:501–513.

Weijer, C. 2000. Benefit-sharing and other protections for communities in genetic research. *Clinical Genetics* 58:367–368.

Winickoff, D. E., and R. N. Winickoff. 2003. The Charitable trust as a model for genomic biobanks. *New England Journal of Medicine* 349:1180–1184.

41

Biobanks and the "social" in social justice

by Sarah Wilson

THE POTENTIAL IMPORTANCE of the relationship between biobanks and social justice is encapsulated within the term "population biobanks," which has become commonplace in discussions of biobank databases. It is the social aspects of life, rather than each individual life, which are key to the success of projects in this field, not only in terms of numbers of participants, but also because of the relevance of social influences such as family, environmental, and lifestyle factors. What then is the relationship between justice as pertaining to the individual and to the social, as these matters relate to biobanks? In particular, is there something unique about population biobanks when it comes to questions of social justice? This paper will consider these questions, and explore them in relation to key themes of social justice, paying particular attention to the tensions that exist between concepts of the individual and the common good, and of "public" and "private" interests. Throughout the paper I will also consider how theories and principles of social justice relate to these discussions, and whether or not they offer support for the development of biobank projects.

Questions of social justice often focus around the distribution of the benefits and burdens of cooperative living, the fair and equitable treatment of members of a society, and the various rights and responsibilities of individuals and states within a social system. Key questions in theories of social justice relate to what is seen as an appropriate balance between the rights and responsibilities of the state and of the individual, as well as the balance between the interests of the individual and the interests of the society. These issues and, indeed, what counts as interests or rights, are a central factor differentiating between theories of social justice. The following discussion considers how these issues arise in relation to population biobanks.

Although I have just identified the concept of balancing conflicting interests as a central element in matters of social justice, it is obviously true that these multiple interests are not always in conflict. In some instances the interests of the individual, of the public, and of the state itself, coincide. This may often be the case when medical advances improve the health of individuals, as this will have general public health benefits and consequently be beneficial to the state. Such reasoning may support the development of biobank projects, as these projects are often promoted on the promise of advances in health care. Therefore, the potential improvements in health care that may come about through the biobank research projects are likely to provide individual health benefits as well as public health benefits, both of which are obviously advantageous to the state. It might also be argued that the potential economic advantage associated with the commercial possibilities of biobanks provide both public and individual benefits as well as benefits for commercial enterprises. These claims to advantage are particularly relevant to a Utilitarian theory of social justice, whereby biobank projects would be justified provided the benefits are calculated as being to the greatest advantage of society. In the previous summary, biobanks are calculated as having the potential to benefit individuals, the public, and the state. However, the same is not necessarily true in the reverse. That is, the possible burdens associated with participation in, or the results of, biobank projects may be more detrimental to specific individuals' interests than to the public interest *per se*.[1]

There are several ways in which biobank projects might have results that are disadvantageous to individual interests, and might therefore be seen as burdens that would need to be assessed against the possible benefits of the projects. These might include the potential "burden" of genetic knowledge on the individual, the possibility of discrimination on genetic grounds (and the resulting implications for fair and equitable treatment), and issues relating to concerns about surveillance and the potential loss of individual privacy. Genetic knowledge may reach the individual either directly, as promised in the Estonian Gene Bank project, or indirectly, for example as a result of it becoming common knowledge that, hypothetically, those with genes for blue eyes also have a genetic propensity for heart disease—or some such unsought information. The question of such unsought information, and of the loss of the right "not to know," has been much discussed in relation to genetics generally, and is particularly pertinent to biobank projects in that they deal with population genetics (see for example Chadwick, Levitt, and Shickle 1997). This aspect of biobanks is also one of the grounds for concern about genetic discrimination, particularly in relation to insurance and employment issues, and this area touches on several matters of social justice. Not only does the issue of discrimination raise questions relating to equitable treatment for individuals (and/or groups) within society, it also brings out the ways in which the different economic or institutional groups within society make claims to have their interests taken into account. This is one sense in which public interests, that is, interests of the public, may clash with private, that is, commercial, interests. A further sense in which public and private interests may be in tension is reflected in more generalized concerns about surveillance and loss of privacy. In this context, the private represents that aspect of people's lives that they view as entitled to protection from state or public interference, and in biobank projects there is a tension between this "private" area of life and the access by many public bodies to personal information. The public and the private are given different emphasis within different theories of social justice. For example, it has been suggested that under a Libertarian rights-based model, individuals should be entitled to the fullest genetic information about themselves it is possible to have. Furthermore, this argument uses Locke's theory of property to justify private ownership of genetic infor-

---

1 At least in the first case—it is arguable whether something that is against the interests of individual members of society may not also inevitably be against the public interest.

mation (Moore 2000). It can be seen how this form of argument might be used to justify the development of commercial genetic biobank projects, and the provision to individuals of their genetic information.

These different understandings of the public and the private are also represented in questions relating to the ownership of biobanks, and to the appropriate relationship between participants and projects. One aspect of this is the claim that biobanks are, or should be treated as, global public goods, and the ensuing claim that ownership should be largely in the public realm. These claims are based largely on arguments relating to the nature of the contents of biobanks: for example that the human genome is part of the common heritage of humanity (HUGO 2002), and/or that biobanks are primarily a tool of information, and that information is a public good (Thorsteinsdóttir et al. 2003). A further claim to public ownership, or at least public interests, lies in the argument that the information contained within biobanks comes largely from publicly funded bodies.[2] This can be seen as in part a matter of compensatory justice, and one response to this has been an emphasis on the importance of benefit sharing, as outlined for example in the HUGO Statement on Benefit Sharing, April 2000. This question of compensatory justice similarly underlies discussion of what might be owed to the participants in biobank projects. Whilst it might not be appropriate to talk of the "burden" of participating in the research, there is an issue relating to the notion of compensation for participation—or at least of non-exploitation. That is, whilst samples may be freely given to the research project, there is some unease about these then being exploited for financial benefit, which accrues only to private companies. Again, benefit sharing has been put forward as one response to this. In general, benefit sharing relates to benefits to *the* public (perhaps donations to health care projects) or to *a* public (in the form of community-specific treatments perhaps), rather than to individual financial benefits. This notion of compensation, or of a reciprocal arrangement between public and private commercial interests, and between individuals and corporate bodies, can be seen as exemplary of one of the key aspects of social justice, as some concept of reciprocity is an essential component of social cooperation.

The preceding instances have shown both where the interests of the individual and the public may coincide, and where the potential disadvantages to individuals might need to be balanced against the social or community good. Some reference has also been made to the competing interest claims of economic or institutional groups, as well as the need to address the compensatory or reciprocal nature of the relationship between these groups. However, the tension between individual interests and an anonymized public or institutional interest is not the only site of conflict. A point commonly made about genetic information is that this information has implications not only for the individual but also potentially for the family and/or known community of which the individual is a member, and that the interests of these are not necessarily the same. The discussion above has already raised the question of biobanks being the source of unsought information, and this may be particularly true in the case of family and community ge-

---

2 For example, the Icelandic Health Sector Database was to make use of pre-existing medical records and genealogical data.

netic links. This adds a further layer of complexity to the issues surrounding the "right not to know," and to concerns relating to genetic discrimination. In the reverse, a family or community may benefit from one individual's participation in a biobank project. As already suggested, the difficulty lies in balancing these various interests.

On the other side of the equation, there are also circumstances in which the potential advantages to individual members of society, or even individual societies, may correspond to a disutility for wider society. For example, the development of health care based on technologically advanced methods brings with it the potential for increasing disparities in access to health care, both nationally and internationally, raising questions of international justice as well as national justice. Egalitarian theories of social justice are pertinent here, particularly those that place emphasis on the meeting of needs. Where a needs-based approach is taken, it may well be argued that resources be concentrated on primary care rather than on expensive, unproven technologies.[3]

The above outline shows that population biobanks raise many issues pertinent to social justice, in particular as they represent tensions between individual and social interests. Some of these issues might be dealt with, though not resolved, by appropriate legislation and regulation. For example, banning the use of genetic information in insurance (as Estonia has done), whilst it may not satisfy the commercial interests of the insurance companies, both respects the privacy of individuals, and the ideal of a non-discriminatory society.

The interesting question then is not so much whether there is something unique in the case of biobanks, but lies in understanding how issues arising from biobanks illuminate matters of social justice, and *vice versa*. Whilst it is true that biobanks as collections of biological samples are not in themselves new, it might be argued that the combination of information and samples held within population biobanks *is* new. Such a claim forms the basis for the argument that biobanks are global public goods, thus linking biobanks with issues of public and private interests and ownership as discussed above. A further claim is that the unique factor in the case of biobanks is that "it is not solely the individual but also the community that needs protection and should be asked for consent" (Sutrop 2004, 7), a claim which brings into focus the complex relationship between individuals, communities, society, and the state.

It may be that the discussion of uniqueness is in fact irrelevant, as what is important is the opportunity presented for rethinking questions of social justice. The above paper has shown how exploring matters of social justice through a focus on biobanks illustrates recurring themes such as the boundaries between the public and the private, and the appropriate balance of rights and responsibilities for individuals, society, and the

---

3 The following statements from the recent WHO report on Genomics and world health illustrate this particularly clearly: "Advances in genomics for global health care must be assessed for their relative value in the practice and delivery of health care compared with the costs and efficacy of current approaches to public health, disease control and the provision of basic preventive medicine and medical care. Conventional, tried and effective approaches to medical research and medical practice must not be neglected while the medical potential of genomics is being explored" (WHO 2004).

state. This interconnection between the individual, the public, and the social echoes the concerns of both social scientists who emphasize the importance of public participation (see for example Weldon 2004), and political philosophers who promote discourse ethics and concepts of dialogic democracy (see for example Benhabib 1992; also Hekman 1995). Rather than attempt to resolve these concerns by fitting them into existing frameworks, an alternative approach would be to open the debate about genetic technologies, and science itself, to the wider public, to break down the barriers dividing "science" from "ethics," the "professionals" from "the public," and to establish acceptable research priorities, social priorities, and overlapping visions of a fair and just society. In this way, the biobank discussion could provide a unique starting point in bringing together the ideals, the theories, and the practice of social justice.

Acknowledgements

This paper was produced as a part of the ELSAGEN project (Ethical, Legal and Social Aspects of Human Genetic Databases: A European Comparison), financed between 2002 and 2004 by the European Commission's 5th Framework Programme, Quality of Life (contract number QLG6-CT-2001-00062). I gratefully acknowledge the support of the European Community. The information provided is the sole responsibility of the author; the Community is not responsible for any use that might be made of data appearing in this publication.

References

Benhabib, S., 1992. *Situating the self: Gender, community, and postmodernism in contemporary ethics*. Cambridge: Polity Press.
Chadwick, R., M. Levitt, and D. Shickle, eds. 1997. *The right to know and the right not to know*. Aldershot: Avebury.
Hekman, S. 1995. *Moral voices, moral selves. Carol Gilligan and feminist moral theory*. Cambridge: Polity Press.
Human Genome Organization (HUGO). 2002. *Statement on human genomic databases*. Prepared by the Ethics Committee of the Human Genome Organization. Available online at: http://www.gene.ucl.ac.uk/hugo/HEC_Dec02.html
Moore, A. D. 2000. Owning genetic information and gene enhancement techniques: Why privacy and property rights may undermine social control of the human genome. *Bioethics* 14:97–119.
Sutrop, M. 2004. Preface. *Trames* 8:5–12.
Thorsteinsdóttir, H., A. S. Daar, R. D. Smith, and P. A. Singer. 2003. Genomics — a global public good? *The Lancet* 361:891–892.
Weldon, S. 2004. "Public Consent" or "Scientific Citizenship"? What counts as public participation in population based DNA collections? In *Genetic databases:*

Socio-ethical issues in the collection and use of DNA, ed. R. Tutton and O. Corrigan. London: Routledge.

World Health Organization (WHO). 2004. *Genomics and world health: Report of the Advisory Committee on Health Research*. Geneva: WHO.

42

Policy for human genetic resources as compared to environmental genetic resources

by Karin Erika Bengtsson

RELATIVE TO EARLIER practice, commercialization of genetic testing has dramatically changed the context of testing in the United States and other countries, taking it out of the clinic and removing the intermediary roles of clinical geneticists and genetic counselors. New technologies are introduced to the market and pressure for their use is increasingly fuelled by well-organized consumer groups, such as the various hereditary cancer groups. Financial agreements are being struck between guards of hospital-based human biological samples collections, and pharmaceutical companies and other funders of genetic research. For example, a consortium of public laboratories that do the banking and private organizations that do the paying is being developed (Kent and Tambuyzer 1999).

The research project "Eurogenbank: Banking of genetic material and data in Europe — legal, ethical and economical issues," funded by the European Union, was concerned with genetic databanking of all sorts of genomic information. There is a clear rationale for this: increasingly, genebanks run on a commercial basis hold genetic data from all sorts of organisms. Among their users are researchers and life science companies. In recent years we have seen major restructuring of transnational companies in this area taking place: one example is the merging of Zeneca, which used to be a plant sciences company, with Astra, formerly a pharmaceutical company.

Access to human genetic resources

The Convention on Biological Diversity has in some circles been described as the facilitator of the commercial development of biotechnology (UNEP 1992). It prescribes that countries shall make genetic resources available, and for their proper conservation establish mechanisms for benefit-sharing with trustees of such resources. At a meeting of the Conference of the Parties (COP) to the Convention in 1995 the item "Access to genetic resources" was on the agenda. Under this item, a section on human genetic resources was included. The secretariat of the Convention had prepared a fairly neutral report, pointing in the direction of the need for further research (UNEP 1995a, 18).

At this COP meeting there was the most heated debate on access to human genetic resources of the whole two-week conference, with about 60 or so countries wanting to make interventions. Eventually it was concluded that human genetic resources have a special status, as expressed in the wording: "[The Conference of the Parties] reaffirms that human genetic resources are *not* included within the framework of the Convention" (UNEP 1995b; emphasis in original).

So these countries did see political motives for keeping human genetic resources separate from other kinds of genetic resources. The arguments put forward by the government delegates acknowledged reasons of human rights and freedoms, including the troubling prospect of nation states eventually exercising power over the use of DNA samples of their citizens.[1] On the other hand, this particular kind of genetic resource could be shared in the form of samples among collaborators, including international collaborators and commercial entities. It was possible for the governments to stop this from happening. Among the particularities that made government delegates arrive at the conclusion that human genetic resources were not fitting subject matter for the CBD was the moral and religious meanings different cultures attached to genetic material, especially when human beings were involved. The European Union expressed the view that access to human genetic resources involved ethics, and that there were other relevant UN bodies for issues of human rights, primarily the UNESCO International Bioethics Committee. Delegations made statements like this: "Genetic material of human origin should not be seen as a natural resource to be exploited by commercial interests."

Commercialization and benefit-sharing

Obviously these matters have to be debated and supervised continually. Knoppers (1999, 7), in a recent publication, poses the question: "Is the biodiversity model applicable to human genetic material used in biotechnology or are there different models?" Arguably, interests in favor of commercialization of genetic information in general will be more likely to support a harmonization of rules for genetic databases with no significant differences accredited to the sources of DNA material.

Following the controversy surrounding DNA sampling, the issues of payment in return for human samples and of equitable sharing of benefits with a community which has provided DNA samples for population research were addressed by the Human Genome Organization Ethics Committee. In a statement it recommended that "inducement through compensation for individual participants, families, and populations should be prohibited. This prohibition does not include agreements with individuals, families, groups, communities or populations that foresee technology transfer, local training, joint ventures, provision of health care or of information, infrastructures, reimbursement of costs, or the possible use of a percentage of any royalties for humanitarian purposes" (HUGO 2000). To my judgment these formulations are aggravating. As the quotation shows, there are reasons for paying attention to the influence commercial interests have on the discussion of benefit-sharing. Phrases like "technology transfer" and

---

1 Notes taken during participant observation by author.

"joint ventures" have clear connections with the commercial world. They suggest commercial activity to a rather worrying extent.

Countries like China and India have put legislation in place to restrict the export and off-shore commercialization of human genetic material in the absence of benefit-sharing (GenEthics News 1997). Can we trust them, and the commercial entities they contract with, to be wise custodians of the (human) rights of their citizens?

Exploitation of genes and genetic information constitutes a process whereby human and nonhuman sources of genes are commodified as mere means to commercial and scientific ends (Hanson 1999). Arguably, the development of genetic databases makes it more difficult to fix the borderline between what are to count as practices in the human field as compared to commercialization of environmental genetic resources.

Environmental genetic resources

The Convention on Biological Diversity (CBD) establishes a new international framework for access to genetic resources and the sharing of benefits from their use. It redefines the principles governing access and benefit-sharing, starting from the principle that parties have sovereign rights over their genetic resources, rather than from the "common heritage" principle. The provisions on genetic resources of the CBD represent the international community's effort to define principles for the use of genetic resources from all sources. This broad scope is consistent with recent developments in technology, which are demonstrating that a growing range of biological materials containing genetic resources have significant value for applications such as pharmaceuticals and biotechnological processes.

Before the CBD came into effect in 1993, environmental genetic resources were freely available. Most of the world would probably have preferred it to stay like that, but the question of access to genetic resources is intimately linked to that of intellectual property rights. The biodiversity-rich countries are mainly Southern, while biological resources have been brought to the North to supply the growing life sciences industry. This has contributed to strained relations between the Northern countries, and advocates of farmers and indigenous peoples.

The Convention recognizes sovereign rights of states over (environmental) genetic resources. Access to genetic resources must be on "mutually agreed terms." For instance, states may wish to ensure that standards take into account the possibility that scientific researchers might later make genetic resources available to third parties for commercial development. States are able to create rules and regulations that dictate how access to genetic resources may be handled. Co-operation agreements may direct specific means for sharing of benefits and/or profits. Relevant types of innovation related to biodiversity with corresponding intellectual property rights systems include: genetic resources as resources for crop breeding and for agricultural and other biotechnologies; informal knowledge as a source for development of products in pharmaceutical, industrial, food and other sectors; biochemicals as sources of pharmaceuticals and industrial applications; and informal knowledge as a source of information for conservation, sustainable management and other applications.

An ideological assumption behind the CBD is that in a market economy, one way to promote sustainable use is to offer sustainably produced products to consumers who prefer to support conservation by buying such goods. So we can say that the biodiversity framework is thoroughly grounded in an ecological order.

Human genetic resources

Let me make the point that not all genetic information of human origin is medical information, though it is certainly biological information. For human genetic banking and databases with human genetic information there are specific considerations that we do not find in the environmental field. These centre around privacy and confidentiality, databases with population-wide coverage, and ethical and social implications of the commercialization of genetic information (Bengtsson 2000).

The problems of privacy and confidentiality, important as they are in themselves, are the surface of a much deeper problem about the social and cultural context of molecular medicine. This problem arises from the human inclination to classify persons into groups and populations with names and numbers. Human genetic research will divide and redistribute populations based on their genetic characteristics. Patient-centered databases may be valuable for local clinicians and helpful in genetic counseling and molecular diagnosis of the disorder for the family members. The conditions for the ethical use of such databases are important to formulate. Our efforts could be aimed at the strengthening and reinforcement of medical confidentiality within the deontological safeguards of the physician-patient relationship.

Databases established on the basis of national or ethnic origin raise questions of power and control in surveillance activities. The existence of a national database may be helpful for the identification of mutations, and encourage regional or local scientists and medical doctors not directly involved in human variation or mutation detection to use this information. A particular genetic locus may have a differential pattern of variation in certain geographic or ethnic groups, due to exposure to different environmental factors or genotypic differences. Therefore, mutation data for that particular locus may have been generated for certain ethnic groups. Genetic characteristics common to an ethnic group may have social implications in terms of stigmatizing population groups or giving rise to adverse behavior in terms of self-confidence.

There are ethical and social implications of the commercialization of genetic information from human genetic resources. A potentially worrying feature of genetic information is the way in which it is perceived as intricately bound up with our identities as persons. This is perhaps most obviously illustrated by databases built up for forensic purposes and used in civil and criminal proceedings. Fears that genetic information might be used to define specific groups of peoples for political or subversive purposes have been raised (Chee Heng et al. 1995). Repressive governmental regimes can enter into agreement with commercial human genetics companies just as well as liberal democracies.

Could it be that intellectual property protection of human genetic data will be an obstacle to disclosure of information, thus challenging ethical discussion? For genetic testing procedures to be validated in a transparent manner there is a need for human genetic information to be publicly available, for example in the form of databases. It would seem that the most important ethical concern with regard to databases with human genetic information open to third persons is to ensure that the information will not be traceable back to particular individuals. I now proceed to place human vulnerability in this field.

Human vulnerability in light of exploitation

Appreciations of the ethical complexities involved in genetic data banking in the human field are difficult to make and we have only just begun revealing layers of ways in which genetic data can have implications for identity and self-determination. "From a holistic point of view, it is, perhaps, precisely the fragmentation of Moore's self, the severing of a constitutive part of his identity, that amount to an affront to dignity." (Kahn 2000, 924). To be asked by a health care professional if one wants to have one's genetic sample stored for unlimited time puts us all in an existential difficulty. To have one's genetic data stored, arguably, is a drive further in this direction of being vulnerable as a human being. The notion of "genetic information" carries with it a strong connotation of being exact and final (Bengtsson 1997). The human condition perhaps makes us all vulnerable to the implications of giving and receiving genetic knowledge about ourselves and our communities. The individual needs to be reassured that social benefits are not a sufficient ground to undermine respect for self-determination by not requiring informed consent for human genetic research (Dolgin 2001). Arguably, social benefits are incommensurable with respect for the integrity of the human being, and cannot simply be weighed against it.

However, human capacities for agency are limited (O'Neill 2001). We can also be exploited when we decide for ourselves. Human dignity and respect for individual freedom is understood in the light of the Kantian notion of self-determination. Human dignity is the most fundamental and absolute principle of human rights (LeBris and Knoppers 1997). Considering the individual as the owner of his or her genetic information could put the concept of human dignity into serious difficulties. The temptations to consent to access in return for financial contributions would be present, should such ownership be absolute. (A concept of partial ownership is preferable.) The prospect of individuals entering into contractual relationships with private entities, in which it is stipulated that genetic data will be stored for it to be accessed by, for example, physicians, allied health professionals, insurers (eventually), and pharmacists, and further genetic testing evaluations carried out, points to a form of surveillance that could become a reality in the not too distant future. The individual might have a presymptomatic test for mutations in the BRCA1 and/or BRCA2 genes involved in familial breast and ovarian cancer. If she were to be put on a regional surveillance program, she would be vulnerable to becoming a consumer (potentially) getting to know ever more genetic information about herself. Who will in fact decide for her as the years pass by? Can she remain an agent, following on from her initial status as an autonomous decision-maker receiving genetic information during a genetic counseling session?

Another example: imagine a group of people who have made their DNA samples and genetic data available to a consortium for genetic research, facilitated by their membership in a national Muscular Dystrophy Association. This group will become a target for the marketing of measures for disease prevention. They will also be vulnerable, both as individuals and as a group. The national circumstances and the health care management system in place will have influence over them, as will the private entities involved in banking and exploiting genetic data about their condition.

The strongest opportunity for exploitation of human genetic resources, arguably, is where financial gain of the individual concerned would be involved. This would most probably be for very rare diseases where examination of samples have led to important new developments (Berg 2001). Individuals with such conditions could become involved in financial agreements.

Benno Müller-Hill discusses how geneticists who have no connection to religion, philosophy, or the enlightenment, have to define their own value system. The biological differences between humans that these geneticists study indicate that equal rights are "impractical and misguided" (Müller-Hill 2000, 581). Müller-Hill warns that this is to think like the Nazis, and that we must guard against deterministic definitions of the human being. One might wonder if human genetic databases will not be test-beds that will come to influence the way in which we view humans.

In examining whether there is a lesser privacy interest in pharmacogenetics, Chadwick (2004, 189) points to the situation in which there is a choice between a drug and *no* medication, giving rise to "the ethical problem of (perceived) abandonment." She discusses the need to regulate equitable and appropriate care for all in an age of molecular genetics, especially for those that are genetically predisposed to disease.

A substantive ethical issue is that of discrimination in the health care setting on the basis of genetic information. Issues centre around our humanity, rationing possibly being facilitated by application of genetic information about the human, and the ethical dilemmas health care personnel will face when handling and acquiring genetic information about individuals (Bengtsson 1999, 36ff.). Indeed, it will probably be difficult to find public funding to support the integration of genetic testing, drugs, etc.—unless those services are marketed as cost saving mechanisms.

In this essay I have tried to show that the ethical issues involved in human genetic databases are very different from the ones involved in the environmental resources field. As Habermas (2003, 31) writes: "Something may, for good moral reasons, be not for us to dispose over and still not be 'inviolable' in the sense of the unrestricted or absolute validity of fundamental rights (which is constitutive for 'human dignity' as defined in Article 1 of the Basic Law)," i.e. the German constitution.

Basic values are independent, autonomous. They are the ones that allow a human being to live authentically and with dignity. The philosophical rationale for banking human genetic data may well rest on quality of life considerations; the rejoinder may well remain an insistent concern for the sanctity of life and a warning against dehumanization in the face of inexorable technology.

To conclude, I emphasize that there are issues unique to human genetics and the banking of genetic material of human origin that warrant careful consideration. Particularly the notion of capitalizing on one's own, or a group's genomic characteristics, is ethically sensitive to a magnitude that justifies the argument that benefit-sharing models as they are taken from the biodiversity arena, are unsuitable for human genetic material. Legal frameworks will have to change to accommodate this.

Acknowledgement

The author was supported by a grant from the Swedish Ethics in Health Care Program.

References

Bengtsson, K. E. 1997. Molecular genetics information put into medical practice. Paper accepted for the conference *Health Care Ethics: Nuremberg 50 Years On*, University of Freiburg, Germany, October.

Bengtsson, K. E. 1999. *Genetic information and counselling in the health care setting: A question of autonomy?* MA thesis, University of Central Lancashire, Preston, U.K.

Bengtsson, K. E. 2000. Principal areas of ethical concern with regard to human genetic banking. Paper presented at the seminar *Human Genetic Banking*, University of Central Lancashire, U.K., 10th February.

Berg, K. 2001. The ethics of benefit sharing. *Clinical Genetics* 59:240–243.

Chadwick, R. 2004. Pharmacogenetics, genetic screening, and health care. In *Genetics and ethics: An interdisciplinary study*, ed. G. Magill. Saint Louis, MO: Saint Louis University Press.

Chee Heng, L., L. El-Hamamsy, J. Fleming, N. Fujiki, G. Keyeux, B. M. Knoppers, and D. Macer. 1995. Bioethics and human population genetics research. *Proceedings of the UNESCO International Bioethics Committee*, Third Session, Volume I: 39–63. Paris: UNESCO. Available online at: http://www.biol.tsukuba.ac.jp/~macer/PG.html

Dolgin, J. L. 2001. Ideologies of discrimination: Personhood and the "genetic group." *Studies in History and Philosophy of Biology and Biomedical Sciences* 32 (4): 705–721.

GenEthics News. 1997. China and India move to control gene export. October/November.

Habermas, J. 2003. *The future of human nature*. Cambridge, U.K.: Polity Press.

Hanson, M. J. 1999. Biotechnology and commodification within health care. *Journal of Medicine and Philosophy* 24:267–287.

Human Genome Organization (HUGO). 2000. *Statement on benefit-sharing*. Prepared by the Ethics Committee of the Human Genome Organization. Available online at: http://www.hugo-international.org/hugo/benefit.html.

Kahn, J. 2000. Biotechnology and the legal constitution of the self: Managing identity in science, the market, and society. *Hastings Law Journal* 51:909–952.

Kent, A., and E. Tambuyzer. 1999 Partnerships between patients, medicine, and the biotechnology industry. *Nature Biotechnology* 17:935.

Knoppers, B. M. 1999. Biotechnology: sovereignty and sharing. In *The commercialization of genetic research: Ethical, legal and policy issues*, ed. T. A. Caulfield and B. Williams-Jones. New York: Kluwer Academic/Plenum Publishers.

LeBris, S., and B. M. Knoppers. 1997. International and comparative concepts of privacy. In *Genetic secrets*, ed. M. Rothstein. New Haven: Yale University Press.

Müller-Hill, B. 2000. Truth, justice, and genetics. *Perspectives in Biology and Medicine* 43 (4): 577–583.

O'Neill, O. 2001. Informed consent and genetic information. *Studies in History and Philosophy of Biology and Biomedical Sciences* 32 (4): 689–704.

United Nations Environment Programme (UNEP). 1992. *Convention on biological diversity*. Available online at: http://www.biodiv.org

United Nations Environment Programme (UNEP). 1995a. *Access to genetic resources and benefit-sharing: Legislation, administrative and policy information.* UNEP/CBD/COP/2/13. Report prepared by the Secretariat of the Convention on Biological Diversity for the second meeting of the Conference of the Parties to the Convention. Available online at: http://www.biodiv.org/convention/cops.asp

United Nations Environment Programme (UNEP). 1995b. Decision II/11: Access to genetic resources. Decision of the second meeting of the Conference of Parties to the Convention on Biological Diversity, Jakarta, 6–17 November. Available online at: http://www.biodiv.org/decisions/default.aspx?m=cop-02

Problems with targeting law reform at genetic discrimination

by Mark J. Taylor

IN THIS PAPER I wish to consider some of the problems associated with targeting law reform specifically and exclusively at discriminatory uses of genetic information: genetic discrimination.

I will suggest that a possible problem with targeting genetic discrimination *per se* is that, for legislation to be workable, it would have to draw a circle around an exemplar of "bad behavior" selected *not* because it represented the paradigm case of *unfair* genetic discrimination, nor even of the unfair use of *genetic* information, but rather because it utilized a particular "type" of (genetic) information.

While I would not argue against the regulation of *unfair*[1] genetic discrimination, and indeed would strongly argue in favor of its proscription, I question whether legislation targeted specifically and exclusively at a particular kind of (genetic) information represents the most appropriate way of effecting such reform.[2] Indeed, I suggest that such an approach may commit regulators to a choice between regulatory frameworks ranged along a spectrum from the "ineffectively narrow" in scope to the "impractically broad."

Justification for this view may, I believe, be found through sustained consideration of the kinds of "information" that any law targeting "genetic discrimination" might attempt to regulate. I will suggest that, broadly speaking, the term may be understood to apply to one of three kinds of discriminatory conduct, and I refer to these as "primary," "secondary," and "tertiary" genetic discrimination.

Three "kinds" of genetic discrimination

I do not seek here to adjudicate between the various definitions of genetic discrimination that have already been offered, nor to add to them, but rather to show how three

---

1 To describe something as "unfair" presupposes a particular perspective within normative moral theory. The alternatives are many and I do not seek here to try and defend a choice between them. While selection of relevant perspective is clearly of great importance, my concern here is rather to demonstrate how inappropriate it may be to locate regulation around a particular definition of "genetic information."
2 I am not alone in expressing doubt: see for example Lemmens 2000.

"kinds" of conduct may be identified that any "conventional" definition of genetic discrimination must relate to in some way.[3] Recognizing that significant problems may be associated with law reform either ignoring *or* acknowledging (and policing) the boundaries between these "kinds" of discrimination helps illustrate some of the problems with targeting *genetic* discrimination as such.

Primary

Primary genetic discrimination may be described as discrimination deliberately practiced on the grounds of a perceived genetic characteristic. It is the deliberate expression of preference for known genetic variation.

If an insurance company deliberately seeks to raise premiums for those who have tested positively for the possession of BRCA1 or BRCA2, then it commits *primary* genetic discrimination. It has identified a genetic variation, ascribed some value to possession of that variation, and then used information about an individual's possession of that variation to inform its decision making process.

Before expressing a preference for a perceived genetic variation in this way, it is obvious that the discriminator must have acquired genetic information about the individual concerned: they must have been *informed* of something about an individual's genetic profile. They must then have *used* this information when making a decision (about the individual).

Secondary

Secondary genetic discrimination may be described as discrimination practiced on the grounds of a property/characteristic that is *associated* with genetic variation. It is not the presence or absence of genetic variation *per se* that informs the decision, but rather the presence or absence of this other variable.

Some of the clearest examples of secondary genetic discrimination may occur in circumstances where genetic information *could* be gathered through a "genetic interpretation" of data that might not ordinarily be recognized as yielding it. An individual seeking to discriminate on the grounds of biological sex is one example. Another would be discrimination practiced on the ground that an individual possessed a "genetic" disorder diagnosed through non-genetic test (e.g. color blindness). Although the discriminator may not be actively seeking to differentiate *on the grounds of the genetic variation as such* the characteristic selected for (or against) relates to genetic variation and genetic information *could* be inferred from awareness of the characteristic's genetic pedigree.

There may be a qualitative difference in the *kind* of genetic information associated with secondary genetic discrimination when compared with primary, but genetic information is nonetheless available to a discriminator possessing an appropriately informed interpretive framework.[4]

---

3 For the purposes of establishing a boundary around the "possible alternatives," I have assumed only that "to discriminate" is "to distinguish between and treat differently." This allows for the possibility of the "fair" or "unfair," "justified" or "unjustified," use of genetic information when differentiating between individuals.

4 The genetic information gathered might, for example, be purely negative in character: if an

Tertiary

Tertiary genetic discrimination occurs when a decision making process is informed by the presence or absence of a property that is not uniformly shared across all genetic variations. The nature of the relationship between the property selected for and genetic variation may be statistically weak, and it may not be of a causal nature at all, but discrimination practiced on the basis of this property will nevertheless have a disproportionate impact upon groups possessing particular genetic variations. Discriminating on the grounds of "strength" might provide one example of tertiary genetic discrimination. Various genetic profiles may be more or less closely associated with physical strength and disproportionately affected by selection for it.

The weak nature of the relationship between the characteristic and genetic variation means that very little genetic information could be gathered from information about the characteristic itself. Such genetic information as could be gathered would be highly speculative in nature.

Problems of distinction and definition

It would be wrong to imply that regulators would have to choose between the three "kinds" of discrimination outlined in a mutually exclusive fashion. Rather, the categories may be thought of as sequentially tracing a boundary around a paradigm case of genetic discrimination in an increasingly expansive fashion. If regulation were to target genetic discrimination *per se*, then it would have to establish how far its concept of genetic information extended.

There is, I suggest, good reason for *not* attempting to regulate *only* primary genetic discrimination. But, if you expand your definition to include "secondary" then there is good reason for not attempting to exclude tertiary discrimination from regulation. If you expand your definition of genetic discrimination to include tertiary genetic discrimination then there is a danger that you may be attempting to regulate *any* kind of discrimination at all. There is, in fact, no good reason for believing that legislating to the boundary of any one of these categories would be particularly effective at preventing (unfair) (genetic) discrimination.

From primary to secondary

It must be remembered that primary genetic discrimination only takes place when an individual knowingly and deliberately selects for, or against, a genetic variation *per se*. If the property preferred is simply *associated* with a particular genetic characteristic then the discrimination will be secondary, rather than primary, genetic discrimination.

One of the practical problems associated with drawing a legislative line around *primary* genetic discrimination would be in establishing that it has taken place. If an insurance company discriminates against an applicant due to their possession of either the BRCA1 or BRCA2 gene, then it appears fairly straightforward that it is the genetic data

---

individual is free of a particular genetic disorder, then one might assume that they do not possess genetic variations associated with it.

themselves that have informed the company's decision. How though might the company come to perceive that an individual possesses these gene variants?

In many cases, genetic tests do not rely upon *direct* analysis of an individual's DNA. The *term* "genetic test" is generally *defined* to include techniques that seek the elucidation of an individual's genetic architecture through the analysis of things other than their DNA. "Genetic tests" can look for particular gene products, such as specific proteins or enzymes, or they can test for properties found further along the biochemical pathway toward phenotypic expression, perhaps even within phenotypic expression itself. As soon as you allow for "genetic tests" searching for characteristics that are *associated* with particular genetic variations, without necessarily *directly* revealing genetic architecture, a difficulty in enforcing a distinction between primary and secondary discrimination becomes apparent.

Would-be discriminators do not seek genetic information because they wish to express a preference for a particular genetic characteristic *unless that genetic characteristic has itself been associated* with something that they value. Genetic information only has instrumental value to a discriminator due to its perceived ability to act as an *indicator* of an individual's possession of, or susceptibility to, some other characteristic or trait. These "valued" characteristics or traits are likely to be found many biochemical steps from the "raw" DNA itself. The pathways linking the DNA to the valued characteristic may then present many opportunities for "tests" at *interim* points.

Regulating *only* primary genetic discrimination would drive discriminators to rely upon indicators that, while "backwards" indicative of genetic variation, were also "forwards" indicative of the "valued" characteristic. That is, they would select indicators that only informed of genetic variation given a particular interpretation. While undeniably *capable of* providing "genetic information," preference for these characteristics would only constitute *secondary* genetic discrimination.[5]

Any attempt to preserve the distinction between primary and secondary genetic discrimination by inquiring into the state of mind of the discriminator would clearly be problematic. It may never be possible to establish in practice whether an individual is drawing, *and relying upon*, genetic inferences from "non genetic" data presented to them. Indeed, regulating only *primary* genetic discrimination would probably have the effect of genuinely reducing overall awareness that the properties tested for have genetic significance at all.

While it may currently be in vogue for things to be marketed with associations to DNA technology emphasized, regulation of only *primary* genetic discrimination would likely reverse this trend. Tests would probably be labeled according to the actual property tested for or the characteristic valued. Without being expressly informed of the genetic basis of the tests, individuals may not appreciate the genetic significance of the information they receive (as many women using "pregnancy testing kits" may remain unaware of the information about hormone levels that these kits are able to provide).

---

5 Regulating the use of information *about the detail of genetic characteristics* may drive individuals to discriminate at a different level of resolution, even if this results in a degradation in the predictive value of the information. Ironically, this may lead to less justifiable discrimination being unregulated.

Regulating primary, but not secondary, genetic discrimination would not discourage the use of data *capable* of yielding genetic information, but only the overt use of genetic information itself. The only genetic discrimination that may be effectively regulated would be that which could *only* be informed by direct DNA analysis. I would suggest that this would be ineffective at achieving the objectives sought through legislative reform and, so far as it had any effect at all, would certainly fail to effectively target the most unfair discrimination.

From secondary to tertiary

Adopting a definition of genetic discrimination that captured acts of secondary genetic discrimination as well as primary would not, however, constitute a panacea. While good reasons may exist for not limiting regulation to purely primary genetic discrimination, any extension of regulation to include secondary, but not tertiary, genetic discrimination would encounter its own problems.

Secondary genetic discrimination occurs when an individual discriminates on the grounds of a characteristic that has a particular genetic pedigree. It does not matter whether the individual is aware that *genetic* information could be gathered from knowledge of the characteristic. Tertiary genetic discrimination occurs when the comparator selected lacks that level of association with genetic variation but, nevertheless, the expression of a preference for the comparator has a disproportionately high risk of affecting those possessing particular genetic variations.

One of the difficulties in separating tertiary from secondary genetic discrimination is in establishing the extent to which any given characteristic is actually related to genetic variation. The degree of association between very many characteristics and genetic variation remains hotly contested even by the scientific community. Without some degree of consensus on this point it would be impossible to distinguish in practice between secondary and tertiary genetic discrimination with the certainty and consistency required for effective regulation. All of a court's time would be spent attempting to establish whether the relevant genetic pedigree could be established for the characteristic in question.

Beyond tertiary?

The obvious response may appear to be to widen the definition to include tertiary genetic discrimination. Any legal reform targeted at tertiary genetic discrimination must face the possibility that it is attempting to regulate discrimination on the grounds of absolutely any characteristic. All human characteristics and/or traits are likely to be disproportionately associated with certain genetic variations.

A point worth emphasizing here is that, even at this level, it *could be* the *perception* of a link between genetic variation and characteristic driving an individual to discriminate in a particular way. Genetic (mis)information may motivate even tertiary genetic discrimination. In this case, it would seem perverse that it should be the lack of an *objectively* demonstrated link that *removed* the discrimination from the jurisdiction of law.

Conclusion

Genetic discrimination may be unfair. Unfair genetic discrimination ought to be prohibited. Genetic discrimination is, however, practiced due to genes being attributed with the propensity to indicate possession of *other* characteristics and traits. If individuals are prevented from seeking this information at one level of "genetic resolution," then they may seek it at another. There is nothing to support the thesis that as you move away from the paradigm case of *primary* genetic discrimination discriminatory practices will be any less unfair, or indeed, any less likely to be informed by data capable, or perceived to be capable, of yielding genetic information.

Even *if* the use of 'genetic information' *per se* could be shown to be uniquely related to unethical conduct, locating legislative reform around the concept itself may condemn the regulator to an undesirable choice between the effective regulation of a "sub-set" of genetic information—defined by its pedigree and not the risks it poses—and the impractical task of regulating all data that could be perceived to yield genetic information.

If we wish to proscribe the worst examples of *unfair* genetic discrimination, then we might do better by tackling the worst examples of unfair discrimination. We should perhaps turn from attempting to characterize those discriminatory practices to be regulated according to the (perceived) pedigree of the comparator used (which may be impossible to determine), and identify them instead by the nature of the threat that they pose to individual rights and fundamental freedoms. In this way we may focus upon preventing *unfair* discrimination and not fixating upon the perceived nature of the comparator.

Acknowledgements

I am very grateful to Dr. Shaun Pattinson and Mr. Peter Odell for their comments on a draft of this paper.

References

Lemmens, T. 2000. Selective justice, genetic discrimination, and insurance: Should we single out genes in our laws. *McGill Law Journal* 45:347–412.